Power Management Integrated Circuits

Devices, Circuits, and Systems

Series Editor
Krzysztof Iniewski
Emerging Technologies CMOS Inc.
Vancouver, British Columbia, Canada

PUBLISHED TITLES:

PUBLISHED TITLES:

Micro- and Nanoelectronics: Emerging Device Challenges and Solutions
Tomasz Brozek

Microfluidics and Nanotechnology: Biosensing to the Single Molecule Limit
Eric Lagally

MIMO Power Line Communications: Narrow and Broadband Standards, EMC, and Advanced Processing
Lars Torsten Berger, Andreas Schwager, Pascal Pagani, and Daniel Schneider

Mixed-Signal Circuits
Thomas Noulis

Mobile Point-of-Care Monitors and Diagnostic Device Design
Walter Karlen

Multisensor Attitude Estimation: Fundamental Concepts and Applications
Hassen Fourati and Djamel Eddine Chouaib Belkhiat

Multisensor Data Fusion: From Algorithm and Architecture Design to Applications
Hassen Fourati

MRI: Physics, Image Reconstruction, and Analysis
Angshul Majumdar and Rabab Ward

Nano-Semiconductors: Devices and Technology
Krzysztof Iniewski

Nanoelectronic Device Applications Handbook
James E. Morris and Krzysztof Iniewski

Nanomaterials: A Guide to Fabrication and Applications
Sivashankar Krishnamoorthy

Nanopatterning and Nanoscale Devices for Biological Applications
Šeila Selimovic´

Nanoplasmonics: Advanced Device Applications
James W. M. Chon and Krzysztof Iniewski

Nanoscale Semiconductor Memories: Technology and Applications
Santosh K. Kurinec and Krzysztof Iniewski

Novel Advances in Microsystems Technologies and Their Applications
Laurent A. Francis and Krzysztof Iniewski

Optical, Acoustic, Magnetic, and Mechanical Sensor Technologies
Krzysztof Iniewski

Optical Fiber Sensors: Advanced Techniques and Applications
Ginu Rajan

Optical Imaging Devices: New Technologies and Applications
Ajit Khosla and Dongsoo Kim

PUBLISHED TITLES:

Organic Solar Cells: Materials, Devices, Interfaces, and Modeling
Qiquan Qiao

Physical Design for 3D Integrated Circuits
Aida Todri-Sanial and Chuan Seng Tan

Power Management Integrated Circuits and Technologies
Mona M. Hella and Patrick Mercier

Radiation Detectors for Medical Imaging
Jan S. Iwanczyk

Radiation Effects in Semiconductors
Krzysztof Iniewski

Reconfigurable Logic: Architecture, Tools, and Applications
Pierre-Emmanuel Gaillardon

Semiconductor Radiation Detection Systems
Krzysztof Iniewski

Smart Grids: Clouds, Communications, Open Source, and Automation
David Bakken

Smart Sensors for Industrial Applications
Krzysztof Iniewski

Soft Errors: From Particles to Circuits
Jean-Luc Autran and Daniela Munteanu

Solid-State Radiation Detectors: Technology and Applications
Salah Awadalla

Structural Health Monitoring of Composite Structures Using Fiber Optic Methods
Ginu Rajan and Gangadhara Prusty

Technologies for Smart Sensors and Sensor Fusion
Kevin Yallup and Krzysztof Iniewski

Telecommunication Networks
Eugenio Iannone

Testing for Small-Delay Defects in Nanoscale CMOS Integrated Circuits
Sandeep K. Goel and Krishnendu Chakrabarty

Tunable RF Components and Circuits: Applications in Mobile Handsets
Jeffrey L. Hilbert

VLSI: Circuits for Emerging Applications
Tomasz Wojcicki

Wireless Medical Systems and Algorithms: Design and Applications
Pietro Salvo and Miguel Hernandez-Silveira

PUBLISHED TITLES:

Wireless Technologies: Circuits, Systems, and Devices
Krzysztof Iniewski

Wireless Transceiver Circuits: System Perspectives and Design Aspects
Woogeun Rhee

FORTHCOMING TITLES:

Advances in Imaging and Sensing
Shuo Tang and Daryoosh Saeedkia

Introduction to Smart eHealth and eCare Technologies
Sari Merilampi, Krzysztof Iniewski, and Andrew Sirkka

Magnetic Sensors: Technologies and Applications
Laurent A. Francis and Kirill Poletkin

Nanoelectronics: Devices, Circuits, and Systems
Nikos Konofaos

Radio Frequency Integrated Circuit Design
Sebastian Magierowski

Semiconductor Devices in Harsh Conditions
Kirsten Weide-Zaage and Malgorzata Chrzanowska-Jeske

X-Ray Diffraction Imaging: Technology and Applications
Joel Greenberg and Krzysztof Iniewski

Power Management Integrated Circuits

Edited by
Mona M. Hella
Rensselaer Polytechnic Institute, New York, USA

Patrick Mercier
University of California, San Diego, USA

Krzysztof Iniewski MANAGING EDITOR
Emerging Technologies CMOS Services Inc.
Vancouver, British Columbia, Canada

CRC Press
Taylor & Francis Group
Boca Raton London New York

CRC Press is an imprint of the
Taylor & Francis Group, an **informa** business

MATLAB® is a trademark of The MathWorks, Inc. and is used with permission. The MathWorks does not warrant the accuracy of the text or exercises in this book. This book's use or discussion of MATLAB® software or related products does not constitute endorsement or sponsorship by The MathWorks of a particular pedagogical approach or particular use of the MATLAB® software.

CRC Press
Taylor & Francis Group
6000 Broken Sound Parkway NW, Suite 300
Boca Raton, FL 33487-2742

Frist issued in paperback 2018

© 2016 by Taylor & Francis Group, LLC
CRC Press is an imprint of Taylor & Francis Group, an Informa business

No claim to original U.S. Government works

ISBN-13: 978-1-4822-2893-9 (hbk)
ISBN-13: 978-1-138-58603-1 (pbk)

This book contains information obtained from authentic and highly regarded sources. Reasonable efforts have been made to publish reliable data and information, but the author and publisher cannot assume responsibility for the validity of all materials or the consequences of their use. The authors and publishers have attempted to trace the copyright holders of all material reproduced in this publication and apologize to copyright holders if permission to publish in this form has not been obtained. If any copyright material has not been acknowledged please write and let us know so we may rectify in any future reprint.

Library of Congress Cataloging-in-Publication Data

Names: Hella, Mona Mostafa, editor. | Mercier, Patrick, editor.
Title: Power management integrated circuits / editors, Mona M. Hella and Patrick Mercier.
Description: Boca Raton : Taylor & Francis Group, 2016. | Series: Devices, circuits, and systems | Includes bibliographical references and index.
Identifiers: LCCN 2015043099 | ISBN 9781482228939 (alk. paper)
Subjects: LCSH: Low voltage integrated circuits. | Electric power--Conservation.
Classification: LCC TK7874.66 .P74 2016 | DDC 621.3815/37--dc23
LC record available at http://lccn.loc.gov/2015043099

Visit the Taylor & Francis Web site at
http://www.taylorandfrancis.com

and the CRC Press Web site at
http://www.crcpress.com

Contents

Contents

Preface

Big data generated by the Internet of Things (IoT), healthcare, and the world wide web (WWW) are changing our lifestyle and our society. Small chips are enabling this change through data sensing, gathering, processing, storing and networking through wireless and wired connections. This explosive growth of electronic devices and their deployment in new applications have sparked an urgency to address their environmentally benign and sustainable energy needs. The spread of mobile computing and the IoT devices is limited by both battery life and form factor. Research in the field of power management circuits and systems in the last 5–10 years has explored integrated power management units with a small form factor, increased power density, and efficient performance over a wide range of output power in the quest for replacing and/or more efficiently operating with conventional rechargeable batteries and carbon-based sources.

The book begins with a comparison between inductive and capacitive dc–dc converters in terms of their passive devices, amenability to integration, and efficiency at various load conditions. A hybrid inductive–capacitive converter is proposed for wide-range dynamic voltage scaling, with details on the static, dynamic performances and discussion of different loss mechanisms. Next, the design of single inductor dual output (SIDO) and single inductor multiple output (SIMO) converters are covered in detail in Chapters 2 and 3, including the presentation of different design goals such as reducing the number of power switches and their associated power losses, extending the output current range, reducing ripple, and improving dynamic performance. Various control techniques are discussed to meet the different design goals with an emphasis on adaptive pseudo-continuous conduction mode (PCCM) detailed in Chapter 3. Design aspects of switched capacitor (SC) dc–dc converters are given in Chapters 4 through 6. While advantageous in terms of their integration potential, switched capacitor converters have been limited to lower power density applications in addition to lower efficiencies as the load moves away from optimum design conditions. Techniques such as quasi-SC converters, soft-charging or resonant SC converters, and recursive SC converters are detailed to address some of the aforementioned limitations. Chapters 7 and 8 present a different perspective on power management units through the use of GaAs pHEMTs for efficient high-frequency switching converters. The details of device design, high-quality factor passives, and circuit design techniques tailored to GaAs technology are discussed in addition to reconfigurable output passive networks to maintain the high efficiency of GaAs converters over wider voltage and current ranges. Some of the circuit techniques such as resonant gate drivers are compared in both GaAs and CMOS technologies as in Chapter 8. While many of the chapters are focused around several key publications in the field, rather than republishing the original papers, the authors have expanded the material to provide more background and breadth than the original publications. As such, the book would complement a graduate level course on power electronics integrated circuits.

We hope you find this book useful in your exploration of power management integrated circuits and systems. There are many unique challenges to working with integrated circuits whether it is in standard nanometer scale silicon technologies or in III–V technologies. However, there are also many rewards to reap from having such a "power system-on-chip" (PSOC) platform. As the editors, we would like to thank the contributors to this book, including the graduate students and the contributing authors who have worked tirelessly to share their insights with you in this book.

Mona M. Hella
Patrick P. Mercier

MATLAB® is a registered trademark of The MathWorks, Inc. For product information, please contact:

The MathWorks, Inc.
3 Apple Hill Drive
Natick, MA 01760-2098 USA
Tel: 508-647-7000
Fax: 508-647-7001
E-mail: info@mathworks.com
Web: www.mathworks.com

Series Editor

Krzysztof (Kris) Iniewski is managing R&D at Redlen Technologies Inc., a start-up company in Vancouver, Columbia, Canada. Redlen's revolutionary production process for advanced semiconductor materials enables a new generation of more accurate, all-digital, radiation-based imaging solutions. Kris is also a founder of Emerging Technologies CMOS Inc. (www.etcmos.com), an organization of high-tech events covering communications, microsystems, optoelectronics, and sensors. In his career, Dr. Iniewski held numerous faculty and management positions at the University of Toronto, University of Alberta, SFU, and PMC-Sierra Inc. He has published over 100 research papers in international journals and conferences. He holds 18 international patents granted in the USA, Canada, France, Germany, and Japan. He is a frequent invited speaker and has consulted for multiple organizations internationally. He has written and edited several books for CRC Press, Cambridge University Press, IEEE Press, Wiley, McGraw-Hill, Artech House, and Springer. His personal goal is to contribute to healthy living and sustainability through innovative engineering solutions. In his leisurely time, Kris can be found hiking, sailing, skiing, or biking in beautiful British Columbia. He can be reached at kris.iniewski@gmail.com.

Editors

Mona Mostafa Hella received a BSc and an MSc with honors from Ain Shams University, Cairo, Egypt, in 1993 and 1996, respectively, and a PhD in 2001 from The Ohio State University, Columbus, Ohio, all in electrical engineering. She is currently an associate professor in the electrical, computer, and systems engineering department at Rensselaer Polytechnic Institute. Prior to that, she has held positions at several companies, including RF Micro Devices and Spirea AB. Dr. Hella was the recipient of the Egyptian Government Award of Excellence (1993), the Micrys Fellowship (1997–1998), and the Texas Instrument Fellowship (1999–2000). She was an associate editor of the *IEEE Transactions on Very Large Scale Integration (VLSI) Systems* from 2011 to 2014. She has served on the technical program committees for ISCAS, GLS-VLSI and RFIC symposium. She has been a member of the administrative committee of the microwave theory and technique society from 2007 to 2009. She has been a trust leader for the NSF-funded engineering research center on "smart lighting" since 2010 and a Fulbright scholar in 2015. Her research interests include the areas of high frequency circuit design and mixed signal design for energy harvesting and biomedical applications.

Patrick P. Mercier received a BSc in electrical and computer engineering from the University of Alberta, Edmonton, Alberta, Canada, in 2006, and an SM and a PhD in electrical engineering and computer science from the Massachusetts Institute of Technology (MIT), Cambridge, Massachusett, in 2008 and 2012, respectively.

He is currently an assistant professor in electrical and computer engineering at the University of California, San Diego (UCSD), where he is also the co-director of the Center for Wearable Sensors. His research interests include the design of energy-efficient microsystems, focusing on the design of RF circuits, power converters, and sensor interfaces for miniaturized systems and biomedical applications.

Prof. Mercier received a Natural Sciences and Engineering Council of Canada (NSERC) Julie Payette fellowship in 2006, NSERC Postgraduate Scholarships in 2007 and 2009, an Intel PhD fellowship in 2009, the 2009 ISSCC Jack Kilby Award for Outstanding Student Paper at ISSCC in 2010, a Graduate Teaching Award in Electrical and Computer Engineering at UCSD in 2013, the Hellman Fellowship Award in 2014, the Beckman Young Investigator Award in 2015, and the DARPA Young Faculty Award in 2015. He currently serves as an associate editor of the *IEEE Transactions on Biomedical Circuits and Systems* and the IEEE *Transactions on Very Large Scale Integration* and is a coeditor of *Ultra-Low-Power Short-Range Radios* (Springer, 2015).

Contributors

Saurabh Chaubey
Department of Electrical and Computer
Engineering
University of Minnesota
Minneapolis, Minnesota

Ke-Horng Chen
Institute of Electrical Control
Engineering
National Chiao Tung University
Hsinchu, Taiwan, Republic of China

T. Paul Chow
Rensselar Polytechnic Institute
Troy, New York

Ramesh Harjani
Department of Electrical and Computer
Engineering
University of Minnesota
Minneapolis, Minnesota

Mona Mostafa Hella
Department of Electrical, Computer,
and Systems Engineering
Rensselaer Polytechnic Institute
Troy, New York

Sudhir S. Kudva
Department of Electrical and Computer
Engineering
University of Minnesota
Minneapolis, Minnesota

Yutian Lei
Department of Electrical and Computer
Engineering
University of Illinois at
Urbana–Champaign
Urbana, Illinois

Zemin Liu
Department of Electrical, Computer and
Systems Engineering
Rensselaer Polytechnic Institute
Troy, New York

D. Brian Ma
Department of Electrical Engineering
The University of Texas at Dallas
Richardson, Texas

Patrick P. Mercier
Department of Electrical and Computer
Engineering
University of California, San Diego
San Diego, California

Vipindas Pala
Department of SiC Power Devices
Wolfspeed, A Cree Company,
Research Triangle Park, North Carolina

Han Peng
Department of High frequency Power
Electronic Lab
GE Global Research
Niskayuna, New York

Robert Pilawa-Podgurski
Department of Electrical and Computer
   Engineering
University of Illinois at
   Urbana–Champaign
Urbana, Illinois

Loai G. Salem
Department of Electrical and Computer
   Engineering
University of California, San Diego
San Diego, California

Jason Stauth
Dartmouth College
Thayer School of Engineering
Hanover, New Hampshire

Yi Zhang
Department of Electrical Engineering
The University of Texas at Dallas
Richardson, Texas

1 Efficient On-Chip Power Management Using Fully Integrated DC–DC Converters

Saurabh Chaubey, Sudhir S. Kudva, and Ramesh Harjani

CONTENTS

1.1 INTRODUCTION

Device scaling has resulted in the implementation of entire systems on a single chip. The increased level of integration has reduced system costs and has proved advantageous from a signal integrity point of view by reducing the number of high-speed signals that need to be routed off-chip. Unfortunately, this higher level of integration has resulted in increased power dissipation in both mobile and stationary devices. Battery life of mobile devices and increased package and cooling costs for stationary devices are the driving forces behind the new focus on methods to reign power dissipation [1]. The supply voltage is one of the primary levers available to control power dissipation [2]. As shown in Equation 1.4, power dissipation in a digital system is approximately proportional to third power of the supply voltage (V_{DD}^3). We can consider this by viewing Equations 1.1 through 1.4. The active power dissipation of a digital circuit is given by Equation 1.1, where C_{tot} is the total digital capacitor switched, V_{DD} is the supply voltage, and f is the frequency of switching. On-chip power conversion is a complex process especially in battery-operated devices. Figure 1.1a shows a typical teardown of a smartphone that indicates the requirement of different power domains, making it a complete system. It should be noted that the majority of the blocks are of digital or mixed signal type. The demand for small form factor and more features resulted in stacked integrated circuits (ICs) and double stacked printed circuit boards (PCBs). Chemical energy–based batteries are the primary power source in handheld devices. State-of-the-art batteries occupy 20% by volume and 25% by weight. Typical volume density of Li–ion batteries is 400 J/mL and a typical weight energy density of 900 J/g. According to Ref. [3], one word on 3G SMS will cost about 1 J and talking on 3G for 1 minute will cost about 30 J, which means, a best state-of-the-art system solution can provide 30 minutes of 3G talk per gram of lithium battery. These numbers throw light on the problem of energy density of such batteries. Another important aspect of Li–ion is the variation of its voltage with time as depicted in Figure 1.1b. Given a time-variant input voltage profile, we need to provide fairly time-constant voltages involving both buck and boost conversions. So the power conversion system not only requires good static performances (steady-state efficiencies) but excellent dynamic behavior (fast feedback at input and output side).

  Linear regulators, inductive converters, and switched-capacitor (SC) converters are the primary DC–DC converter arsenal to tackle this problem. Inductive converters

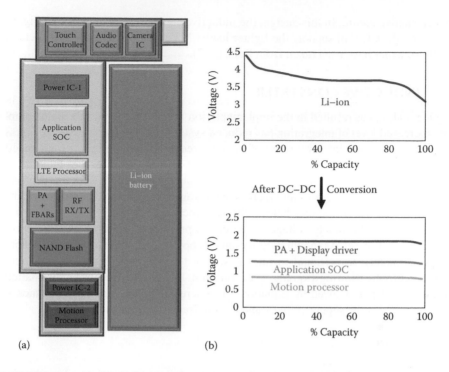

(a)   (b)

FIGURE 1.1 (a) Typical smartphone teardown showing different power domains. (b) Discharge profile of Li–ion battery and requirements of different voltage levels by different components.

require high-Q passives that are not always conveniently available in low-cost CMOS processes. Capacitive converters (CCs) are most easily implemented in modern CMOS processes as they primarily only need switches and capacitors. This ability to be easily integrated, finer device geometries, and newer technologies like deep trench capacitors have increased the interest in SC converters. The conversion efficiency of SC converters has traditionally not matched that of inductive converters, but a better understanding of the different topologies and their parasitic losses has slowly increased their efficiencies [4].

Inductive converters are more efficient at higher loads but show droop in efficiency at lower loads due to switching losses. On the other hand, CCs are easier to implement, perform better at lower loads, but fair poorly at higher loads due to increasing conduction losses. As the final part of this chapter presents, for a wide output range of dynamic voltage scaling (DVS)-based load, a hybrid of inductive and CC is required.

This chapter starts by discussing the design details of inductive converter and its static and dynamic performances. The design that we will discuss is a wide output inductive converter, suitable for DVS-based applications achieving 78% efficiency. Next, we will dwell on the capacitive conversion part of DC–DC power conversion. We will discuss the various loss mechanisms involved in the power conversions. Finally, we will look into the possibility of using a hybrid converter of both inductive and capacitive nature. In this design, the inductive converter will support the higher loads and the CC will support the lighter loads. Major technical and design details for this chapter have been taken from Refs. [5–7].

1.2   INDUCTIVE CONVERTER

Device scaling has resulted in the implementation of entire systems on a single chip. The increased level of integration has reduced system costs and has proved advantageous from a signal integrity point of view by reducing the number of high-speed signals that need to be routed off-chip. Unfortunately, this higher level of integration has resulted in increased power dissipation in both mobile and stationary devices. Battery life of mobile devices and increased package and cooling costs for stationary devices are the driving forces behind the new focus on methods to reign in power dissipation [1]. The supply voltage is one of the primary levers available to control power dissipation [2]. As shown in Equation 1.4, power dissipation in a digital system is approximately proportional to third power of the supply voltage (V_{DD}^3). We can consider this by viewing Equations 1.1 through 1.4. The active power dissipation of a digital circuit is given by Equation 1.1, where C_{tot} is the total digital capacitor switched, V_{DD} is the supply voltage, and f is the frequency of switching

$$P = C_{tot}V_{DD}^2 f \tag{1.1}$$

Likewise, the switching frequency is given by the following equation, where I is the current supplied to the circuit:

$$f = \frac{I}{C_{tot}V_{DD}} \tag{1.2}$$

The *on* current for the transistor is given by

$$I \propto (V_{DD} - V_T)^{-1-2} \tag{1.3}$$

where V_T is the threshold voltage. The exponential power term decreases as the short channel effects increase for smaller device dimensions [8]. Substituting Equations 1.2 and 1.3 into Equation 1.1, we arrive at

$$P = C_{tot}V_{DD}^2 \frac{I}{C_{tot}V_{DD}} \propto V_{DD}^{-2-3} \tag{1.4}$$

If the supply voltage is reduced further to operate the circuit in the subthreshold region of operation, further savings can be achieved as the current decreases exponentially in this region as shown by the following equation [8]:

$$I = I_s e^{\frac{V_{GS}}{nkT/q}} \left(1 - e^{-\frac{V_{DS}}{kT/q}}\right)(1 + \lambda V_{DS}) \propto e^{V_{DD}} \tag{1.5}$$

where

    k is the Boltzmann constant
    T is the absolute temperature in Kelvin
    I_S and n are empirical parameters

It is this strong dependence on V_{DD} that is utilized in DVS-based systems where the supply voltage is dynamically varied depending on the load being executed by the system. When running an application that can be run at a slower speed, the supply voltage and the frequency of operation are scaled down to reduce power dissipation. The advent of multicore and application-specific cores presents the next level of challenges in power management and distribution. Maximum power saving is possible when each of the cores forms separate voltage domains and DVS is applied to them individually [9–11]. This level of fine-grain power control is only possible with individual power regulators for each voltage domain. Off-chip voltage regulators require individual power pins to interface the regulated power to the on-chip voltage domains and a few additional pins to interface with the off-chip power converter. The increased pin-count and multiple board-level regulators increase the system cost. Additionally, routing of the multiple separate regulated supply voltages increases the top-level system routing complexity and leads to increased losses in the power delivery network. For DVS to be successful, a fully integrated on-chip power converter appears as an optimal solution, which is the focus of this chapter.

High-efficiency, fully integrated power converters have been demonstrated [12]. However, DVS places additional constraints on the converter. The power consumption of a digital block to which DVS is applied may vary over a wide range, as shown in Figure 1.2a (reconstructed plot using the data in Ref. [13]) for a low-voltage motion estimation accelerator. In this example, the power varies more than four orders of magnitude when the voltage is varied from 0.25 to 1.4 V. As the maximum performance of such application-specific cores is not always required, significant power savings can be achieved if the block voltage can be adapted to the load requirements dynamically. Ultra-DVS [14] using subthreshold operation has demonstrated significant energy savings. The feasibility of subthreshold operation has been shown in a wide variety of circuits [15,16], making it an important component in DVS systems. However, this necessitates a fully integrated DC–DC power converter that operates efficiently over a wide output power range. For testing purposes, we have chosen a ring oscillator (RO) as a representative digital circuit whose V–I profile is shown in Figure 1.2b. This matches the profile of variation of power with voltage of Figure 1.2a. We will use this V–I profile to load the converter, which will be explained in detail in Section 1.2.3.1.

There are different options available for implementing an on-chip voltage regulator, including linear, inductive switching and CCs. The efficiency of an inductive switching regulator depends only on the parasitics of its components, unlike the linear and CCs whose efficiency depends primarily on the conversion ratio [17]. Hence, inductive switching regulators or buck converters can achieve very high efficiencies and thus have been selected for this design. However, their efficiency decreases for lower-output powers. We have made additional improvements to the typical buck converter architecture to overcome these limitations and achieve high efficiency over a wide power range that is necessary for a DVS system.

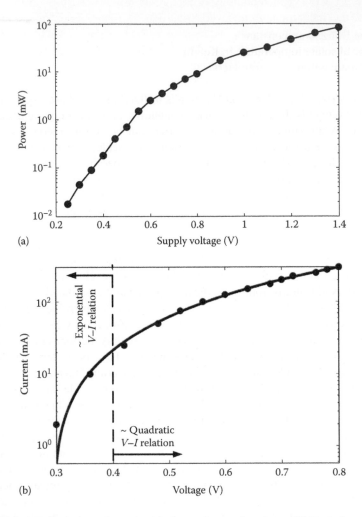

FIGURE 1.2 (a) Variation of power with change in supply voltage. (b) Variation of current with change in supply voltage.

The rest of this chapter is organized as follows. Section 1.2 focuses on the individual components and modifications made to the traditional buck converter architecture and summarizes the changes necessary for the integrated converter. Section 1.3 describes the additional changes made to improve the transient response and achieve automatic mode control. Measurement results are presented in Section 1.4, followed by comparison with other works in Section 1.5.

1.2.1 SWITCH SCALING AND FREQUENCY SCALING ARCHITECTURE

The block-level circuit diagram of a typical buck converter is shown in Figure 1.3. For a fully integrated implementation, the size of the passives that can be implemented

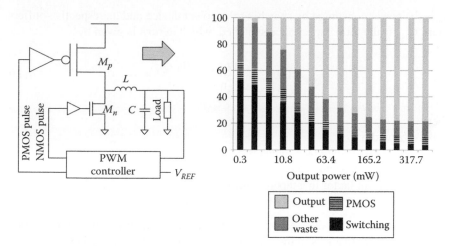

FIGURE 1.3 A typical buck converter and % power dissipation versus load power for a pulse width modulated buck.

on-chip is limited due to the limited chip area available. The ripple on the output voltage is given by the following equation [18], where L and C are the inductance and filter capacitance, D is the duty cycle for generating the required output voltage, and f_{sw} is the switching frequency of the buck converter:

$$\Delta V = \frac{D(1-D)V_{DD}}{8LCf_{sw}^2} \tag{1.6}$$

To mitigate the problem of the small size of the passive components, the switching frequency f_{sw} needs to be increased to meet the output ripple voltage specifications. However, increasing the operating frequency results in increased switching losses in the NMOS and PMOS power devices and their corresponding drivers. In Figure 1.3, we plot the simulated percentage of power dissipated in various components versus the power level for an integrated pulse width modulated (PWM) buck converter. The different components of input power P_{in} are as follows:

$$P_{in} = P_{out} + P_{sw} + P_{cond,PMOS} + P_{otherloss} \tag{1.7}$$

where P_{out} is the power supplied to the load

$$P_{sw} = C_{gate}V_{DD}^2 f_{sw} \tag{1.8}$$

P_{sw} is the power consumed by switching the power device and its respective buffers, where C_{gate} is the total capacitance switched, which in turn is given by

$$C_{gate} = WLC_{ox} + \frac{WLC_{ox}}{\eta} + \frac{WLC_{ox}}{\eta^2} + \cdots \tag{1.9}$$

where
  W is the width
  L is the length of the power devices being switched
  C_{ox} is the gate capacitance per unit area
  η is the fan-out factor in a tapered buffer design

The PMOS and NMOS devices are in triode region of operation during conduction. Hence, the conductive loss in the PMOS device can be approximated by

$$P_{cond,PMOS} = \frac{I_{PMOS}^2}{\mu_p C_{ox}(W_p/L)(V_{DD} - V_{Tp})} \tag{1.10}$$

where
  μ_p is the hole mobility in PMOS
  W_p is the width of the PMOS power device
  V_{Tp} is the threshold voltage of the PMOS device
  I_{PMOS} is the RMS current in the PMOS power device

The mean square value, I_{PMOS}^2, is given by $I_{PMOS}^2 = D^2 I_{load}^2 + I_{PMOS,rms}^2$, where $I_{PMOS,rms}$ is the RMS value of $I_{PMOS,ripple}$. $I_{PMOS,ripple}$, in turn, is given by

$$I_{PMOS,ripple} = \begin{cases} -\dfrac{D(1-D)V_{DD}}{2Lf_{sw}} + (t-nT)\dfrac{(1-D)V_{DD}}{L} & \text{for } nT \leq t \leq (nT+DT) \\ 0 & \text{for } (nT+DT) \leq t \leq (n+1)T \end{cases} \tag{1.11}$$

$P_{otherloss}$ is the power lost in the rest of the circuit:

$$P_{otherloss} = P_{cond,NMOS} + P_{cond,L} + P_{sc} \tag{1.12}$$

where

$$P_{cond,NMOS} = \frac{I_{NMOS}^2}{\mu_n C_{ox}(W_n/L)(V_{DD} - V_{Tn})} \tag{1.13}$$

is the conductive losses in the NMOS power device, with an electron mobility of μ_n, width of W_n, and I_{NMOS}. The RMS current through the NMOS power device, which is the sum of $(1 - D)I_{load}$ and $I_{NMOS,ripple}$ is given by

$$I_{NMOS,ripple} = \begin{cases} 0 & \text{for } nT \leq t \leq (nT + DT) \\ \dfrac{D(1-D)V_{DD}}{2Lf_{sw}} - (t - (n+D)T)\dfrac{DV_{DD}}{L} & \text{for } (nT + DT) \leq t \leq (n+1)T \end{cases}$$

$$\text{(1.14)}$$

$$P_{cond,L} = R_s I_{ind}^2 \tag{1.15}$$

is the loss in the inductor series resistance R_s, where I_{ind} is the RMS value of current through the inductor, and

$$P_{sc} = V_{DD}I_{sc} \tag{1.16}$$

is the loss due to the direct current flowing when the PMOS and NMOS devices are simultaneously *on*.

Different techniques have been implemented in this design to reduce each of these wasteful components of power, but special attention is paid to reducing the switching power losses as it forms a significant portion of this wasteful power. At low-output powers, more than 50% of the total input power is dissipated in switching the power devices and their associated buffers. The switching power losses can be reduced by either reducing the capacitance being switched or by reducing the frequency of operation. Both techniques will be applied to our converter.

1.2.1.1   Switch Scaling

The size of the switching power device is selected based on the maximum load current supported by the converter. When the load current decreases, using a smaller power device increases the efficiency by decreasing the switching power losses. In order to reduce the capacitance being switched, the power device along with its drivers is split into multiple parts. Depending on the load current requirement, an appropriately sized power device is switched and the remaining power devices and its drivers are completely turned off. By scaling the switch size [19,20], we reduce the switching losses at lower powers, increasing the overall efficiency. In this design, we have only scaled the PMOS power device and its corresponding drivers because the PMOS power devices are much larger than the NMOS power devices because of their smaller current per unit width. Also, in a DVS-based system, higher-output voltages correspond to larger load currents. Higher-output voltages in a buck converter are achieved by turning on the PMOS power device for a longer portion of the switching cycle, which necessitates a larger PMOS power device to reduce the conductive losses. Simulations showed that scaling the NMOS power device provides limited increase in efficiency and only adds to the design complexity. The optimal sizing of the PMOS power device can be calculated by considering both the switching power losses and the resistive losses in the PMOS device:

$$P_{sw\_PMOS} = P_{sw} + P_{cond,PMOS} \tag{1.17}$$

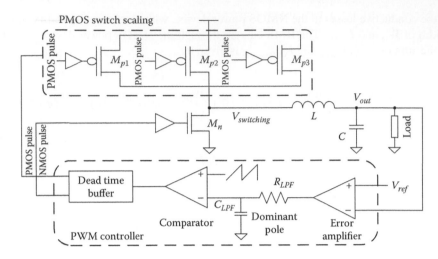

FIGURE 1.4   Constant frequency pulse width modulation mode with switch scaling.

Differentiating P_{sw_PMOS} with respect to W_p and setting $dP_{sw\_PMOS}/dW_p = 0$, the optimum switching transistor width can be calculated as

$$W_{p,opt} = \frac{I_{PMOS}}{V_{DD}C_{ox}} \sqrt{\frac{1}{\mu_p(V_{DD} - V_{Tp})f\left(1 + \dfrac{1}{\eta} + \dfrac{1}{\eta^2} + \cdots\right)}} \tag{1.18}$$

From Equation 1.18, it is evident that to minimize switching losses and PMOS conductive losses, the width of the PMOS power device needs to be scaled proportionally to the current through the PMOS power device. The current through the PMOS power device is in turn proportional to load current (and to the duty cycle). In order to cover the entire output load current range, the PMOS power device is split into three parts with widths 2, 6, and 12 mm. Using these devices, effective drive sizes of 2 mm (1×), 8 mm (4×), 14 mm (7×), and 20 mm (10×) are achievable. The NMOS power device is a single 2 mm wide device.

The switch scaled mode uses a fixed frequency-based PWM controller as shown in Figure 1.4. The switching frequency of the PWM-based controller is 300 MHz. The PWM controller uses the dominant pole-based compensation technique with the open-loop dominant pole set at approximately 40 kHz, achieved by an RC low-pass filter. Poles and zeros of the feedback loop path were extracted via circuit simulations and then used to create a MATLAB® model of the system to examine system stability [18]. The system simulations showed a phase margin of 90° at a gain crossover frequency of 1.95 MHz as shown in Figure 1.5. A large phase margin was used in this design to ensure stability of the system even after process variations. The dead time buffer following the comparator is used to reduce the P_{sc} component of the wasteful power. A tapered buffer design has been used to drive the NMOS and PMOS power devices with a fan-out factor of 8.

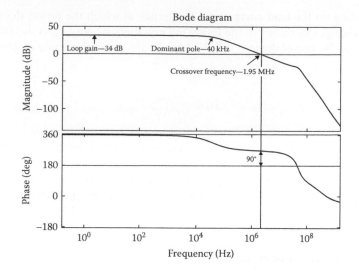

FIGURE 1.5   Bode plot for full system MATLAB® simulation of the converter.

1.2.1.2   Frequency Scaling

System-level simulations show that though switch scaling achieved efficiency improvements at high and medium output powers, the conversion efficiency was still low at
lower-output powers. To combat this problem, we perform automatic frequency scaling at the lowest output powers to further increase the efficiency. Frequency scaling is
implemented by operating the converter in pulse frequency modulation (PFM) mode.
In this mode, when the output voltage dips below the reference voltage, the PMOS
power device is turned *on* to charge the filter capacitor, which provides the load current. The NMOS power device is turned *on* for a short period of time after turning *off*
the PMOS device to discharge the inductor. The feedback path uses a clocked comparator to sample the difference between the output voltage and the reference voltage
and generates the switching pulses based on this difference as shown in Figure 1.6.

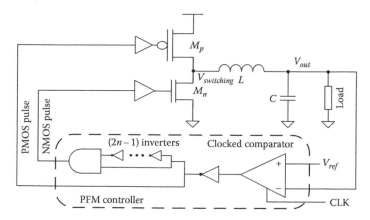

FIGURE 1.6   Frequency scaling using pulse frequency modulated controller.

Depending on the load current, the frequency at which the power devices are switched changes as shown in the following equation, where f_{PFM} is the switching frequency of the power devices in PWM mode and ΔV is the ripple on the output voltage:

$$f_{PFM} = \frac{I_{load}}{C\Delta V} \tag{1.19}$$

The converter here operates in discontinuous conduction mode where both the PMOS and NMOS switches are *off* simultaneously for a part of the clock period. This mode of operation can only support low load currents because of the small size of the filter capacitor that provides the load current when both the PMOS and NMOS power devices are *off*. During this mode of operation, the width of the PMOS and NMOS devices are fixed at 2 mm each.

1.2.1.3 Integrated Converter

In order to obtain high efficiency over the entire output power range, the proposed converter operates in single-phase PWM mode at high-output powers and in PFM mode at low-output powers [21]. A block diagram for the multimode integrated converter is shown in Figure 1.7. A state machine, discussed in more detail in Section 1.2.2, selects the appropriate mode of operation—PWM or PFM and the appropriate size PMOS

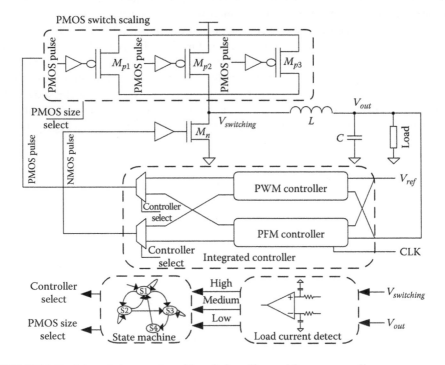

FIGURE 1.7 Integrated converter with switch scaling and frequency scaling.

power device in that particular mode. When the output power is high, all the three switches are operational in the PWM mode (PWM-10X). At medium-output powers, the 12 mm switch along with its driver chain is switched *off* and the other two operational switches, with a combined size of 8 mm (PWM-4X), provide the output power. At low-output powers, only the 2 mm switch is functional in the PFM mode (PFM-1X).

In order to conserve power when one of the controllers is operational, the other controller is completely turned *off*. The PWM controller is turned *off* by shutting off the current sources and PFM mode is turned *off* by gating the clock to the clocked comparator.

1.2.2　Load Current Detection and State Machine

As shown in Equation 1.18, the optimum size of the PMOS switch is decided by the load current. Hence, in order to select the optimum size of the PMOS power device, a method to sense the current in the load is necessary. Addition of any resistance in the load current path for the purposes of current sensing results in additional losses and reduces efficiency. Hence, we make use of the parasitic series resistance of the inductor to detect the current as shown in Figure 1.8a. The DC voltage drop across the inductor is directly proportional to the load current. The voltage across the inductor is filtered using a RC low-pass filter to calculate the average DC voltage across the inductor. Based on this measurement, current consumption is deduced and the detector logic generates a current-level signal that takes on three discrete values—high, medium, or low—to be used by the state machine. Temperature and process variations can cause the inductor DC resistance to vary, leading to erroneous current-level signals. We compensate for this variation by altering the variable offset in the comparator generating the current-level signals. This variable offset signal can be generated with the help of on-chip temperature sensors [22].

In this prototype chip, a simple one-hot encoding-based state machine, shown in Figure 1.8b, operates the converter in three states: PWM-10X, PWM-4X, and PFM-1X modes. (For testing purposes, the state machine can be bypassed to operate the converter in additional states: PWM-7X and PWM-1X.) The state machine operates at a low-frequency dissipating minimal power.

1.2.2.1　PWM Transient Speedup

The PWM mode controller uses dominant pole compensation in the feedback loop. Because of the small bandwidth required to meet the stability criteria, the transient response is very slow. In order to speed up the transient response, current sources are added to charge the low-pass filter node when a large difference between the reference voltage and the output voltage is detected as shown in Figure 1.9. Under normal condition, the current sources are turned off and do not affect the controller. The algorithm implemented in the decision circuit is also shown in Figure 1.9.

1.2.2.2　Inductive Converter Passives

In our design, all the passive components required in the converter are implemented on-chip. These components occupy a large area and also affect the overall efficiency of the converter. Hence, on-chip passives need to be custom-designed to reduce the

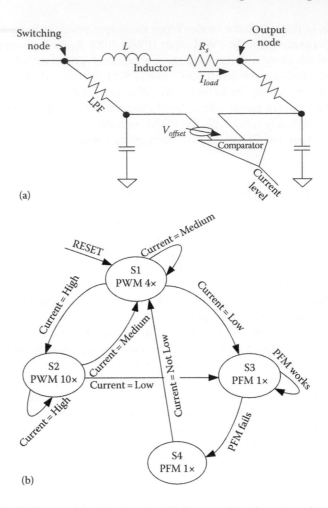

FIGURE 1.8   (a) Current detection circuitry. (b) State machine for automatic mode change.

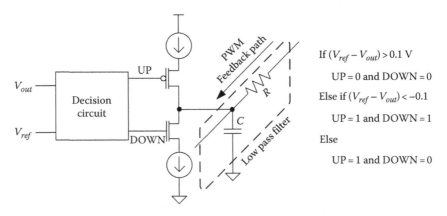

FIGURE 1.9   Pulse width modulated speedup circuit and decision circuit algorithm.

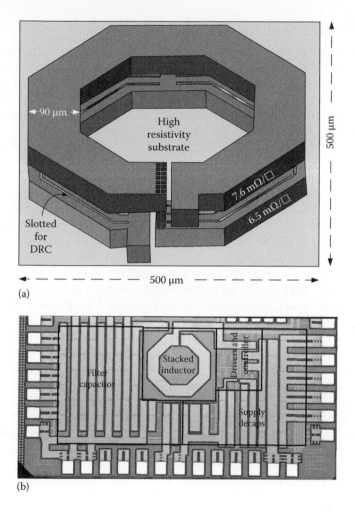

FIGURE 1.10 (a) Layout of the stacked inductor fabricated. (b) Chip micrograph of the wide output range DC–DC converter.

area as well as to minimize the associated parasitics, which reduce the efficiency. The $P_{cond,L}$ component of the wasteful power is directly proportional to the series resistance of the inductor (and the square of the inductor RMS current) and needs to be minimized. A custom-stacked inductor using the top two low-resistivity metal layers was designed as shown in Figure 1.10a. The inductor occupies an area of 500 μm × 500 μm.

Simulations of the inductor (with 90 μm wide metals) placed over high-resistivity substrate in ADS momentum show an inductance of 2 nH and series resistance of 0.245 Ω at DC [23]. Only the bottom metal needed to be slotted to meet CMP DRC rules. The top metal was not slotted to reduce series resistance. The filter capacitor is of size 5 nF and effective series resistance (ESR) of 74 mΩ and is constructed using dual-MIMcaps and MOScaps stacked to conserve area.

1.2.3 MEASUREMENT RESULTS

The prototype design was implemented in IBM 130 nm CMOS process. Figure 1.10b shows the die microphotograph of the wide output range DC–DC converter. The converter core occupies an area of 1.13 mm², and the total design area including the decoupling capacitors is 1.59 mm². A supply decoupling capacitor was added conservatively to this prototype due to the large package bondwire inductors and can be eliminated for low-inductance packages including flip-chip designs.

1.2.3.1 Efficiency

Figure 1.11 plots the measured efficiency of the converter in the PWM mode with a switching frequency of 300 MHz and in the PFM mode for varying output currents at different output voltages. A maximum efficiency of 74.45% is obtained at an output voltage of 860 mV and a load current of 125 mA while the system was operating in the PWM-4X mode. The maximum power supplied by the converter is 266 mW. In the PFM mode, the efficiency varies from 60.8% to 42.8%. The clock generation block (a RO) used during the PFM mode consumes 490 μW of power. Our efficiency estimate excludes this power as we assume the presence of a system clock. However, if we include this power, the low-end efficiency reduces from 42.8% to 32.3% but it has minimal effect at higher powers.

The system was designed for a switching frequency of 300 MHz. With a few adjustments in the reference current, the system was also made to operate in the PWM mode with switching frequencies of 250 and 200 MHz for evaluation purposes. The efficiency of the converter for an output voltage of 0.86 V and varying load currents is shown in Figure 1.12 for different operating frequencies and ambient temperature conditions. The maximum efficiency at 200 MHz and 27°C ambient temperature is 76.4%. The maximum efficiency increased to 77% when the ambient temperature around the chip was reduced to 8°C. (The effect of temperature and its significance on the design is explained in Section 1.2.3.3.) The improved efficiency at 200 MHz is largely a result of a reduction in switching losses.

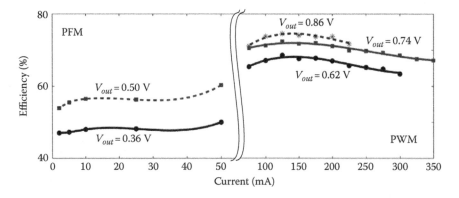

FIGURE 1.11 Efficiency of the converter in pulse width modulated and pulse frequency modulated modes for different output voltages.

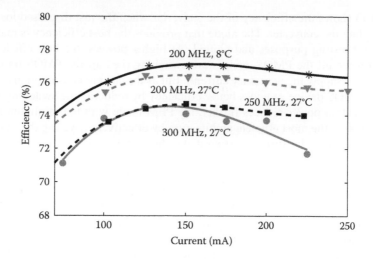

FIGURE 1.12   Converter efficiency for different switching frequencies V_{out} = 860 mV.

As discussed earlier, one of the motivations for this converter architecture was to supply power to digital DVS-based systems. In the case of digital systems, the current has a quadratic relation with supply voltage in strong inversion and an exponential relation with supply voltage in the subthreshold region of operation. Figure 1.13 shows the current consumption profile for a group of ROs, which is chosen as a representative digital circuit, when the supply voltage is varied. The region of operation when the RO circuit operates in the subthreshold region has been zoomed-in and plotted on a logarithmic scale. The power consumption varies from 0.6 mW at 0.3 V to 240 mW at 0.8 V, that is, a variation of 400× for a supply voltage variation of 2.6×. Here, both the output voltage and load current of the converter are being simultaneously varied, which results in the variation of power over this wide range.

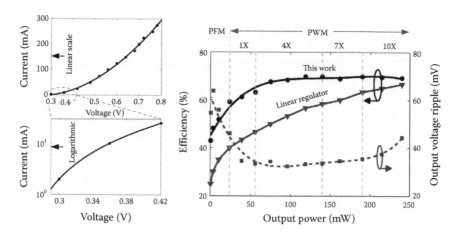

FIGURE 1.13   V–I profile of ring oscillator and efficiency of the converter for the V–I profile shown.

Figure 1.13 shows the efficiency of the converter when the just discussed load pro-file is loading the converter. The mode that provides the best efficiency is manually selected for testing purposes and plotted. At higher powers, the best efficiency is obtained when all the PMOS devices are switching (i.e., all the PMOS transistors are needed to supply the necessary current). But as the current consumption reduces, a smaller PMOS device provides better efficiency by reducing the switching losses. When the output power reduces further, the PFM mode with the minimum PMOS device becomes the most efficient architecture by effectively lowering the switching frequency, and thereby further cutting down the switching losses. For comparison purposes, the theoretical efficiency for a linear regulator is also plotted. Our con-verter performs better than the linear regulator at all output powers. Ripple in the output voltage is also shown at the corresponding output powers. We note that the ripple is fairly constant in the PWM mode and rises slightly as we reduce the load in the PFM mode due to the effective reduction in the switching frequency.

Ripple on the output voltage in the PWM mode at 300 MHz is shown in Figure 1.14a. We suspect that the high-frequency artifacts seen on the top graph of the PWM mode

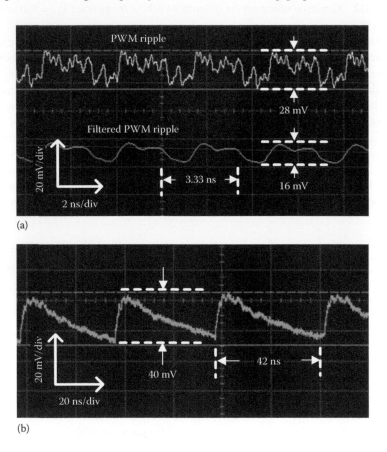

(a)

(b)

FIGURE 1.14   (a) Output voltage ripple in pulse width modulation. (b) Output voltage ripple in pulse frequency modulation.

are due to the reference square wave signal that is used in the ramp generation, which is buffered using a chain of inverters that is attached to the converter power supply. The current drawn by these buffers at different time instances, due to the individual inverter delays, during their transition results in high-frequency noise on the supply, which rides on top of the actual PWM ripple. When this high-frequency content is filtered out using a 600 MHz filter, the effective ripple on the output voltage reduces to 16 mV. The ripple on the output voltage in the PFM mode is shown in Figure 1.14b. As can be seen in this figure, the high-frequency ripple artifacts are absent in the PFM mode as the ramp generation unit is turned off. The output voltage is 750 mV for the PWM mode and 375 mV in the PFM mode. To get an enlarged view of the ripple on the output voltage, the signal was AC coupled to the oscilloscope and hence the lack of DC information. Also, please note that the timescale in the PFM mode is much larger than the timescale in the PWM mode because of the decrease in switching frequency with reduced load. As shown in Figure 1.12, higher efficiency can be achieved by reducing the switching frequency but reduction in switching frequency also results in increased ripple in the output voltage. Measured unfiltered ripple voltages at 200, 250, and 300 MHz are 46.2, 37.2, and 28.5 mV, respectively. Likewise, measured filtered (using a 600 MHz filter) ripple voltages at 200, 250, and 300 MHz are 31.8, 24.7, and 16 mV, respectively. The filtered ripple voltage values match the predicted values (Equation 1.6) reasonably well. The deviation of the unfiltered ripple voltage values can be attributed to high-frequency content injected by the buffers in the ramp signal generator.

Table 1.1 shows the switching + controller power in the different modes of operation. When operating in the PWM mode, the switching (+ controller) power component varies only with the size of the PMOS transistor and the associated driver chain. But in the PFM mode as the switching frequency changes, the switching power component changes even though the PMOS switch size is constant.

1.2.3.2   Transient Response

The transient response when switching from one reference voltage to another in the PWM mode with transient speedup turned off is shown in Figure 1.15a. Transition time for the output voltage to switch from 0.6 to 0.85 V is measured to be 1.4 μs.

TABLE 1.1
Switching + Controller Power in Different Modes

Mode	PMOS Transistor Size (mm)	Switching + Controller Power Loss in mW	
		Simulated	Measured
PWM	20 (10×)	17.74	24.5
PWM	14 (7×)	12.6	17.4
PWM	8 (4×)	8.44	11.7
PWM	2 (1×)	3.84	4.32
PFM	2 (1×)	0.43–2.077	0.6–4.7

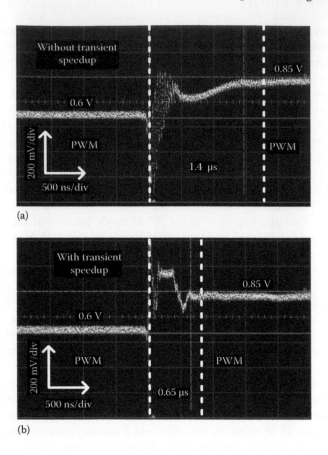

(a)

(b)

FIGURE 1.15 Transient response in pulse width modulated → pulse frequency modulated mode with and without speedup.

The transition time with speedup enabled was measured to be 0.65 µs as shown in Figure 1.15b, achieving an effective speedup of >2X. The transition from 0.85 to 0.6 V was measured to be 0.9 µs for both the cases of with speedup and without speedup. As can be seen, the high → low transition is much faster than the low → high transition, which is the reason that turning on the speedup mechanism had little effect. We see some overshoot and undershoot in the transient response owing to the simple nature of the control system used in this design.

The transient response and adaptive mechanisms were further verified by switching from one extreme state to the other, that is, PWM-10X to PFM-1X and vice versa. Figure 1.16a shows the output voltage switch from 374 to 717 mV. The output current at 374 mV is 30 mA, and the output current at 717 mV is 152.5 mA. The converter is in PFM-1x PMOS for the lower-output power and switches to PWM-10x PMOS for the higher-output power. The time taken to make this transition is 5.27 µs. Figure 1.16b shows the transient response in the reverse direction. The time taken to make this transition is 30.73 µs. The longer transition time for the down transition is because at the lower-output voltage the converter in the PWM mode fails to track the reference

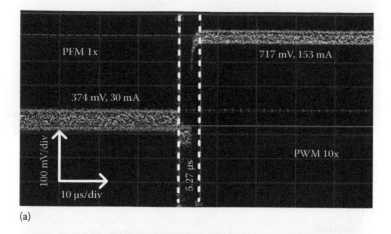

(a)

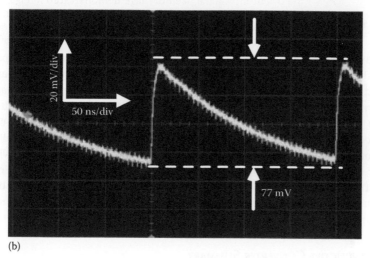

(b)

FIGURE 1.16 (a) Transient voltage response to output current change from 30 mA to 152.5 mA. (b) Transient voltage response to output current change from 152.5 mA to 30 mA.

voltage, which delays the state machine from making a decision in favor of PFM-1X. Finally, when the converter enters the PFM-1X mode, the converter tracks the reference voltage. This is a particular problem due to an implementation issue in our prototype design that is easily fixed with a small change to the PWM feedback system.

1.2.3.3 Effect of Temperature

The converter is a fully integrated implementation intended to supply power to digital circuits. Particularly, in the DVS mode, the temperature under which the converter has to operate may be significantly higher than the ambient temperature because of the heat dissipated by the digital circuitry surrounding the converter. To study the effect of temperature on the efficiency, the ambient temperature around the chip was raised by blowing air at 100°C using the Temptronic Thermostream thermal-inducing

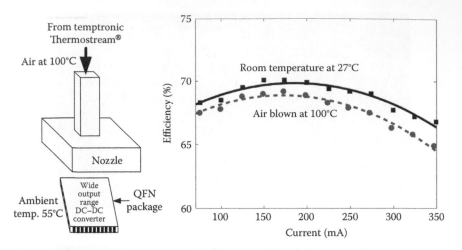

FIGURE 1.17   Efficiency variation with temperature V_{out} = 760 mV.

system as shown in Figure 1.17. The independently measured ambient temperature around the chip was 55°C. The efficiency reduces with an increase in the temperature, and this reduction in efficiency is greater at higher load currents than at lower load currents as shown in Figure 1.17. An increase in the temperature results in an increase in series resistance of the inductor as well as an increase in the channel resistance of the PMOS and NMOS devices. This increase in the resistance causes larger conductive losses at higher load currents than at lower load currents when the resistive loss forms only a small fraction of wasteful power. The efficiency reduces by about 3% at 350 mA load current as compared to 1% at 75 mA of load current. The reduced efficiency at elevated temperature needs to be taken into account when calculating the total system efficiency in real-world operating conditions.

1.2.4   INDUCTIVE CONVERTER SUMMARY

In this chapter, we have presented a wide output range, fully integrated on-chip power converter for DVS-based applications. To obtain high efficiency over the entire range of output powers, the converter switches between different modes of operation. At higher-output powers, switch scaling with a constant frequency PWM-based mode of operation is used. At low-output powers, a constant switch width but variable frequency-based PFM control is used. Switching between the different modes happens automatically with the converter tracking the output current. The prototype chip supplies output power from 0.6 to 266 mW. The converter achieves a peak efficiency of 77% under reduced temperature and a maximum efficiency of 74.45% under normal operating conditions. The efficiency varies between 42.8% and 74.45% over the entire wide power range, which, to the best of our knowledge, is the highest reported range for a fully integrated on-chip design.

From this discussion, we conclude that we can achieve highly efficient power conversion with fairly less voltage ripple for higher loads. But at lower loads (for DVS application, this means lower voltage), in order to maintain the same absolute voltage ripple, frequency of operation has to be increased, thus increasing the switching losses. But for lighter loads like around 1 mA and less, the CC performs better. So next we will see the details and design of a capacitive DC–DC converter, highlighting the just discussed observation.

1.3 CAPACITIVE CONVERTER

1.3.1 INTRODUCTION

Reduced power dissipation in integrated systems has become a critical design goal. One of the most effective methods to reduce power dissipation in digital system is to dynamically scale the supply voltage (DVS) based on the load conditions. Further savings in power can be obtained by using multiple independent voltage domains with DVS applied to each one of them. Designing for multiple voltage domains has necessitated that the power converter be fully integrated on-chip to overcome the constraints imposed by limited pin count and routing. Fully integrated inductive converters [6,12], linear regulators, and CCs [17] have been demonstrated.

Among these, CCs are particularly well suited for digital CMOS processes as they require only capacitors and MOS transistors as switches. Additionally, CCs can be designed to be highly efficient with their theoretical maximum efficiency given by $\eta = V_{out}/V_m$ [17], where V_m is the maximum unloaded voltage of the converter. Hence, maximum efficiency is achieved when V_{out} is close to V_m. However, the maximum load current supported by the converter is $I_{max} \propto C_b f_s (V_m - V_{out})$. In order to support the load current, $C_b * f_s$ has to be appropriately selected. The value of $C_b * f_s$ increases with an increase in the max load current to be supported by the converter as shown in Figure 1.18a. This results in an increase in the converter losses. The overall converter losses include switching losses (P_{SW}), losses in the converter core (P_{C_loss}), and fixed loses due to biasing (P_{bias}) as shown in Figure 1.18b. These losses can be modeled to calculate the overall efficiency as has been done for a 1:1 converter in Figure 1.19, where the thin dashed lines indicate the efficiency of the converter core and the thicker lines, the overall efficiency including the switching and bias loss in a 130 nm technology. The efficiency reduction for output voltages close to V_m is much higher than that for voltages away from V_m where the efficiency remains flat for varying I_{max}. Hence, the converter components are selected to optimize the overall efficiency and maintain a flat profile for a wide range of I_{max} values. For this design even though 0.95 V has a higher core efficiency, the overall efficiency is lower at higher load currents. However, selecting $V_{out} = 0.85$ V results in a fairly constant and higher efficiency over the entire range. The loss in efficiency from an ideal converter depends on the converter configuration, including factors such as number of switches in series with the bucket capacitor, the effective overdrive on the gate of switches, etc., and may be more severe for multimode designs. Only the switching component of the loss scales with device technology and is proportional to L^2, where L is the smallest feature size. For the same load conditions and a smaller technology, the shape of the curves remains similar but the y values increase:

$$\Delta V_{ripple} = \frac{I_{load}}{C_{tank} f_s} \tag{1.20}$$

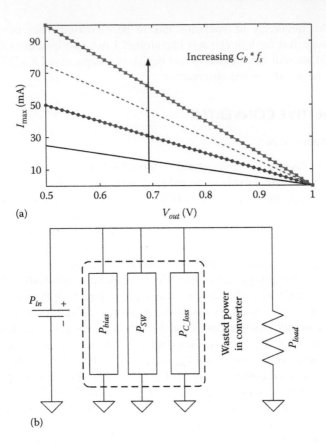

(a)

(b)

FIGURE 1.18 (a) Variation of I_{max} with output voltage (V_o). (b) Power components breakup for the converter.

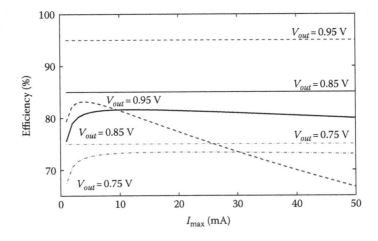

FIGURE 1.19 Overall efficiency of the converter for different loads.

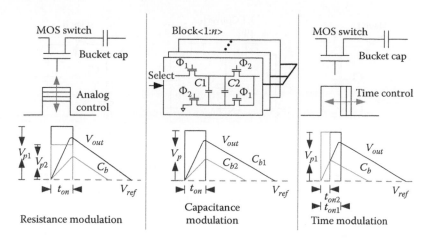

FIGURE 1.20 Catalogue of ripple control techniques.

All SC converters have output voltage ripple. However, fully integrated CCs suffer from the presence of significant ripple in the output due to the area-limited size of the tank (C_{tank}) or the decoupling capacitor as shown in Equation 1.20. Here for a fixed ripple voltage and a smaller tank capacitor, the switching frequency has to increase. Unfortunately, increasing the switching frequency increases the switching losses and reduces the overall efficiency. For fully integrated CCs, there is a clear chip area versus efficiency trade-off.

The root cause of the problem is that the bucket capacitor ($C1$ in Figure 1.21) values are selected depending on the maximum load current at the highest output voltage. For other operating conditions, the size of the bucket capacitor is larger than what is actually required. Therefore, when the output voltage or the load current decreases, the tank capacitor is overcharged by the bucket capacitors, resulting in increased ripple. The ripple can be reduced by reducing the amount of charge transferred to the tank capacitor. This can be achieved by modulating the resistance of switches, the size of the bucket capacitor, or the charge/discharge time of the bucket capacitors as shown in Figure 1.20. The first method of modulating the switch resistance requires analog amplifiers and hence is not very convenient from a scaling perspective. In this design, we use capacitance modulation for coarse ripple control and introduce digital bucket capacitor charge/discharge time modulation for fine ripple control both of which are digital in nature and hence suitable for easily scaling from one technology to other.

The power converter is targeted for near threshold to subthreshold voltage operation [14–16] and provides an output voltage in the 0.3–0.5 V range from a 1.2 V input.

1.3.2 PARTIAL CHARGING/DISCHARGING AND EFFICIENCY

The effect of partial charging/discharging of the bucket capacitors on the efficiency of the converter is illustrated using a simple 1:1 converter in Figure 1.21. Assuming a constant output voltage, the efficiency of the energy transfer in two stages is evaluated,

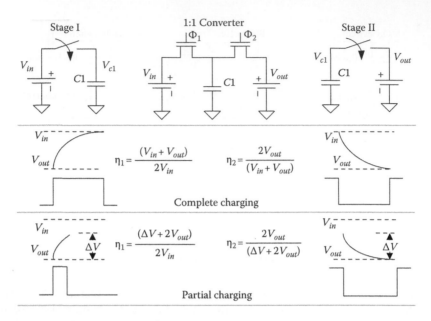

FIGURE 1.21 Effect of partial charging on efficiency.

that is, from input to the bucket capacitor in stage I and from the bucket capacitor to output in stage II. The total converter efficiency is the product of efficiencies in stages I and II. For the fully charged/discharged case, the bucket capacitor is charged to V_{in} in stage I, the energy stored in the bucket capacitor in this phase is $E_{cap1} = C_b(V_{in}^2 - V_{out}^2)/2$, and the energy supplied by the input source is $E_{in} = C_bV_{in}(V_{in} - V_{out})$. In stage II, the bucket capacitor that was charged to V_{in} is discharged to V_{out}, the energy supplied by the bucket capacitor is $E_{cap2} = C_b(V_{in}^2 - V_{out}^2)/2$, and the energy delivered to the load is $E_{out} = C_bV_{out}(V_{in} - V_{out})$. Hence, the efficiency of the converter is $\eta_1 = E_{cap1}/E_{in} = (V_{in} + V_{out})/2V_{in}$ and $\eta_2 = E_{out}/E_{cap2} = 2V_{out}/(V_{in} + V_{out})$ in stages I and II, respectively, which results in a overall efficiency of $\eta = \eta_1\eta_2 = V_{out}/V_{in}$.

The same exercise can be repeated for partial charging where in stage I the bucket capacitor is partially charged to $V_{out} + \Delta V$ instead of V_{in}. The energy stored in the bucket capacitor in stage I is $E_{cap1} = C_b(\Delta V^2 + 2V_{out}\Delta V)/2$, and the energy delivered by the input supply is $E_{in} = C_bV_{in}\Delta V$. For stage II, the energy supplied by the bucket capacitor is $E_{cap2} = C_b(\Delta V^2 + 2V_{out}\Delta V)/2$, and the energy that is delivered to the output is $E_{out} = C_bV_{out}\Delta V$. The overall converter efficiency is $\eta = \eta_1\eta_2 = V_{out}/V_{in}$, where $\eta_1 = (\Delta V + 2V_{out})/2V_{in}$ and $\eta_2 = 2V_{out}/(\Delta V + 2V_{out})$.

The overall efficiency of the converter remains the same in both cases, but the intermediate efficiencies vary. In partial charging, the efficiency of the charging stage is less than that for complete charging. However, the efficiency in the discharging stage for partially charging is higher than that for complete charging. *Hence, by utilizing partial charging/discharging, the overall efficiency of the converter does not change but the output ripple voltage is reduced.* Effectively, partial charging behaves like a smaller capacitor allowing for larger range of "capacitive modulation."

1.3.3 IMPLEMENTATION

Figure 1.22 shows the block diagram of the fully integrated CC. The converter core uses two phases offset by 180°. Additionally, two CC modes of operation ($V_m = V_{dd}/2$ and $V_m = V_{dd}/3$) are used to achieve high efficiency over the entire output voltage range [17]. In the $V_m = V_{dd}/2$ mode, the bucket capacitor is connected between the input and output in phase 1 and in parallel with the output in phase 2. Whereas in the $V_m = V_{dd}/3$ mode, two capacitors are connected in series in phase 1 and then both these capacitors are connected in parallel across the output in phase 2. The switches are arranged so as to reuse the same bucket capacitors in both modes of operations. Two copies of the converter core are connected in parallel. Either both the converter cores are used or only one of them is used to supply the output load. Regulation of the output voltage is achieved by a single bound hysteretic controller that forms the primary control loop. The primary loop changes the frequency of switching of the converter depending on the output voltage and the load current using a clocked converter to compare the output voltage to the reference voltage and initiates switching action when output dips below reference voltage. The switching frequency modifies the effective $R_{out} = 1/(f_s \times C_b)$ of the converter, thereby achieving output voltage regulation.

A secondary loop is responsible for controlling the ripple on the output voltage. The frequency of oscillation measured using a counter is an indirect measure of the ripple on the output voltage, that is, the ripple increases with reduced frequency, and forms the input to the secondary loop. If the switching frequency is equal to the reference clock

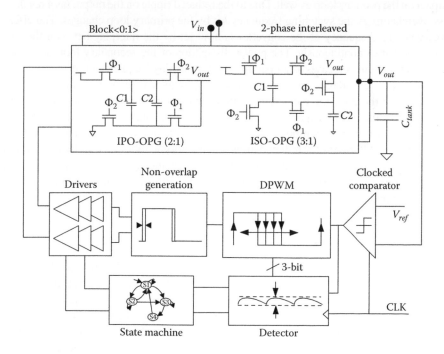

FIGURE 1.22 Fully integrated capacitive converter.

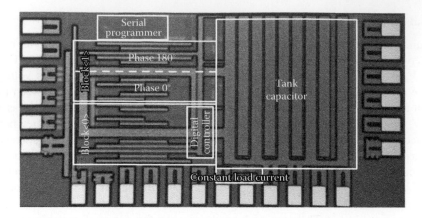

FIGURE 1.23   Die photograph of fully integrated capacitive converter.

frequency, then both the converter cores are put in use and the pulse width is maximized for both charging/discharging of the bucket capacitors. If the switching frequency is less than 1/6 of the reference frequency, then only one of the converter cores is used and the minimum pulse width is used for charging/discharging the bucket capacitors. For any frequency in between these two limits, the number of converter cores used is retained from the previous cycle and only the charge/discharge time is modified based on the switching frequency. The changes instituted by the secondary loop do have second-order impact on the primary loop as well. Due to the reduced ripple on the output (as a result of less overcharging), the switching frequency set by the primary loop changes. The effective switching frequency of the converter is now set by the interplay between the two loops and varies continuously. The power dissipation of the secondary loop decreases as the load current decreases as it operates at the switching frequency of the converter.

The bucket capacitors are constructed using dual-MIMcaps in order to reduce losses due to the parasitic capacitances in which switches and control circuitry are placed below the bucket capacitors. The total size of the bucket capacitor was 936 pF reused by both converter modes. A 5 nF tank capacitor is built using both dual-MIMcaps and MOScaps for reduced area. The prototype fabricated in IBM's 130 nm CMOS process occupied a total area of 0.97 mm² including the tank capacitors and the decoupling cap on the input supply. The die photo of the fully integrated converter is shown in Figure 1.23.

1.3.4   CC MEASUREMENT

The efficiency of the converter at fixed output voltages and different load currents is shown in Figure 1.24. The converter achieves a maximum efficiency of 70% for V_{in} = 1.3 V and V_{out} = 0.5 V. The max efficiency represents the efficiency of the converter without charge/discharge time modulation and full capacitance switching. When the state machine is enabled, depending on the load conditions an appropriate number of converter cores and charge/discharge time of the bucket capacitors are modulated to reduce the overcharging of the output node. Due to reduced

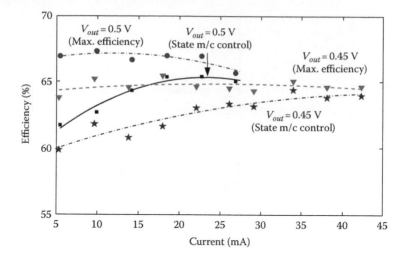

FIGURE 1.24 Overall efficiency of the converter.

overcharging, the frequency of switching increases slightly as compared to the case where the state machine is disabled. This results in a slight decrease in efficiency due to increased switching loses. With a smaller technology, switching losses will have a lower impact on the overall efficiency and the efficiency decrease due to state machine action is expected to be minimal.

The converter can also be manually controlled to set the number of converter cores to be used and the time for charging/discharging the bucket capacitors can be set to the desired value. Figure 1.25 shows the efficiency of the converter core neglecting the power consumed in the controller and drivers, when charge/discharge pulse width

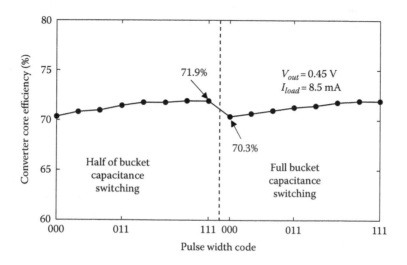

FIGURE 1.25 Core efficiency of the converter.

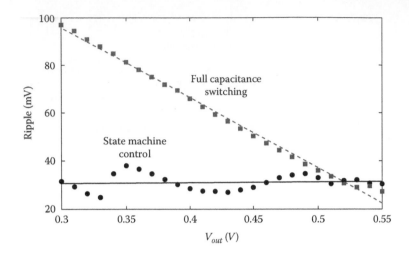

FIGURE 1.26 Variation of ripple with V_{out} for $I_{load} = 4$ mA.

and the bucket capacitance size is varied. Code 111 represents no time modulation and code 000 represents minimum pulse width for charging/discharging. The efficiency of the converter core remains fairly constant and varies between 70.3% and 71.9% over the entire range of pulse width modulation, which confirms our theory presented in Section 1.3.2. Figure 1.26 shows the variation of the ripple when V_{out} is varied from 0.3 to 0.55 V. When the full capacitance is switching without ripple control, the ripple varies nearly linearly with the output voltage. When the ripple control is enabled, the ripple is almost constant and is nearly independent of the output voltage.

Figure 1.27 shows the transient measurement of the output voltage when $V_{out} = 0.4$ V and $I_{load} = 2$ mA. The output voltage is AC coupled and hence contains no DC information. The ripple on the output voltage is 77 mV when all the converter cores are switching and no time modulation of charge/discharge pulse (Figure 1.27a). When only one of the converter cores is used for energy transfer, the ripple reduces to 45 mV (Figure 1.27b). In Figure 1.27c, the state machine decides the mode of operation based on the ripple. The ripple on the output voltage reduces to 27 mV, which is a 65% reduction compared to the case in Figure 1.27a. As discussed earlier, the transient measurement results show the increase in the switching frequency when the overcharging on the output is reduced.

1.3.5 CC SUMMARY

In this chapter, we have demonstrated a fully integrated CC in IBM CMOS 130 nm technology. The integrated converter uses two loops, primary for regulation and secondary for ripple control. The secondary loop in turn utilizes a dual pronged approach, which is all digital in nature to reduce ripple on the output voltage. This technique utilizes capacitance modulation to achieve coarse ripple reduction and uses partial charging/discharging of the bucket capacitors for further ripple mitigation to achieve a 65% ripple reduction at 0.4 V and 2 mA. The CC achieves a maximum efficiency of 70%.

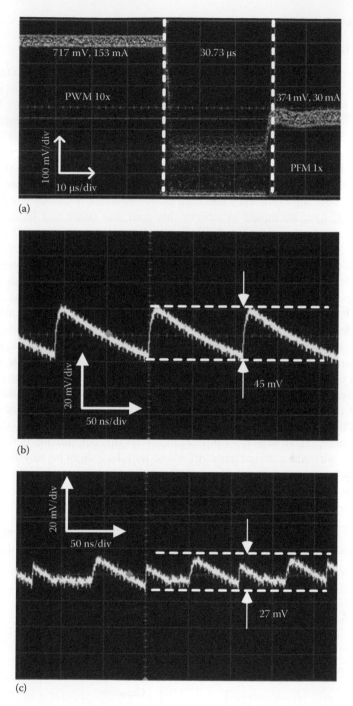

(a)

(b)

(c)

FIGURE 1.27 (a) Transient ripple response when all converter cores are working and are in no-time modulation mode. (b) Output voltage ripple when only one converter core is on. (c) State machine controlled output voltage ripple.

1.4 COMBINED HYBRID CONVERTER

1.4.1 INTRODUCTION

Power dissipation has become a major concern for device scaling and integration, particularly for battery-operated systems [1]. DVS has been one of the most efficient techniques to tackle this issue by using voltage as the lever to reduce power by up to 10^4 X [13]. Increased device leakage and the sluggish increase in battery energy density exasperates this problem. Multiple cores involving independent voltage domains address the leakage problem but require efficient and fully integrated DC–DC power conversion. Figure 1.28 shows the primary motivation for finding a fully integrated DC–DC conversion solution that is able to cater to a high-output power range while maintaining both high-efficiency and high-power densities. Previous works in this area [6,12,24] have attempted to achieve this goal, but either had limited output power ranges or displayed degraded efficiency at lower-output voltages. This chapter presents a parallel hybrid inductive and CC that is able to achieve high efficiencies over a wide power range for DVS systems where the CC caters to the lower-output powers and the inductive converter caters to the higher-output power needs.

We propose to reuse the area under the inductor for digital loads as only the top metal layers are used for the inductor. To check the feasibility of area-reuse concept, a separate test chip as a part of the internal research, fabricated in 130 nm bulk, with ring oscillators as representative digital circuits underneath the inductor was compared to an inductor only design on a high resistivity substrate to study the impact of digital circuits on the inductor and vice versa. The coupling from the inductor to digital circuits is less than −40 dB at 300 MHz, which is less than active circuits on the same substrate. The coupling from the digital to inductor is lower than −50 dB. The presence of digital circuits under the inductor slightly decreases the inductance [25]. However, for power converters this degradation is not significant. Hence, we make a design trade-off where we take a slight hit on efficiency for a considerable increase in area efficiency.

1.4.2 ARCHITECTURE

Figure 1.29 provides the motivation for combining the two converters to achieve higher efficiency across a wide output power range. Note that by changing the voltage

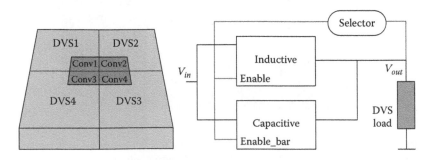

FIGURE 1.28 DC–DC converter per power domain for best efficiency.

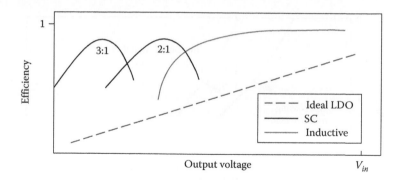

FIGURE 1.29 Desired efficiency profile for each dynamic voltage scaling system.

conversion ratio of CC, higher efficiency at low-power operation can be achieved. But the increase in capacitance value for higher-power outputs puts practical limitations on CCs for higher loads in DVS-based systems. On the other hand, the inductive converter can solve this problem for higher loads. Thus, a hybrid of both these converters can achieve a wide output power range.

The details of the hybrid converter are provided later (Figure 1.31). However, both converters are connected in parallel but only one is switched on at any one time. The switching between the converters is done when the efficiency achieved by one betters that achieved by the other. The switching between the converters is done on the basis of voltage sensing to optimize efficiency. This transition voltage can be estimated for a given load condition and is modeled in the following analysis. The switchover point between the two converters can be decided by considering when the efficiencies of both converters are equal. We know that at the time of switching $V_{out,cap} = V_{out,ind}$, thus the ratio of converter efficiencies, α, is given by

$$\alpha = \frac{\eta_{ind}}{\eta_{cap}} = \frac{I_{in,cap}}{I_{in,ind}} \tag{1.21}$$

From Equation 1.21, the switch from capacitive to inductive converter is done when the ratio α is greater than one and from inductive to capacitive when it is less than one. We can identify the transition voltage, V_{sw}, if we plug in the values of the two converter currents [5,6] in Equation 1.21:

$$\alpha = \frac{I_{in,cap}}{I_{in,ind}} = \frac{\dfrac{1}{k} \cdot \dfrac{V_{max} - V_{out}}{R_{out}}}{\dfrac{V_{out} \cdot I_{out}}{V_{in}} + \dfrac{I_{out}^2 \cdot (R_L + R_{Switch})}{V_{in}}} \tag{1.22}$$

This relation can be used to estimate V_{sw} for any kind of load. Figure 1.30 shows the variation of $\alpha = \eta_{ind}/\eta_{cap}$ with output voltage for our design for a 10 Ω load on the left and a digital load (DVS 25 mA at 0.5 V) on the right. A dotted line shows when $\alpha = 1$.

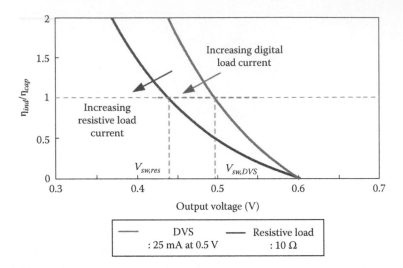

FIGURE 1.30 Transition voltage for digital and resistive loads based on efficiency cross-over criteria.

The transition voltage for the digital load in this plot is 0.5 V, and the transition voltage for the 10 Ω is 0.44 V. Increasing the resistive load current or the digital load current would move the optimal transition voltage lower as indicated in the figure.

1.4.2.1 Transition Circuits

The overall converter is shown in Figure 1.31 with the inductive on top and the capacitive on the bottom. The transition between the inductive converter and CCs is decided digitally (as explained in more detail later), and the effective circuits during the transition are shown in Figure 1.32. In capacitive mode, the filter capacitor of inductive part acts as tank capacitor, and in the inductive mode, the bucket capacitors of CC act as additional filter capacitor. We shall discuss the inductive converter next.

1.4.3 INDUCTIVE CONVERTER

The inductive converter is operational when the output voltage that the converter needs to support is high. The clocks to the digital controller are enabled and the switching action resumes. The inductive converter consists of the core converter unit and the controller that regulates the output voltage to the desired value as shown in Figure 1.31. The controller is a digital PWM (DPWM)-based controller with a time-to-digital conversion–based analog-to-digital converter (ADC) to digitize the error between the reference voltage and the output voltage. The error is then digitally integrated in an accumulator, the output of which forms the code for generation of the appropriate duty-cycle pulses in the DPWM generator. The details of the implementation control loop are discussed as follows.

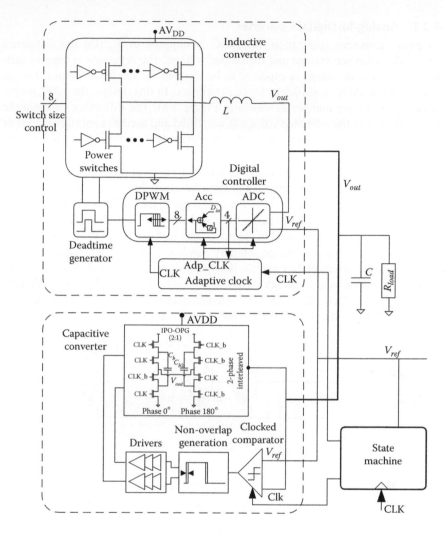

FIGURE 1.31 Overall hybrid capacitive/inductive converter.

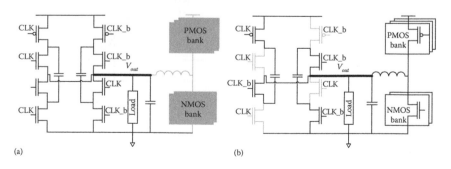

FIGURE 1.32 Equivalent structures of the hybrid converter in (a) capacitive mode and (b) inductive mode.

1.4.3.1 Analog-to-Digital Converter

In a power converter application, the ADC is required to digitize the difference between the reference voltage and the output voltage. The reference voltage is variable, and the output voltage is expected to be close to the reference voltage. Hence, a low-resolution ADC is sufficient for our purposes. In this design, the ADC output is 4-bit signed binary number as shown in Figure 1.33. The difference between the output voltage and the reference voltage is amplified and used to control the delay of two variable delay lines whose input is the reference clock signal. The delay lines are composed of a chain of current starved inverters whose delay is controlled by changing the current in these inverters. The output of one of the delay lines is used as a clock signal for D-flip-flops to take the snapshot of the other delay line. The other delay line is tapped at 16-equidistant nodes to form the data input to the D-flip-flops. Based on the delay difference between the two delay lines, a thermometric code is captured in the D-flip-flops, which is proportional to the difference in the voltage. This thermometric code is then converted to a 4-bit signed binary number. This technique requires a delay matching between the clock line and the data line when the voltage difference is zero. The data line nodes have some extra capacitive loading due to D-flip-flop, which are not present on the clock delay line nodes. Hence, dummy loads are added on this line to match the delay in the two lines when the control input is the same.

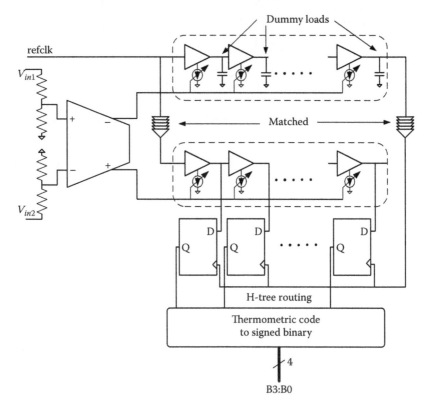

FIGURE 1.33 Analog-to-digital converter details used in the inductive control loop.

1.4.3.2 Accumulator

The ADC output is integrated in a digital integrator built using an accumulator shown in Figure 1.34 with overflow prevention logic. The 4-bit output of the ADC is sign extended to a 10-bit signed binary number and added to the previous data using a ripple carry summer.

1.4.3.3 Digital Pulse Width Modulator

The DPWM generation block receives the MSB 8-bits of the 10-bit accumulator output. The DPWM circuits consist of a D-flip-flop and a chain of delay cells to generate the appropriate delay as shown in Figure 1.4. The D-flip-flop is positive edge triggered. Here, the D-flip-flop data input is connected to V_{dd}. The same edge is propagated through a delay chain whose delay is controlled by the DPWM control word from the accumulator. The delayed edge from the delay line is then converted to a pulse and used to reset the D-flip flop to zero, thereby changing the duty cycle of the clock signal according to the DPWM control word. The delay cell consists of an inverter loaded with binary weighted capacitors connected to a switch to set the appropriate loading on the delay line.

1.4.3.4 Efficiency Improvement Techniques

Two important techniques have been used for efficiency improvement:

1. *Switch size scaling*: In order to reduce the switching losses, the size of the power switches is scaled according to the load conditions. The PMOS switch consists of six equal-sized cells each of width 2 mm and the corresponding drivers preceded by the mux. Hence, the switch size of the PMOS switches is varied from 2 to 12 mm in steps of 2 mm. Similarly, the NMOS switch is divided into two parallel cells of size 2 mm each, that is, the size of the NMOS switches is varied from 2 to 4 mm in steps of 2 mm.

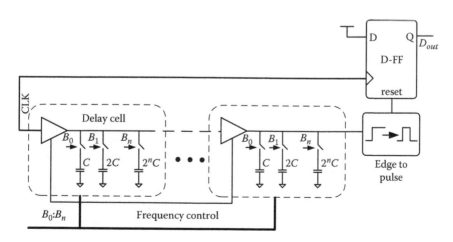

FIGURE 1.34 Accumulator used in control loop.

2. *Adaptive clock frequency*: The clock frequency of the ADC and the accumulator is reduced in the steady state with the DPWM switching frequency remaining unaltered. This is possible because in steady state the reference voltage and the output voltage are not expected to change significantly. The steady-state condition is detected by checking the ADC output, which is the difference between the reference voltage and the output voltage. When the error is less than a certain value, a clock whose frequency is 1/4th the frequency of the reference clock is selected with the normal clock frequency being selected when there are transients occurring that are likely to result in larger errors.

1.4.4 CAPACITIVE CONVERTER

For our DVS load, the switched CC is enabled by the state machine when the reference voltage is lower than 0.5 V. In this prototype, a single-mode (IPO-OPG) [5] CC, which is two-phase interleaved, has been implemented. The IPO-OPG configuration achieves a maximum conversion ratio of 2:1 ($V_{out} = V_{in}/2$). A single bound hysteretic controller achieves the regulation. In this type of control, the output voltage is compared with the reference voltage, and when the output voltage dips below the reference voltage, a switching action is initiated to transfer a quanta of charge from the supply to the output. In this implementation, single CC mode has been implemented but multiple modes can be easily implemented and the appropriate mode selected by the state machine based on the reference voltage.

1.4.5 HYBRID CONVERTER DESIGN

A state machine selects between the inductive converter and the CC based on the reference voltage with the inductive converter being selected when the reference voltage is greater than 0.5 V and CC otherwise. The state machine also turns off the converter that is not in use by gating off the clock to that particular converter. The combined converter is shown in Figure 1.31. The hybrid converter occupies a total area 0.4 mm² including the area occupied by the decoupling capacitors on the input supply. The filter capacitor is shared between the two converters. The input supply voltage of the converter is 1.2 V, and the converter supplies a maximum power of 140 mW.

1.4.6 HYBRID CONVERTER MEASUREMENTS

For steady-state efficiency measurements, the hybrid converter was stressed with two kinds of loads, resistive and digital, in both open- and closed-loop mode. On-chip MOS transistor banks were used as resistive loads, and a group of eight ROs under inductor acted as the representative digital load. Figure 1.35 shows the efficiency for resistive loads for both the converters. The CC achieves 71% closed-loop peak efficiency at 0.52 V, while the inductive converter shows 76.4% peak efficiency at 0.9 V for a resistive load of 20 Ω. V_{dd} was 1.2 V. The efficiency profile for the DVS-based system is characterized by two sets of measurements, one with

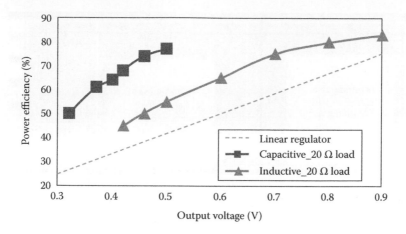

FIGURE 1.35   Efficiency versus output voltage for resistive loads.

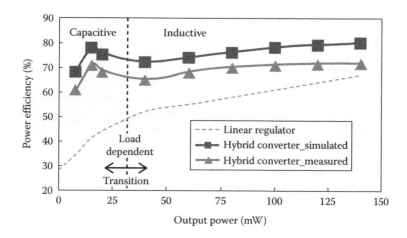

FIGURE 1.36   Efficiency versus power consumption for digital loads.

digital under inductor and the other without. Figure 1.36 shows the measured and measured efficiencies versus power from 2 to 140 mW (digital load). Two separate designs were included on-chip, one with digital under inductor and one without. The measured peak efficiency for the inductive converter was 74.3% peak without digital under inductor and 70.3% for digital under inductor. The slight decrease in efficiency can be attributed to the coupling losses between digital circuits and the inductor. Figure 1.37 shows the die photograph. The converter achieves a power density of 4.1 W/mm². Reusing the area occupied by the inductor by placing digital circuits underneath the inductor increases the power density of the converter to 4.1 W/mm² as shown in Table 1.2. The power density using this technique is the highest achieved to date.

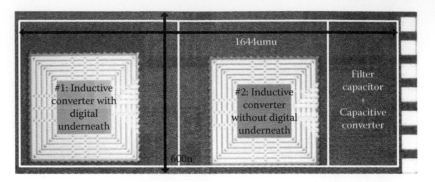

FIGURE 1.37 Die photograph.

TABLE 1.2
Design Summary

Parameter	Value
Technology	IBM 130 nm CMOS
Core area	1.13 mm^2
Area with decaps	1.59 mm^2
Peak achievable efficiency	77% (at 200 MHz and 8°C)
Peak operating efficiency	74.5% (at 300 MHz and room temp.)
Max o/p power	266 mW
Min o/p power	0.6 mW
Input voltage	1.2 V
Max output voltage	0.88 V
Min output voltage	0.3 V
Output voltage ripple at 750 mV (unfiltered)	28 mV
Output voltage ripple at 750 mV (filtered at 600 MHz)	16 mV
O/p power range	443x
Filter capacitor	5 nF
Inductor	2 nH

1.4.7 Hybrid Converter Summary

Using CCs to improve the efficiency at lower voltages enables 70X output power range (Table 1.3) while maintaining the efficiency above 61% even at loads as low as 2-3 mW. The efficiency at extremely small loads could easily be further improved by implementing multiple modes for the CC [5]. Reuse of area under the inductor boosts power density from 0.387 to 4.1 W/mm^2 at the cost of slight degradation in efficiency (<5%). Addition of a single thick metal layer optimized for inductors eliminates any potential power/ground bus congestion and is cheaper than off-chip magnetics.

TABLE 1.3
Performance Summary and Comparison Table

Parameter	Ref. [3]	Ref. [4]	This Work
Technology node	65 nm bulk	130 nm bulk	32 nm SOI
Converter type	Capacitive	Inductive	Hybrid
Power efficiency	73	77.9	76.4
Power density	0.19	0.21	$0.387 \rightarrow 4.10$
Ripple (mV)	<40	40	<30
Power range (mW)	253	315	2–140

1.5   CONCLUSIONS

In this chapter, we described three different kinds of switched-based converters. First, we looked at inductive converter that could achieve high-power efficiencies but fairs poorly at lower loads. Second, we dwelled on CC, which performs better at lighter loads but fairs poorly at higher loads due to increasing conduction losses. Finally, we presented a hybrid of inductive and CC for wide-range DVS-based applications. This converter maintained good efficiency across the both lighter and heavier loads.

REFERENCES

1. International Technology Roadmap for Semiconductors, Executive Summary, 2009.
2. T. D. Burd, T. A. Pering, A. J. Stratakos, and R. W. Brodersen, A dynamic voltage scaled microprocessor system, *IEEE Journal of Solid-State Circuits*, November 2000, 35(11), 342–351.
3. G. Perrucci, F. Fitzek, and J. Widmer, Survey on energy consumption entities on the smartphone platform, in *2011 IEEE 73rd Vehicular Technology Conference (VTC Spring)*, Budapest, Hungary, May 2011, pp. 1–6.
4. R. Harjani and S. Chaubey, A unified framework for capacitive series-parallel dc–dc converter design, in *2014 IEEE Proceedings of the Custom Integrated Circuits Conference (CICC)*, San Jose, CA, September 15–17, 2014, pp. 1–8.
5. S. Kudva and R. Harjani, Fully integrated capacitive converter with all digital ripple mitigation, in *2012 IEEE Custom Integrated Circuits Conference (CICC)*, San Jose, CA, September 9–12, 2012, pp. 1–4.
6. S. Kudva and R. Harjani, Fully-integrated on-chip dc-dc converter with a 450× output range, *IEEE Journal of Solid-State Circuits*, August 2011, 46(8), 1940–1951.
7. S. Kudva, S. Chaubey, and R. Harjani, High power-density, hybrid inductive/capacitive converter with area reuse for multi-domain DVS, in *2014 IEEE Proceedings of the Custom Integrated Circuits Conference (CICC)*, San Jose, CA, September 15–17, 2014, pp. 1–4.
8. J. M. Rabaey, A. Chandrakasan, and B. Nikolic, *Digital Integrated Circuits*. Prentice Hall of India Private Limited, Delhi, India, 2003.
9. T. Hattori, T. Irita, M. Ito, E. Yamamoto, H. Kato, G. Sado, Y. Yamada et al., A power management scheme controlling 20 power domains for a single-chip mobile processor, in *2006 IEEE International Solid-State Circuits Conference (ISSCC 2006)*, San Francisco, CA, February 6–9, 2006, pp. 2210–2219.

10. C. Isci, A. Buyuktosunoglu, C.-Y. Cher, P. Bose, and M. Martonosi, An analysis of efficient multi-core global power management policies: Maximizing performance for a given power budget, in *39th Annual IEEE/ACM International Symposium on Microarchitecture*, Orlando, FL, December 9–13, 2006, pp. 347–358.

11. R. Islam, A. Sabbavarapu, and R. Patel, Power reduction schemes in next generation Intel® ATOM™ processor based SoC for handheld applications, in *IEEE Symposium on VLSI Circuits (VLSIC)*, Honolulu, HI, June 16–18, 2010, pp. 173–174.

12. J. Wibben and R. Harjani, A high efficiency DC-DC converter using 2 nH integrated inductors, *IEEE Journal of Solid-State Circuits*, April 2008, 43(4), 844–854.

13. H. Kaul, M. Anders, S. Mathew, S. Hsu, R. Krishnamurthy, and S. Borkar, A 320 mV 56 μW 411GOPS/watt ultra-low voltage motion estimation accelerator in 65 nm CMOS, in *IEEE International Solid State Circuits Conference*, San Francisco, CA, 2008, pp. 316–317, 616.

14. B. Calhoun and A. P. Chandrakasan, Ultra-dynamic voltage scaling (UDVS) using sub-threshold operation and local voltage dithering, *IEEE Journal of Solid State Circuits*, January 2006, 41(1), 238–245.

15. Y. Pu, J. P. D. Gyvez, H. Corporaal, and Y. Ha, An ultra-low-energy multi-standard JPEG co-processor in 65 nm CMOS with sub/near threshold supply voltage, *IEEE Journal of Solid State Circuits*, March 2010, 45(3), 668–680.

16. I. J. Chang, J.-J. Kim, S. P. Park, and K. Roy, A 32 kb 10T sub-threshold SRAM array with bit-interleaving and differential read scheme in 90 nm CMOS, *IEEE Journal of Solid State Circuits*, February 2009, 44(2), 650–658.

17. R. Balczewski and R. Harjani, Capacitive voltage multipliers: A high efficiency method to generate multiple on-chip supply voltages, in *The 2001 IEEE International Symposium on Circuits and Systems*, Sydney, New South Wales, Australia, May 6–9, 2001, pp. 508–511.

18. R. Erickson and D. Maksimovic, *Fundamentals of Power Electronics*. Springer, Berlin, Germany, 2001.

19. S. Musunuri and P. L. Chapman, Optimization of CMOS transistors for low power DC-DC converters, in *IEEE 36th Power Electronics Specialists Conference*, Recife, Brazil, June 16, 2005, pp. 2151–2157.

20. D. Ma, W.-H. Ki, and C.-Y. Tsui, An integrated one-cycle control buck converter with adaptive output and dual loops for output error correction, *IEEE Journal of Solid State Circuits*, January 2004, 39(1), 140–149.

21. J. Xiao, A. V. Peterchev, J. Zhang, and S. R. Sanders, A 4 μA quiescent-current dual-mode digitally controlled buck converter IC for cellular phone applications, *IEEE Journal of Solid State Circuits*, December 2004, 39(12), 2342–2348.

22. A. Naveh, E. Rotem, A. Mendelson, S. Gochman, R. Chabukswar, K. Krishnan, and A. Kumar, Power and thermal management in the Intel® Core™ Duo processor, *Intel Technology Journal*, May 2006, 10(2), 110–122.

23. MOSIS wafer acceptance tests, http://www.mosis.com/cgi-bin/cgiwrap/umosis/swp/params/ibm-013/v02b_8rf_8l m_dm-params.txt.

24. H.-P. Le, S. Sanders, and E. Alon, Design techniques for fully integrated switched-capacitor dc-dc converters, *IEEE Journal of Solid-State Circuits*, September 2011, 46(9), 2120–2131.

25. C. Yue and S. Wong, On-chip spiral inductors with patterned ground shields for Si-based RF ICs, *IEEE Journal of Solid-State Circuits*, May 1998, 33(5), 743–752.

2 Single-Inductor Multiple-Output DC–DC Buck Converter

Ke-Horng Chen

CONTENTS

2.1 BASIC TOPOLOGY

Battery-operated electronic products have the trend of compact size, longer battery usage, and thinner form factor. It leads to the development of system-on-a-chip (SoC) because the SoC can merge different kinds of digital and analog circuits into one fabricated chip to enhance circuit performance and to reduce the printed-circuit board (PCB) area.

Generally speaking, the SoC is composed of many functional blocks including microprocessor, digital signal processor, random access memory, read-only memory, analog-to-digital and/or digital-to-analog converter, input/output control circuits, and today's key energy controlling module—power management unit (PMU) [1–22].

Conventional PMU merely provides only one constant supplying voltage to power the main functional circuit. Although basic power management uses on/off technique to effectively shut down any power loss, it has limitations because of constant supplying voltage. Instead of basic power management, distributive supplying voltage and current sources are requested to further reduce power loss and thus improve power conversion efficiency due to the progress of advanced fabrication technology. In particular, miniature smart mobile phones, digital cameras, MP3 players, and tablets require specified supplying voltage and current sources in different power-efficient functional submodules, which are designed by complex circuits with increased number of external components. Owing to the introduction of advanced PMU, most of the targeted requirements of battery-operated electronic products can be easily satisfied. However, design complexity in versatile distributive voltage and current sources and increased number of external passive components set the challenge of miniature form factor.

Figure 2.1 depicts the simplest design methodology to generate different supplying voltage sources where a switching regulator (SWR) operated as the first stage to provide the advantage of high-energy conversion efficiency [1–5]. If considering compact size and small volume, the second stage is composed of parallel low dropout (LDO) regulators to provide many distributive voltage sources [17–20]. Owing to the usage of LDO regulators, the PMU does not need to increase the number of external components but the power conversion efficiency is sacrificed. The efficiency becomes worse because large power dissipated on the power metal-oxide-semiconductor field-effect transistors (MOSFETs) if the voltage drop across the

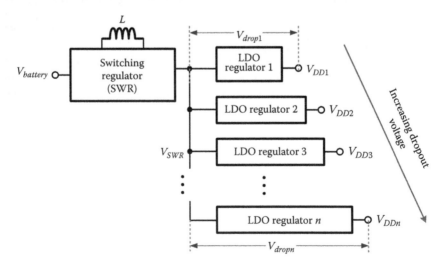

FIGURE 2.1 Conventional power management unit can provide distributive supplying voltage levels but power conversion efficiency is sacrificed.

second stage becomes large. In Figure 2.1, the dropout voltage V_{drop1} of the output voltage V_{DD1} continuously increases to V_{dropn} of the output voltage V_{DDn}, where V_{DD1} is larger than V_{DDn}. The battery usage time is seriously influenced by such a parallel LDO structure, although it can generate different output voltage levels easily.

Since the LDO regulator results in large power dissipation on its pass transistor, the power conversion efficiency of conventional PMU is hard to be improved. That is, to properly enhance the power conversion efficiency of the PMU for SoC applications, multiple SWRs need to be used for high-voltage conversion to get much improvement in power conversion efficiency compared to the counterpart LDO designs. High power conversion efficiency (PCE) can be guaranteed by the utilization of multiple SWRs if multi-output function from the input battery voltage $V_{battery}$ is implemented, as shown in Figure 2.2. However, in commercial and product views, three disadvantages are obvious. The first critical drawback is the cost of multiple off-chip inductors. This issue will cause the failure of this PMU idea. The second one is the large occupied PCB area due to multiple off-chip inductors. The last one is the serious effect of electromagnetic interference (EMI) caused by multiple SWRs. In conclusion, it is impossible to use this PMU idea in Figure 2.2 if the cost, the size, and the EMI problem of the inductor are not solved. That is why the commercial products have not provided such a design in battery-powered electronics because this methodology is not suitable for SoC integration.

The single-inductor dual-output (SIDO) converter and the single-inductor multiple-output (SIMO) converter have been proposed to provide dual or multiple outputs in the PMU with both high-power conversion efficiency and only one off-chip inductor utilization [23–36]. The basic structure of the SIMO DC–DC converter is shown in Figure 2.3. The single inductor can be charged through the distinct energy delivery paths from the input battery voltage $V_{battery}$. Then, the stored energy will be allocated to each distinct output according to their instantaneous energy demands. Different energy control techniques can decide the energy storage and delivery policy

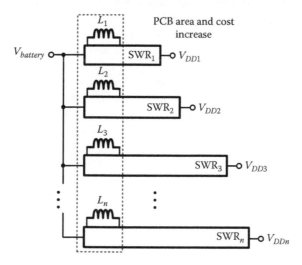

FIGURE 2.2 The idea of power management unit implemented by parallel switching regulators.

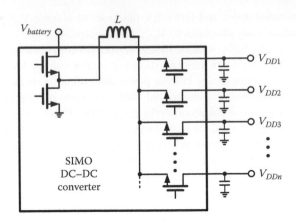

FIGURE 2.3   Basic structure of the single-inductor multiple-output DC–DC converter.

to ensure output voltage regulations through the use of only one off-chip inductor. High efficiency is maintained by the switching operation, not the LDO regulator, while area-efficient PMU is derived by the smart usage of only one external inductor. The SIMO DC–DC converter is one of the good candidates for the PMU designs in SoC applications.

2.1.1   Simplified Single-Inductor Dual-Output DC–DC Converter in the SoC

If the submodules in the SoC can be simply classified into analog and digital circuits, the simplified SIDO converter can provide suitable and independent supplying voltages to those submodules. However, if the demanded supplying voltage range is wide, the SIMO converter may be more suitable than the SIDO converter. Similarly, the reduction of one inductor is area-efficient and the PCB area is effectively reduced since conventional PMU requires two off-chip inductors to generate dual SWR output voltages [37–40]. Low-output voltage ripple, minimized cross-regulation, and high-power conversion efficiency are essential design issues for one SIDO converter. Figure 2.4 shows the proposed SIDO step-down converter in the SoC integration. The embedded power switches in the SIDO converter can deliver energy from $V_{battery}$ to both off-chip capacitors C_{OA} and C_{OB} through an off-chip inductor. Thus, the SIDO converter generates two output voltages, V_{OA} and V_{OB}, to supply the analog and the digital parts, respectively. That is to say, the PMU implemented by the SIDO converter in the SoC system can be properly guaranteed. However, some design challenges in the SIDO/SIMO converters should be carefully considered in order not to deteriorate the SoC performance. The well-known design challenge is the reduction of cross-regulation. When any of the outputs has sudden loading current change, cross-regulation occurs in the rest of the outputs because the accumulated inductor current occurs in the single off-chip inductor. Before reducing the cross-regulation, we need to realize why cross-regulation occurs in steady-state and transient response period.

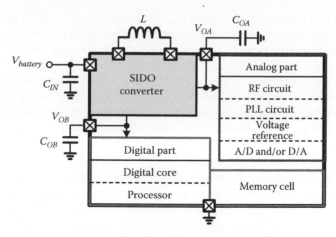

FIGURE 2.4   Brief illustration of the single-inductor dual-output step-down converter in system-on-a-chip integration can provide distinct analog and digital power supplies to analog and digital parts, respectively.

2.1.2   TRANSIENT AND STEADY-STATE CROSS-REGULATION

Typical cross-regulation problem in the SIDO converter is illustrated in Figure 2.5. Cross-regulation may occur both in transient and steady-state periods because of the accumulated energy in the single off-chip inductor. That is, in case of a loading current step ΔI_{OA} at the output voltage V_{OA}, for example, the V_{OA} naturally derives a voltage variation and needs time to be recovered back to its regulated value. However, unintended voltage variation also occurs at the other output voltage V_{OB}

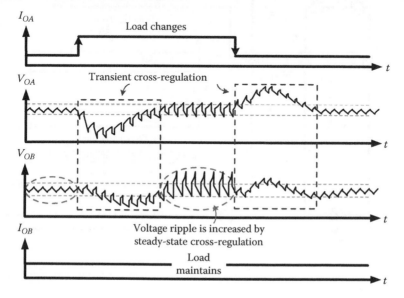

FIGURE 2.5   Illustration of cross-regulation in both transient response period and steady-state condition of the single-inductor dual-output converter.

if it maintains a constant load current I_{OB}. This occurrence is called the transient cross-regulation because the voltage variation at the victim output occurs during the transient response period. The transient cross-regulation resulted from the arranged energy delivery scheme during the load transient response. That is to say, the instantaneous load current variation breaks the balanced energy delivery scheme in steady state. The stored energy in the inductor no longer satisfies the sudden energy demand of the outputs since it is impossible to raise the inductor current level to its new balanced level within a short period of time. Therefore, the voltage drops occur in both output voltages. The voltage variation in the victim output will feedback the unbalance energy to the inductor later. It causes the chain effect in both outputs. Although this feedback can be controlled, the settling time of both outputs becomes longer than before. Transient cross-regulation may cause abnormal operation in the circuits supplied by the victim output voltage. The design challenge is to minimize the original cross-regulation source to enhance the supply quality in the PMU.

Moreover, the steady-state cross-regulation can be clearly observed when the two outputs experience a large loading current difference. When one of the two outputs operates at light loads and the other has to provide a large loading current, the light-load output would be overcharged if no suitable energy delivery scheme is applied on it. Furthermore, owing to discontinuous inductor current received at both outputs, the induced output voltage ripple increases due

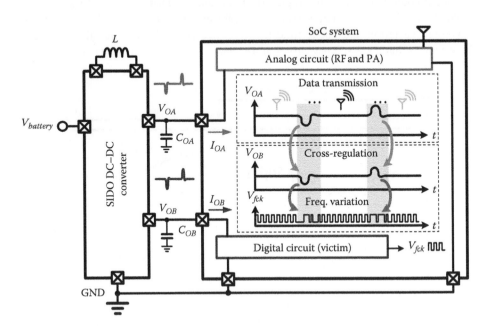

FIGURE 2.6 Detail description of the transient cross-regulation effect in the single-inductor dual-output converter and the performance deterioration in system-on-a-chip applications.

to the existence of equivalent series resistance (ESR) on each output capacitor. More seriously, the functions of some noise-sensitive circuits will be drastically affected. It is important to minimize cross-regulation to ensure low-voltage ripple and decreased transient voltage variation for low-voltage and high-performance SoC applications.

The transient cross-regulation in the SIDO converter will affect the PMU as well as the performance of the powered circuits in SoC applications. Figure 2.6 shows the transient cross-regulation effect of one step-down SIDO converter in SoC applications. The analog supplying voltage source V_{OA} powers the RF and PA circuits while the digital supplying voltage source V_{OB} supplies the digital circuits. During the data transmission period, large power requirement causes a sudden loading current increase at the I_{OA} so as to result in an expected voltage drop at V_{OA}. Simultaneously, the V_{OB} also suffers from the voltage variation even under constant loading conditions since the only inductor is disturbed and used to deliver energy to the loading change at the I_{OA}. The unwilling sudden voltage variation at the V_{OB} will deteriorate the performance of the digital circuits in SoC. The transient cross-regulation varies the frequency of signal V_{fck} and even induces an abnormal operation in the system processor or other digital circuits. How to minimize the cross-regulation in the SIDO converter becomes a necessary research topic.

2.2 BASIC PERFORMANCE DEFINITION IN PMU AND IN SIDO/SIMO CONVERTER

To fairly evaluate PMU performance, some commonly used design specifications are needed to be addressed in the summary, which contains both transient and steady-state emulative items. After reviewing the design specifications, the corresponding design specifications in the SIDO and SIMO converters are also identified to let the readers realize how to design high-performance SIDO and SIMO converters.

2.2.1 LOAD REGULATION

Load regulation indicates how loading current variation can influence the output voltage accuracy if the output voltage is settled down in steady state. The definition of load regulation is expressed as follows, which is used to evaluate the degree of output voltage variation under different loading current conditions:

$$\text{Load regulation} = \frac{\Delta V_{OUT}}{\Delta I_{OUT}} \text{ (V/A)} \qquad (2.1)$$

Smaller value of the load regulation carries out the higher precision of the output voltage level. The performance of the load regulation is related to the open-loop gain in the closed-loop modulated power management module. Therefore, to get better regulation, higher system loop gain is required if system stability is still guaranteed.

2.2.2 Line Regulation

Line regulation is also a steady-state performance to test the output accuracy under the perturbation of input voltage supply. The definition of line regulation is expressed in the following equation, which represents the dependence of the output voltage related to the input voltage variation:

$$\text{Line regulation} = \frac{\Delta V_{OUT}}{\Delta V_{IN}} \text{ (V/V)} \tag{2.2}$$

The smaller value of line regulation realizes the better voltage immunity of the input voltage. Consequently, better line regulation performance can lead to a robust operation because it stands for the ability against input supply variations. Actually, it becomes more and more important because the input supplying voltage is continuously decreased by the trend of advanced nanometer CMOS process. Any supplying voltage variation will lead to SoC performance deterioration. Thus, the PMU needs to have improved line regulation to guarantee supplying high-quality voltages for SoC applications.

2.2.3 Transient Voltage Variations

In the design of SIDO and SIMO converters in the PMU, only one inductor is used to provide energy to two and multiple outputs. It is important to have improved transient response performance even when distinct operation modes are used to get high performance and high efficiency in SoC applications. That is, the loading current condition for each power management module in the PMU is definitely distinct. A good transient response implies the derivation of small voltage drop and the fast transient settling time in each output voltage in case of any sudden loading current variation. If each output voltage can be rapidly settled down, the inductor can be reused to face another loading current variation from the rest of the outputs. Basically, system bandwidth affects the transient response performance. Wide system bandwidth indicates a shorter transient response time in the control loop. Transient response can be illustrated by frequency domain and time domain. In the frequency domain, the position of the poles and zeros will lead to distinct transient response. If the system dominant pole is set at higher frequencies, it results in larger system bandwidth as well as faster response time. However, it is sometimes hard to ensure the stability of the control loop. On the contrary, once the dominant pole is put at low frequencies, slower transient response time is derived. Nevertheless, a stable operation can be realized easily. The illustration of time-domain analysis in load transient response of one buck converter is shown in Figure 2.7. It contains the buck converter with its controller, off-chip inductor, L; feedback voltage divider, R_1 and R_2; output capacitor, C_{OUT}; and the equivalent series resistor of the capacitor C_{OUT}, R_{ESR}. The output voltage waveform of the load transient response is also shown in Figure 2.8. When the load transient response occurs in case of light-to-heavy loading current change, the transient period of t_1 is mainly determined by the system bandwidth, and the drop voltage V_{drop} is related to the value of R_{ESR}. The response of the system

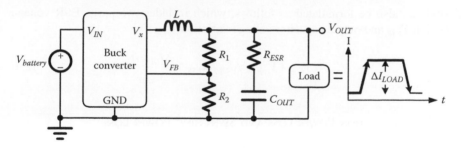

FIGURE 2.7   Illustration of time-domain analysis for the load transient response in a buck converter.

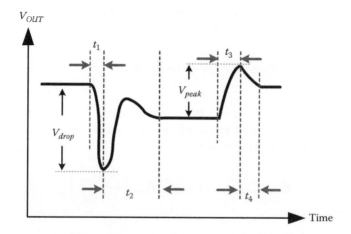

FIGURE 2.8   Output voltage waveform in the load transient response of one buck converter.

control loop is slower than the speed of loading current increase, so that the duty cycle of the buck converter cannot be extended immediately. Insufficient energy will lead to a sudden large voltage drop. The output capacitor then acts as one current source to sustain the request of the output load current. The voltage drop V_{drop} in the load transient response is formulated as

$$V_{drop} = V_{ESR} + V_{cap} = \Delta I_{OUT} R_{ESR} + \frac{\Delta I_{OUT} \cdot t_1}{C_{OUT}} \qquad (2.3)$$

The transient period of t_2 is determined by the time that the high-side power switch would charge the output capacitor to its regulated voltage. It also affected the transient settling time, which is related to the phase margin of the control loop. The total response time including t_1 and t_2 is known as the transient recovery time. Similarly, the voltage variation V_{peak} is derived when the output load is suddenly changed from heavy to light loading current condition. The determination for the transient period of t_3 and t_4 is similar to that of t_1 and t_2, respectively. The voltage V_{peak} in the transient response period can also be formulated as follows, which includes two parts, ESR voltage variation V_{ESR} and output capacitor voltage variation V_{cap}:

$$V_{peak} = V_{ESR} + V_{cap} = \Delta I_{OUT} R_{ESR} + \frac{\Delta I_{OUT} \cdot t_3}{C_{OUT}} \tag{2.4}$$

2.2.4 CONDUCTION POWER LOSS AND SWITCHING POWER LOSS

Due to the large driving current at the power stage in the PMU, the conduction power loss $P_{conduction\_loss\_H}$ obviously affects the power conversion efficiency at heavy loads. As we know, the on-resistance of the power switch is inversely proportional to the size of the power transistor. The conduction power loss of the high-side power switch in the buck converter can be expressed as follows, where the on-resistance R_{onp} is decided by the high-side power switch:

$$P_{conduction\_loss\_H} = (I_{OUT})^2 \cdot R_{onp} \cdot \frac{V_{OUT}}{V_{IN}} \tag{2.5}$$

Similarly, the conduction power loss $P_{conduction\_loss\_L}$ of the low-side power switch in the buck converter is expressed in the following equation, where the on-resistance R_{onn} is decided by the low-side power switch:

$$P_{conduction\_loss\_L} = (I_{OUT})^2 \cdot R_{onn} \cdot \left(1 - \frac{V_{OUT}}{V_{IN}}\right) \tag{2.6}$$

One trade-off exists between the conduction power loss and the active silicon area. Since the utilization of large power switches can directly reduce the conduction power loss at the power stage, the cost is also increased due to the occupation of the large silicon area.

Switching power loss results from power dissipation during the switching interval of the power switches. As depicted in Figure 2.9, the switching loss is proportional to

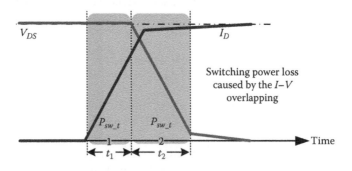

FIGURE 2.9 Illustration of the switching power loss of the power switch during the *I–V* overlapping region.

the input voltage $V_{battery}$ and the loading current I_{OUT} delivered by the buck converter. When the power switches are turned on, the driving current gradually increases while the voltage across the drain and source of the power switch decreases correspondingly. Consequently, the intersection of the conducting current and the voltage drop across the power switch causes the switching power loss during the switching period. During the period of t_1 in the switching operation of the power switch, the switching power loss P_{sw_t1} can be obtained as

$$P_{sw\_t1} = \frac{1}{2}(V_{IN} \cdot I_{LOAD}) \cdot t_1 \qquad (2.7)$$

When the conducting current rises to its target value, the current can be sustained while the drop voltage across the power switch starts to decrease. Similarly, the switching loss can be derived during the period of t_2 in the switching operation as

$$P_{sw\_t2} = \frac{1}{2}(V_{IN} \cdot I_{LOAD}) \cdot t_2 \qquad (2.8)$$

In summary, total switching power loss P_{sw} in one switching operation of one power switch is the summation of P_{sw_t1} and P_{sw_t2}. If the switching frequency f_{sw} is constant, P_{sw} is expressed by the product of the battery voltage $V_{battery}$ and the output loading current I_{LOAD} in the following equation, which is scaling by the summation of t_1 and t_2:

$$P_{sw} = P_{sw\_t1} + P_{sw\_t2} = (V_{battery} \cdot I_{LOAD}) \cdot (t_1 + t_2) \cdot f_{sw} \qquad (2.9)$$

Generally speaking, smaller summation of t_1 and t_2 can result in smaller switching power loss but it will cause a large EMI effect, which is not acceptable in today's commercial products. Another trade-off exists between switching power loss and the EMI effect.

2.2.5 POWER CONVERSION EFFICIENCY

Power conversion efficiency is the most important design issue especially in the PMU designs for battery-operated electronics. The power conversion efficiency formulated as follows is defined as the ratio of the providing output power to the total power received from the input battery voltage source:

$$\eta = \frac{P_{OUT}}{P_{IN}} = \frac{P_{OUT}}{P_{OUT} + P_{conduction} + P_{sw} + P_Q + P_{others}} \qquad (2.10)$$

The total power received from the input battery voltage source contains the following items including the output power P_{OUT}, the conduction power loss at the power stage $P_{conduction}$, the switching power loss at the power stage, P_{sw}, the quiescent current of the control circuit P_Q, and the leakage power loss P_{others} due to parasitic elements in realistic silicon fabrication. Therefore, any power reduction technique for one or more power loss items can improve the overall power conversion efficiency. Especially, it is important to present high-power conversion efficiency in SIDO and SIMO converters in the PMU designs for the SoC. In the following subsection, the design goal of the SIDO/SIMO converter is presented to show how to get a good design. After reviewing the goal, we can check the state of the art to see their advantages and disadvantages simultaneously according to the summarized design goals.

2.3   DESIGN GOAL OF THE SIDO/SIMO CONVERTER

As the mentioned characteristics of SIDO and SIMO converters show, the major advantage of those converter in the PMU design for the SoC is the compact size due to the usage of only one external off-chip inductor. However, if considering other characteristics, it is easy to conclude that the PMU with multiple SWRs has better performance than that with SIDO and SIMO converters. Therefore, it becomes more important to improve those characteristics of SIDO and SIMO converters. We can review each design specification on how to get the improvement.

2.3.1   REDUCTION OF THE NUMBER OF POWER SWITCHES

The numbers of the power switches should be minimized to get the cost reduction in the silicon area since a large number of power switches occupy a large silicon area. If considering the reduction of both conduction loss and switching loss at the power stage to enhance the power conversion efficiency, the energy delivery paths at the power stage of the SIDO converter need to be well arranged. That is to say, minimizing the number of the power switches simply for cost reduction may cause an increase in power loss. A trade-off may exist between the cost reduction in the silicon area and the reduction of power loss. The readers should consider both at the same time. One freewheel power switch is sometimes used for easy stability consideration [23]. However, it causes the increase of both the cost and the power loss. In other words, the reduction of freewheel power switch can reduce the cost and the power loss. It is meaningful to do the reduction of the number of power switches.

2.3.2   REDUCTION CONDUCTION POWER LOSS AND SWITCHING POWER LOSS

Power switches at the power stage are responsible for delivering the energy to dual or multiple outputs in the SIDO/SIMO converter. Since the current flowing through these power switches is large, the size of the power switches needs to be large, too. Nevertheless, the equivalent resistance of the power switches can result in the unwilling conduction power loss. It is related to the size of the power switch and the current level of the inductor. Moreover, the switching power loss due to charging/discharging the power switches occurs every switching cycle. Due to the switching type power converter, the power switches can be turned on/off repeatedly in order to realize distinct energy delivery paths in the SIDO/SIMO converter. The switching power loss is derived at the instantaneous transition period of power switches. As a result, to enhance the full power conversion efficiency in the SIDO/SIMO converter, both conduction power loss and the switching power loss need to be reduced by proper utilization of the energy delivery paths at the power stage.

2.3.3 Extension of Output Loading Current Range

The PMU has to cover all the possible loading current ranges to guarantee correct operation functions in SoC applications. That is, the restriction of the output loading between dual and multiple outputs must be eliminated. Thus, the SIDO/SIMO converter has to correctly operate for delivering the energy to dual or multiple outputs over a wide loading current range of any output. Furthermore, the operation of the SIDO/SIMO converter still needs to be ensured even if there is a large loading current difference between any two or more outputs.

2.3.4 Improvement of Load Transient Response

Load transient response is the basic requirement of the power management design. Since the output loading current may be changed in each of the outputs suddenly, the SIDO/SIMO converter is demanded to have the capability to recover the output voltages from the overshoot or undershoot voltage conditions. In addition, the load transient response is used to demonstrate the system stability to further guarantee a stable operation. Thus, the load transient is also an important design goal in SIDO and SIMO converter designs.

2.3.5 Reduction of Output Voltage Ripple

Since the SIDO/SIMO converter has to power the subcircuits in the SoC system, the quality of the output voltages needs to be guaranteed. Because of the noise-sensitive subcircuits in the SoC system, the output voltage ripple of the power management module is concerned. Consequently, the smaller output voltage ripple can help realize the higher supply quality of the SIDO/SIMO converter.

2.3.6 Reduction of Transient and Steady-State Cross-Regulations

Cross-regulation is a specific problem that can only be derived at the SIDO/SIMO converter structure. It is because the utilization of a single inductor has to achieve the energy delivery operation for the dual outputs. However, since the inductor current condition cannot be changed rapidly, the sudden variation at one output of the SIDO/SIMO converter can result in the interference of the energy delivery functions for the rest of the outputs. That is, obvious voltage variation appears at the output even if its output loading current condition is unchanged. Therefore, the SIDO/SIMO converter has to use advanced control methodology so as to realize the independent output voltages.

2.3.7 IMPROVEMENT OF POWER CONVERSION EFFICIENCY

Power conversion efficiency is always the most important design issue in the power management module design. Since the DC–DC converter aims to provide a stable power source for supplying the subcircuits in SoC applications, the power dissipations need to be reduced for obtaining higher-power conversion efficiency. Moreover, as the trend of battery-operated portable electronics and green energy becomes important, power conversion efficiency is a critical design issue in the PMU designs.

2.4 STATE-OF-THE-ART SIDO/SIMO CONTROL METHODOLOGIES

SIDO or SIMO converters can provide two or multiple output supplying voltage sources for the SoC system by utilizing only one off-chip inductor. However, the energy delivery schemes at the power stage need to be considered carefully in order to guarantee an adequate energy supply for the dual and multiple outputs so as to minimize cross-regulation. Furthermore, the dc level of the inductor current has to satisfy the total loading conditions of the outputs. There are some prior representative design topologies for the SIDO/SIMO converter, which aim to achieve proper operation functions. Here, we discuss and compare state-of-the-art design methodologies to realize their advantages and disadvantages.

2.4.1 PSEUDO-CONTINUOUS CONDUCTION MODE

To minimize the cross-regulation in the SIDO/SIMO converter, the energy delivery function of each output must be designed independently. The discontinuous conduction mode (DCM) control method can ensure the inductor current to be set to zero at the end of the pulse width modulation (PWM) switching cycle. Similarly, in the pseudo-continuous conduction mode (PCCM) operation [23], the inductor current returns back to the dc level I_{dc}, which is derived in the beginning of the PWM switching cycle. Figure 2.10 shows a brief illustration of the classic PCCM control method in the SIDO/SIMO DC–DC converter. With the implementation of freewheel switch, S_f, freewheel period is always activated at the end of each PWM switching cycle. That is, the inductor will be charged in the beginning of each PWM switching cycle. In sequence, the stored energy is delivered to one of the multiple outputs by the controller. After sufficient

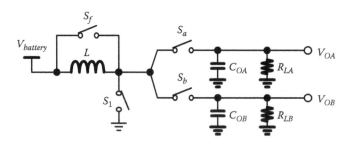

FIGURE 2.10 Brief illustration of the pseudo-continuous conduction mode control technique in single-inductor dual-output/single-inductor multiple-output DC–DC converter. (From Ma, D. et al., *IEEE J. Solid-State Circuits*, 38(6), 1007, 2003.)

energy has been transferred from the charged inductor to the specific output, the free-wheel stage can be activated by turning on the freewheeling switch S_f in order not to deliver extra energy to cause overshoot voltage.

The freewheel stage can reset the inductor current to a predefined dc level, which can be viewed as one constant inductor current level. Owing to the disappearance of the inductor current information at the end of each switching period, the system order can be reduced to one similar to the DCM operation. Consequently, the compensation for the PCCM operation becomes simpler. Proportional compensator or proportional-integral (PI) compensator can be used to increase system stability. The cross-regulation can be minimized because the inductor current is always reset to the I_{dc} so as not to affect the rest of the outputs. Figure 2.11 shows the timing diagram of the PCCM control in the SIDO/SIMO DC–DC converter where the I_{dc} is predefined if considering the heavy loading current condition. Besides, the I_{dc} cannot be adjusted dynamically similar to that in the CCM operation.

Some obvious disadvantages can be found. The first crucial drawback is the high conduction power loss caused by the setting of I_{dc}. In other words, the advantage of easy compensation contributed by the I_{dc} incurs the big problem in the efficiency. Actually, the value I_{dc} should be high enough to meet the heavy loading current condition and thus the system stability can be guaranteed for the whole loading current range. High I_{dc} value seriously deteriorates the efficiency because the inductor current will flow through each power switch and thus the conduction power loss increases. We can conclude that the efficiency is greatly sacrificed at light loads if we simply want to simplify the compensation network.

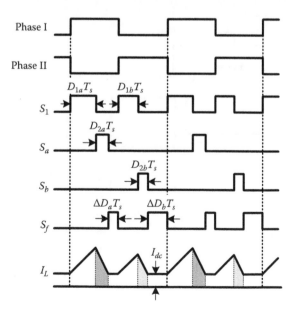

FIGURE 2.11 Timing diagram of the pseudo-continuous conduction mode control in single-inductor dual-output/single-inductor multiple-output DC–DC converter. (From Ma, D. et al., *IEEE J. Solid-State Circuits*, 38(6), 1007, 2003.)

Furthermore, the size of freewheeling power switch is proportional to the I_{dc}. As a result, the size of the freewheeling power switch is large and the silicon cost increases drastically. Some prior arts said that its size can be one-twentieth of the main power switches. However, it was not correct because the I_{dc} will decrease to zero during the freewheeling period. The PCCM operation degenerates to the DCM operation but the efficiency also degrades. The output voltage ripple of the PCCM is similar to that of the DCM because the peak inductor current is determined by the loading current. Consequently, the output voltage ripple increases drastically at heavy loads because the inductor current ripple can be kept constant similar to that in the CCM.

In conclusion, the PCCM has large power loss with the inadequate power conversion efficiency that should be improved in order to meet the requirement for portable electronics nowadays. The freewheel stage needs to be removed and the inductor needs to be lowered down for high efficiency while the compensation skill still needs to be simplified.

2.4.2 ORDERED POWER-DISTRIBUTIVE CONTROL

To simplify the energy delivery control for multiple outputs, the ordered power-distributive control technique has been proposed in the SIMO converter to achieve the proper energy distribution without using the freewheeling power switch. Figure 2.12 shows the structure of the ordered power-distributive control in the SIMO converter. In this example, there are five distinct output voltages in the SIMO converter. Four outputs, V_{o1}–V_{o4}, provide four different positive output voltages while the output V_{oN} generates

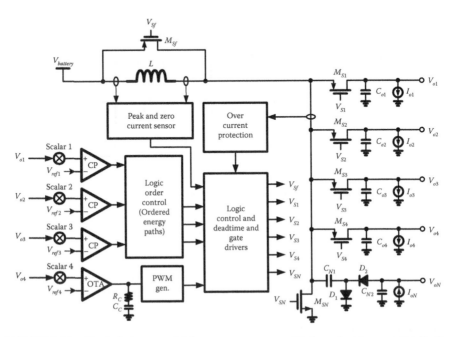

FIGURE 2.12 Single-inductor multiple-output converter with the ordered power-distributive control technique. (From Le, H.-P. et al., *IEEE J. Solid-State Circuits*, 42(12), 2076, 2007.)

one negative output voltage with the help of one charge pump circuit. Also, the power switch S_X is used to provide the energy charging path for the inductor in the beginning of each PWM switching cycle. The switches, S_1 to S_4, can deliver the energy to the output of V_{o1} to V_{o4}, respectively. The negative output voltage is realized with the help of one charge pump structure, which is achieved along with the energy delivery paths of the SIMO converter [24]. Freewheel stage operation can also be activated by turning on the switch S_f to guarantee the stable operation of the SIMO converter. The output voltage of V_{o1} is set as the highest one that is used to drive the buck voltages of the p-type power switches. The output voltage regulations of V_{o1} to V_{o3} are achieved by using three comparators C_P, and the closed-loop error modulation is only left with the output of V_{o4}. This control methodology can achieve the multi-output energy distribution in the SIMO converter with five distinct output voltage levels. Also, the compensation network is reduced to only one similar to the original single-output buck converter.

The illustration of the ordered power-distributive control is shown in Figure 2.13. Because of the PWM operation, the inductor current will be charged to the peak value, which is determined by the linear closed loop in the beginning of each switching cycle. When the inductor current reaches the peak level, the inductor starts to discharge by the comparator-based control. The energy will be delivered to each output voltage, V_{o1} to V_{o4}, in sequence. The rest of the energy transferring to V_{o4} will affect its voltage level and thus the linear closed loop can be established for accurate energy control. That is, the operational amplifier implemented at the feedback path of V_{o4} can adjust the inductor peak current level at the next PWM switching cycle to ensure the proper energy allocation in this ordered power-distributive energy-controlled SIMO converter. Moreover, the utilization of freewheel stage can generate a buffer period with the different PWM switching cycle for ensuing the stable operation similar to the PCCM. But it was not designed large to tolerate large inductor current.

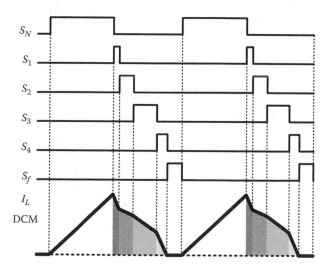

FIGURE 2.13 Illustration of the ordered power-distributive control in single-inductor multiple-output DC–DC converter. (From Le, H.-P. et al., *IEEE J. Solid-State Circuits*, 42(12), 2076, 2007.)

The ordered power-distributive control technique improved some drawbacks in the PCCM. Obviously, it introduces some side effects in the performance degradation. The first shortcoming is the large output voltage ripple caused by the comparator-based control technique, although the compensation network is simplified. The second critical problem is the high inductor current because the charging period of this technique is established by turning on the low-side n-type power switch. There is no energy delivery path during this period. Thus, it results in a high inductor current level. Without addressing the drawbacks of high inductor current, the readers can realize that the demand of reducing inductor current is an important challenge in SIDO and SIMO converters. Furthermore, the driving capability is not large since the inductor current cannot offer all of the heavy load condition with the five outputs simultaneously. As a result, the ordered power-distributive control technique cannot be utilized in SoC applications because of its inevitable restrictions.

2.4.3 Charge-Reservation Control

Another control scheme for the SIDO/SIMO DC–DC converter is the charge-reservation methodology. A small capacitor can be implemented on-chip to monitor the charge condition of output capacitor [25]. Figure 2.14 shows the structure of the SIDO converter with the charge-reservation control. This SIDO converter achieves

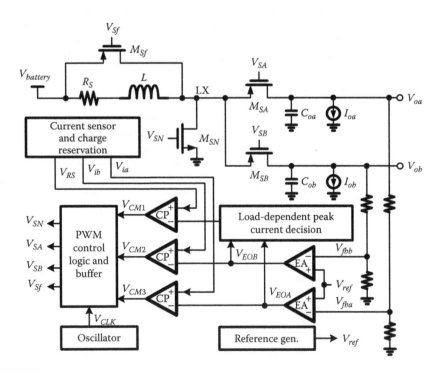

FIGURE 2.14 Single-inductor dual-output converter with the arranged energy paths and the charge-reservation control technique for one buck and one boost output voltages. (From Huang, M.-H. and Chen, K.-H., *IEEE J. Solid-State Circuits*, 44(4), 1099, 2009.)

dual buck and boost outputs with only one single inductor utilization. In order to minimize the number of power switches for obtaining the silicon cost saving, the energy delivery paths for dual outputs must be arranged carefully. There are three main power switches and one freewheel switch in the power stage. Both output conditions can be fed back to the controller by error amplifiers. The full-range current sensing circuit is used to derive the complete inductor current information to achieve the duty cycle modulation with the charge-reservation methodology. The control logic can generate the control signals for the power switches. This technique implements the similar current mode control for low-output voltage ripple. Although both two feedback paths need a PI compensator, the current-mode PI compensator can be used to simplify and reduce the design and the cost.

Figure 2.15 shows the energy delivery paths of the inductor current to deliver energy from input battery voltage source to both outputs. Path 1 delivers energy to the buck output and the inductor current slope is positive. After it reaches the satisfied current level for the buck output, energy path will switch to path 3 to charge the inductor current to the desired peak value, which is determined by both feedback paths. Thus, the inductor is always kept below the calculated peak value. After path 3, the stored energy starts to transfer to the boost output through path 2. Finally, the freewheel operation appears at the end of the switching cycle until the next PWM switching cycle starts. Therefore, the energy distribution for the dual outputs can be well arranged. On the other hand, although this structure can minimize the number of power switches, at least the boost output needs to have a larger loading current compared to the buck output for getting the negative inductor current slope. That is to say, the boost operation is used to release the inductor current when the discharging operation is activated. However, this topology is not suitable in the PMU designs for the SoC because the loading current criteria cannot be met all the time. In other words, the low-side power switch cannot be removed simply

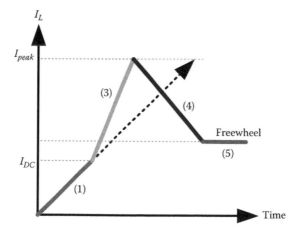

FIGURE 2.15 The energy delivery methodology for the single-inductor dual-output converter with one buck and one boost output voltages. (From Huang, M.-H. and Chen, K.-H., *IEEE J. Solid-State Circuits*, 44(4), 1099, 2009.)

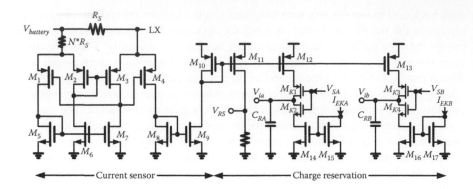

FIGURE 2.16 Implementation of the charge-reservation control methodology. (From Huang, M.-H. and Chen, K.-H., *IEEE J. Solid-State Circuits*, 44(4), 1099, 2009.)

for cost reduction because it will lead to stability problem if considering all over-loading current ranges.

The main idea of charge-reservation control methodology in Figure 2.16 is to determine the value of inductor peak current for getting a precise energy storage in the inductor. It is necessary to sense the full-range inductor current information to the controller by the full-range current sensor. One sensing resistor R_S is used to get the full-range inductor current information. R_S is also included in the freewheeling period in Figure 2.14 to ensure the full-range sensing operation.

In the charge-reservation circuit, the sensed inductor current is used to charge the small on-chip capacitors, which is used to monitor the charge conditions of all output capacitors. For example, when the energy is transferred to the output V_{OA}, the monitoring current is charged to the capacitor C_{RA} to reflect the charge condition for getting the precise energy control at V_{OA}. Similarly, the capacitor C_{RB} is charged as the energy is delivered to the output of V_{OB}. Moreover, the loading current condition can be reflected by the scaled down currents I_{EKA} and I_{EKB} where the gate control signals V_{SA} and V_{SB} are used to decide the charging or discharging for C_{RA} and C_{RB}, respectively.

However, the mismatch of the current sensing operation may lead to the inaccurate duty cycle determination of the charge-reservation control methodology. Furthermore, the effect of the parasitic components at the output energy delivery path cannot be taken into consideration through the charge-reservation control because only C_{RA} and C_{RB} are taken into consideration. Therefore, the precise duty modulation determination for the proper energy delivery scheme is hard to be achieved in this design.

2.4.4 ENERGY CONSERVATION MODE CONTROL

To achieve the power management integration in 65 nm SoC applications, the SIDO converter needs to use 65 nm core devices and to design the low-voltage controller. The SIDO converter with the low-voltage energy distribution controller (LV-EDC) is shown Figure 2.17 [34]. The power switches, which are implemented by the thick oxide devices in 65 nm CMOS technology, can tolerate the voltage range of the

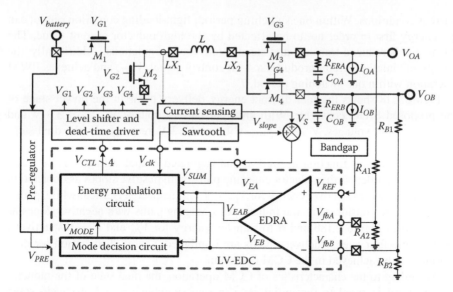

FIGURE 2.17 Single-inductor dual-output step-down DC–DC converter with the low-voltage energy distribution controller. (From Lee, Y.-H. et al., An interleaving energy-conservation mode (IECM) control in single-inductor dual-output (SIDO) step-down converters with 91% peak efficiency, in *Proceedings of IEEE Symposium on VLSI Circuits*, Honolulu, HI, June 16–18, 2010, pp. 57–58.)

battery-powered input voltage source. The voltage dividers are carried out by R_{A1}, R_{A2}, R_{B1}, and R_{B2} to feed output voltage information to the LV-EDC for monitoring output loading current conditions. The current sensing circuit can derive inductor current information by detecting the voltage variation at the switching node LX_1. To achieve a robust current-programmed control, the summing signal V_{SUM} is composed of the slope compensation signal V_{slope} and the current sensing signal V_S to avoid subharmonic oscillation [38].

The advantage of power saving and small silicon area is due to the LV-EDC circuit operating under 1 V low voltage with the use of 65 nm core devices. Low-voltage supply V_{PRE} for the LV-EDC circuit is supplied by the high efficient pre-regulator, which is implemented by a switched-capacitor (SC) circuit cascading with one LDO regulator. The energy distribution regulation amplifier (EDRA), which is used to reflect the output load conditions, generates the error signals V_{EA} and V_{EB} for V_{OA} and V_{OB}, respectively. The EDRA is composed of the compensation enhancement multistage amplifier to ensure low-voltage operation, high-voltage gain, and on-chip compensation [34]. The V_{EAB} generated by the summation of V_{EA} and V_{EB} represents the overall energy demand of dual outputs and thus can indicate the peak inductor current level [34]. The intersections of V_{SUM} with V_{EA}, V_{EB}, as well as V_{EAB}, decide energy modulation by the current-programmed control to guarantee low-output voltage ripple and good transient response because dual outputs can obtain energy in each PWM switching cycle. The mode decision circuit in the LV-EDC helps determine energy operation modes through the signal, V_{MODE}, in accordance with loading current conditions. Within one switching period, light-loading condition output can get energy first in order not to be affected by the high inductor current value. The advantage is reduced transient and steady-state cross-regulations. Additionally, the energy modulation circuit produces a 4-bit control signal, V_{CTL}, and achieves PWM operation within one switching period, T_S.

Figure 2.18 shows the four distinct energy delivery paths in the power stage of the proposed SIDO converter. With an input battery voltage $V_{battery}$ of 3.3 V and nominal output voltages, V_{OA} and V_{OB}, of 1.8 and 1.2 V, respectively, the dual step-down operation is derived. Four distinct energy paths can be composed to guarantee the energy transfer function as well as the regulated output voltages. Path 1 and path 3 are regarded as the inductor charging paths with positive slopes as depicted in Figure 2.18a and c to deliver energy for V_{OA} and V_{OB}, respectively. Meanwhile, path 2 and path 4 are considered as the inductor discharging paths with negative slopes as illustrated in Figure 2.18b and d to transfer energy for V_{OA} and V_{OB}, respectively. When these four energy delivery paths are combined, energy distribution for dual outputs can be achieved in the CCM operation.

According to the characteristics of CCM operation, the final state of the inductor current level is equal to the initial state in one switching period at steady state.

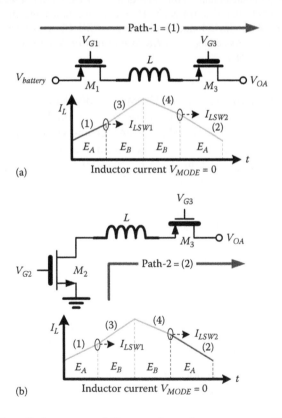

(a)

(b)

FIGURE 2.18 Four distinct energy delivery paths at the power stage. (a) Inductor current charging path for V_{OA}. (b) Inductor current discharging path for V_{OA}. *(Continued)*

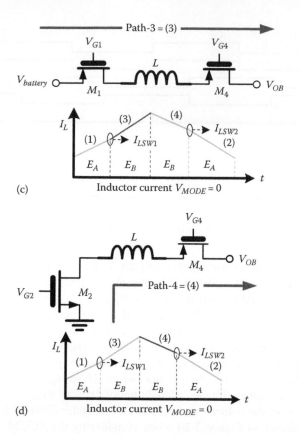

(c)

(d)

FIGURE 2.18 (Continued) Four distinct energy delivery paths at the power stage. (c) Inductor current charging path for V_{OB}. (d) Inductor current discharging path for V_{OB}. (From Lee, Y.-H. et al., An interleaving energy-conservation mode (IECM) control in single-inductor dual-output (SIDO) step-down converters with 91% peak efficiency, in *Proceedings of IEEE Symposium on VLSI Circuits*, Honolulu, HI, June 16–18, 2010, pp. 57–58.)

However, the inductor current waveform, as depicted in Figure 2.18, is the specific one combination of the energy delivery paths in steady state, which is controlled by mode decision signal V_{MODE}. The V_{MODE} signal determines two distant methodologies to achieve the operation in the SIDO converter. As a result, to minimize both transient and steady-state cross-regulations induced by the output loading current steps and the large loading current difference between the dual outputs, respectively, the dual-mode energy delivery methodology can appropriately arrange the energy delivery paths and the inductor current level. Thus, the transient cross-regulation can be reduced by rapidly adjusting the inductor current when the load transient response occurs, and the steady-state cross-regulation can be minimized simultaneously with the output voltage ripple in the SIDO converter.

Figure 2.19 shows the timing diagram of the energy conservation mode (ECM) control with the order of path 1, path 3, path 4, and path 2. This ordered sequence is circulated by the triggering of the system clock, V_{clk}. The error signal V_{EA}

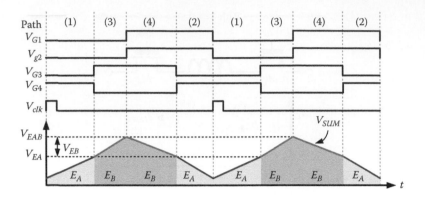

FIGURE 2.19 The timing diagram of the proposed energy conservation mode control.

determines the transitions from path 1 to path 3 and from path 4 to path 2. That is, path 1 and path 2 are decided by V_{SUM} and V_{EA}, respectively, for the output V_{OA}. Similarly, path 3 and path 4 are decided by V_{SUM} and V_{EB}, respectively, which is the difference between V_{EAB} and V_{EA}. The ECM control achieves the separated dual step-down operations by the inductor current superposition scheme. Both error signals as well as the energy demands of the dual output can be modulated within one PWM switching cycle. Thus, each output receives the power form input battery in each switching period. Also, each power switch, M_1 to M_4, would switch twice in one switching cycle time with the ordered energy delivery paths.

A comparison of the inductor current waveforms under different loading current conditions is shown in Figure 2.20 when considering the PCCM and the ECM control methodologies. The combination of two separated step-down operations in the ECM results in that the average inductor current, I_{L,avg_ECM}, is equal to the summation of the two output loads. The value of I_{L,avg_ECM} is much lower than that of I_{L,avg_PCCM} in the PCCM control due to a freewheel stage used to regulate the inductor

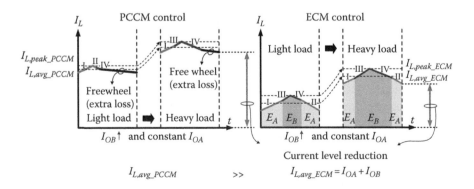

FIGURE 2.20 Comparison of the inductor current waveforms under different load conditions when considering the pseudo-continuous conduction mode and the energy conservation mode control methodologies.

current. The PCCM control has to raise its inductor current level to provide the adequate energy for dual outputs due to the occupation of the energy delivery time for the freewheel stage. In contrast, the ECM control yields a low average inductor current level to reduce conduction power loss and to enhance efficiency by removing the freewheel stage. In the ECM control, if the load current I_{OB} of the V_{OB} increases and the loading current I_{OA} of the V_{OA} is fixed, it leads to an increase in total output power. Simultaneously, the average inductor current of the ECM control I_{L,avg_ECM} arises to reflect the heavy load condition. The total energy E_A for V_{OA} in Figure 2.20 must be constant at both light and heavy loading conditions since I_{OA} is not changed. On the other hand, the total energy E_B for the V_{OB} is increased to provide a sufficient power owing to the increase of I_{OB}. The periods of path 1 and path 2 are reduced, but the periods of path 3 and path 4 are extended because of the increase in I_{L,avg_ECM}. Moreover, the effect from PVT condition can be minimized by high gain error amplifiers [39]. They modulate the error signals properly to ensure correct function in the ECM control. The low value of I_{L,avg_ECM} greatly enhances the power conversion efficiency in the ECM control owing to the use of superposition technique. Furthermore, the derivation of the minimum average inductor current level helps reduce the output voltage ripple to get improved supply quality.

2.5 SUMMARY

SIDO and SIMO converters are used in the PMU design for SoC applications. Design goals include (1) reduction of the number of power switches, (2) reduction of conduction power loss and switching power loss, (3) extension of output loading current range, (4) improvement of load transient response, (5) reduction of output voltage ripple, (6) reduction of transient and steady-state cross-regulations, and finally (7) improvement of power conversion efficiency. Those design goals cannot be all met at the same time. Thus, multidimensional trade-off should be done depending on the required specifications. Most importantly, circuit designers have to weigh in the advantages and disadvantages of each control technique before selecting to use a specific method.

REFERENCES

1. H.-W. Huang, K.-H. Chen, and S.-Y. Kuo, Dithering skip modulation, width and dead time controllers in highly efficient DC-DC converters for system-on-chip applications, *IEEE J. Solid-State Circuits*, 42(11), 2451–2465, November 2007.
2. K.-H. Chen, C.-J. Chang, and T.-H. Liu, Bidirectional current-mode capacitor multipliers for on-chip compensation, *IEEE Trans. Power Electron.*, 23(1), 180–188, January 2008.
3. G. Patounakis, Y. W. Li, and K. L. Shepard, A fully integrated on-chip DC-DC conversion and power management system, *IEEE J. Solid-State Circuits*, 39(3), 443–451, March 2004.
4. M. Alimadadi, S. Sheikhaei, G. Lemieux, S. Mirabbasi, and P. Palmer, A 3GHz switching DC-DC converter using clock-tree charge-recycling in 90 nm CMOS with integrated output filter, in *IEEE International Solid-State Circuits Conference, Digest of Technical Papers*, San Francisco, CA, February 11–15, 2007, pp. 532–533.

5. C.-Y. Hsieh and K.-H. Chen, Adaptive pole-zero position (APZP) technique of regulated power supply for improving SNR, *IEEE Trans. Power Electron.*, 23(6), 2949–2963, November 2008.

6. F.-F. Ma, W.-Z. Chen, and J.-C. Wu, A monolithic current-mode buck converter with advanced control and protection circuit, *IEEE Trans. Power Electron.*, 22(5), 1836–1846, September 2007.

7. M. D. Mulligan, B. Broach, and T. H. Lee, A 3 MHz low-voltage buck converter with improved light load efficiency, in *IEEE International Solid-State Circuits Conference, Digest of Technical Papers*, San Francisco, CA, February 11–15, 2007, pp. 528–529.

8. I. Doms, P. Merken, C. V. Hoof, and R. P. Mertens, Capacitive power management circuit for micropower thermoelectric generators with a 1.4 μA controller, *IEEE J. Solid-State Circuits*, 44(10), 2824–2833, October 2009.

9. Y.-H. Lee, S.-J. Wang, Y.-Y. Yang, K.-L. Zheng, P.-F. Chen, C.-Y. Hsieh, Y.-Z. Ke, K.-H. Chen, Y.-K. Chen, C.-C. Huang, and Y.-H. Lin, A DVS embedded power management for high efficiency integrated SoC in UWB system, in *Proceedings of IEEE Asian Solid-State Circuits Conference*, Taipei, Taiwan, November 16–18, 2009, pp. 321–324.

10. J. Kwong, Y. K. Ramadass, N. Verma, and A. P. Chandrakasan, A 65 nm sub-Vt microcontroller with integrated SRAM and switched capacitor DC-DC converter, *IEEE J. Solid-State Circuits*, 44(1), 115–126, January 2009.

11. M.-H. Huang, P.-C. Fan, and K.-H. Chen, Low-ripple and dual-phase charge pump circuit regulated by switched-capacitor-based bandgap reference, *IEEE Trans. Power Electron.*, 24(5), 1161–1172, May 2009.

12. P. Favrat, P. Deval, and M. J. Declercq, A high-efficiency CMOS voltage doubler, *IEEE J. Solid-State Circuits*, 33(3), 410–416, March 1998.

13. S. Das, D. Roberts, S. Lee, S. Pant, D. Blaauw, T. Austin, K. Flautner, and T. Mudge, A self-tuning DVS processor using delay-error detection and correction, *IEEE J. Solid-State Circuits*, 41(4), 792–804, April 2006.

14. S. Xiao, W. Qiu, G. Miller, T. X. Wu, and I. Batarseh, An active compensator scheme for dynamic voltage scaling of voltage regulators, *IEEE Trans. Power Electron.*, 24(1), 307–311, January 2009.

15. M. E. Sinangil, N. Verma, and A. P. Chandrakasan, A reconfigurable 8T ultra-dynamic voltage scalable (U-DVS) SRAM in 65 nm CMOS, *IEEE J. Solid-State Circuits*, 44(11), 3163–3173, November 2009.

16. P. Hazucha, S. T. Moon, G. Schrom, F. Paillet, D. Gardner, S. Rajapandian, and T. Karnik, High voltage tolerant linear regulator with digital control for biasing of integrated DC-DC converters, *IEEE J. Solid-State Circuits*, 42(1), 66–73, January 2007.

17. Y.-H. Lin, K.-L. Zheng, and K.-H. Chen, Power MOSFET array for smooth pole tracking in LDO regulator compensation, *IEEE Trans. Power Electron.*, 23(5), 2421–2427, September 2008.

18. R. J. Milliken, J. Silva-Martínez, and E. Sanchez-Sinencio, Full on-chip CMOS low-dropout voltage regulator, *IEEE Trans. Circuits Syst. I, Reg. Papers*, 54(9), 1879–1890, September 2007.

19. C.-H. Lin, K.-H. Chen, and H.-W. Huang, Low-dropout regulators with adaptive reference control and dynamic push–pull techniques for enhancing transient performance, *IEEE Trans. Power Electron.*, 24(4), 1016–1022, April 2009.

20. M. Al-Shyoukh, H. Lee, and R. Perez, A transient-enhanced low-quiescent current low-dropout regulator with buffer impedance attenuation, *IEEE J. Solid-State Circuits*, 42(8), 1732–1742, August 2007.

21. J. A. Starzyk, Y.-W. Jan, and F. Qiu, A DC-DC charge pump design based on voltage doublers, *IEEE Trans. Circuits Syst. I*, 48, 350–359, March 2001.

22. Y.-K. Luo, K.-H. Chen, and W.-C. Hsu, A dual-phase charge pump regulator with nano-ampere switched-capacitor CMOS voltage reference for achieving low output ripples, in *Proceedings of IEEE International Conference on Electronics, Circuits and Systems*, Boulder, CO, August 31–September 3, 2008, pp. 446–449.

23. D. Ma, W.-H. Ki, and C.-Y. Tsui, A pseudo-CCM/DCM SIMO switching converter with freewheel switching, *IEEE J. Solid-State Circuits*, 38(6), 1007–1014, June 2003.

24. H.-P. Le, C.-S. Chae, K.-C. Lee, S.-W. Wang, G.-H. Cho, and G.-H. Cho, A single-inductor switching DC-DC converter with five outputs and ordered power-distributive control, *IEEE J. Solid-State Circuits*, 42(12), 2076–2714, December 2007.

25. M.-H. Huang and K.-H. Chen, Single-inductor multi-output (SIMO) DC-DC converters with high light-load efficiency and minimized cross-regulation for portable devices, *IEEE J. Solid-State Circuits*, 44(4), 1099–1111, April 2009.

26. K.-S. Seol, Y.-J. Woo, G.-H. Cho, G.-H. Cho, J.-W. Lee, and S.-I. Kim, Multiple-output step-up/down switching dc-dc converter with vestigial current control, in *IEEE International Solid-State Circuits Conference, Digest of Technical Papers*, New York, February 2009, pp. 442–443.

27. D. Kwon and G. A. Rincón-Mora, Single-inductor-multiple-output switching DC-DC converters, *IEEE Trans. Circuits Syst. II, Exp. Briefs*, 56(8), 614–618, August 2009.

28. E. Bonizzoni, F. Borghetti, P. Malcovati, F. Maloberti, and B. Niessen, A 200 mA 93% peak efficiency single-inductor dual-output DC-DC buck converter, in *IEEE International Solid-State Circuits Conference, Digest of Technical Papers*, San Francisco, CA, February 11–15, 2007, pp. 526–527.

29. X. Jing, P. K. T. Mok, and M. C. Lee, A wide-load-range constant-charge-auto-hopping control single-inductor dual-output boost regulator with minimized cross-regulation, *IEEE J. Solid-State Circuits*, 46(10), 2350–2362, October 2011.

30. Y.-J. Woo, H.-P. Le, G.-H. Cho, G.-H. Cho, and S.-I. Kim, Load-independent control of switching DC-DC converters with freewheeling current feedback, *IEEE J. Solid-State Circuits*, 43(12), 2798–2808, December 2008.

31. W. Xu, Y. Li, X. Gong, Z. Hong, and D. Killat, A dual-mode single-inductor dual-output switching converter with small ripple, *IEEE Trans. Power Electron.*, 25(3), 614–623, March 2010.

32. C.-S. Chae, H.-P. Le, K.-C. Lee, G.-H. Cho, and G.-H. Cho, A single-inductor step-up DC-DC switching converter with bipolar outputs for active matrix OLED mobile display panels, *IEEE J. Solid-State Circuits*, 44(2), 509–524, February 2009.

33. M.-H. Huang and K.-H. Chen, Single-inductor dual buck-boost output (SIDBBO) converter adaptive current control mode (ACCM) and adaptive body switch (ABS) for compact size and long battery life in portable devices, in *Proceedings of IEEE Symposium on VLSI Circuits*, Kyoto, Japan, June 16–18, 2009, pp. 164–165.

34. Y.-H. Lee, K.-H. Chen, Y.-H. Lin, Y.-Y. Yang, S.-J. Wang, Y.-K. Chen, and C.-C. Huang, An interleaving energy-conservation mode (IECM) control in single-inductor dual-output (SIDO) step-down converters with 91% peak efficiency, in *Proceedings of IEEE Symposium on VLSI Circuits*, Honolulu, HI, June 16–18, 2010, pp. 57–58.

35. M. Belloni, E. Bonizzoni, E. Kiseliovas, P. Malcovati, F. Maloberti, T. Peltola, and T. Teppo, A 4-output single-inductor DC-DC buck converter with self-boosted switch drivers and 1.2 A total output current, in *IEEE International Solid-State Circuits Conference, Digest of Technical Papers*, San Francisco, CA, February 3–7, 2008, pp. 444–445.

36. M.-H. Huang, Y.-N. Tsai, and K.-H. Chen, Sub-1V input single-inductor dual-output (SIDO) DC–DC converter with adaptive load-tracking control (ALTC) for single-cell-powered systems, *IEEE Trans. Power Electron.*, 25(7), 1713–1724, June 2010.

37. Y. Qiu, X. Chen, and H. Liu, Digital average current-mode control using current estimation and capacitor charge balance principle for DC-DC converters operating in DCM, *IEEE Trans. Power Electron.*, 25(6), 1537–1545, June 2010.

38. R. W. Erickson and D. Maksimović, *Fundamentals of Power Electronics*. Norwell, MA: Kluwer, 2001.
39. Y.-H. Lee, Y.-Y. Yang, K.-H. Chen, Y.-H. Lin, S.-J. Wang, K.-L. Zheng, P.-F. Chen, C.-Y. Hsieh, Y.-Z. Ke, Y.-K. Chen, and C.-C. Huang, A DVS embedded power management for high efficiency integrated SoC in UWB system, *IEEE J. Solid-State Circuits*, 45(11), 2227–2238, November 2010.
40. S. O. Cannizzaro, A. D. Grasso, R. Mita, G. Palumbo, and S. Pennisi, Design procedures for three-stage CMOS OTAs with nested-miller compensation, *IEEE Trans. Circuits Syst. I, Reg. Papers*, 54(5), 933–940, May 2007.

3 SIMO Power Converters with Adaptive PCCM Operation

Yi Zhang and D. Brian Ma

CONTENTS

3.1 BACKGROUND AND MOTIVATION

Driven by modern very large scale integration (VLSI) systems and their fast-emerging applications, the demands on multiple on-chip power supplies have been ever increasing due to two main reasons. First, as complexity of electronic devices such as cell phones and laptops increases, multiple power supplies are essential to power key functional blocks, such as processors, liquid crystal display, and radio transceivers. Second, as power consumption in semiconductor chips rises, efficient system power management techniques are in high demand. This leads to the high popularity of the so-called dynamic voltage/frequency scaling (DVFS) techniques (Burd et al. 2000, Chang and Pedram 1997, Cheng and Baas 2008, Ma and Bondade 2010, Usami et al., 1998), where multiple-supply-based power management schemes are crucial. By powering distinct functional modules and subsystems with different supply voltages and by adaptively switching their supplies according to instantaneous power needs, power and energy savings achieved through DVFS techniques are significant (Burd et al. 2000, Luo et al. 2007, Ma and Bondade 2010).

Historically, multiple DC power supplies are implemented by transformer-based isolated DC–DC converters or several nonisolated DC–DC converters. However, due to the use of bulky off-chip inductive components and many power switches, system volume, printed circuit board (PCB) footprint, and electromagnetic interference noise are all considerably large. In the meantime, regulation accuracy can be compromised by commonly used "master–slave" regulation approaches (Ma et al. 2001a).

To mitigate these issues, single-inductor multiple-output (SIMO) power converters were proposed (Ki and Ma 2001, Ma et al. 2001a, 2003a,b). By time-sharing a single inductor and some power switches, a SIMO power converter offers multiple power outputs, which can be regulated independently. This leads to a significant reduction in cost, volume, and PCB footprint. Because of their cost-effective features, the application of SIMO power converters has been proliferating in recent years, spanning from hybrid power source units (Huang et al. 2010), energy-harvesting systems (Kim and Rincon-Mora 2009, Sze et al. 2008) to light-emitting diode backlighting (Chen et al. 2012), active matrix OLED display panels (Chae et al. 2009), and electronic paper displays (Lee et al. 2010).

As more SIMO converter-based IC products are being commercialized in the market, cross-regulation effect, as one of the most critical design challenges, has drawn extensive attention. In a SIMO switching converter, because the inductor is time-shared by several subconverters, a duty ratio change in one subconverter may affect the amount of energy transferred to other subconverters even if the operating conditions (load currents, duty ratios, etc.) in other subconverters remain constant. As a result, this may cause voltage variations on the affected outputs. This effect was named as cross-regulation in Ma et al. (2003b) because a local operating condition change in one subconverter has impacted the regulation performances across the others. In worst-case scenarios, a SIMO converter fails to operate due to severe cross-regulation.

In order to avoid the cross-regulation effect, the regulation time slot designated to each subconverter should be completely isolated. For this reason, SIMO converters in the early stage usually operate in discontinuous conduction mode (DCM) (Ki and Ma 2001, Ma et al. 2001a, 2003b). However, in heavy-load scenarios, peak inductor current may rise very high in DCM, causing substantial voltage and current ripples and switching noise. Alternatively, if continuous conduction mode (CCM) is applied to alleviate large peak current, cross-regulation occurs immediately. In order to reduce the heavy current stress while retaining low cross-regulation, a pseudo-continuous conduction mode (PCCM) operation was first proposed in Ma et al. (2002). Instead of returning to zero as in DCM, the inductor current stays constant at a predefined current level through freewheel switching actions. With the same load condition, the peak inductor current in PCCM can be much lower than that in DCM, while still avoiding cross-regulation effect since the regulation actions on any two outputs are isolated by one freewheel switching period. Unfortunately, the PCCM operation also has its own drawbacks. Especially, in the unbalanced load conditions, large conduction power loss is observed in subconverters handling light loads. Moreover, since the freewheel switch turns on/off multiple times in one switching cycle, the overall switching power loss also increases. To overcome these drawbacks, a new control technique is required to reduce both conduction and switching power loss in the freewheel switch.

3.2 DEVELOPMENT OF SIMO POWER CONVERTERS

To better understand the PCCM operation for SIMO converters, we first examine the development of SIMO power converter, as illustrated in Figure 3.1 (Ma et al. 2001a, 2003b). Consider two conventional switching boost converters in DCM operating at the same switching frequency. One possible operation scheme is illustrated in Figure 3.1a. At the beginning of each switching cycle T, the inductor L_1 is charged at a rate of V_{IN}/L_1 until $D_{11}T$ expires, where D_{11} is the duty ratio of the switching power converter 1. During $D_{21}T$, the inductor L_1 is discharged at a rate of $-(V_{IN} - V_{O1})/L_1$ and

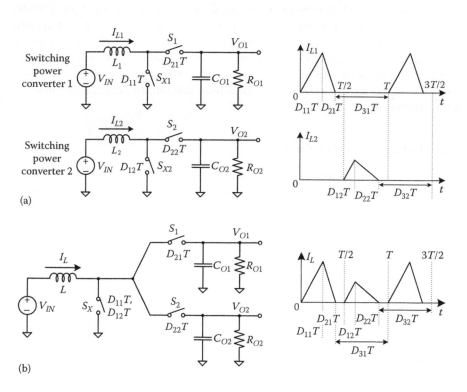

FIGURE 3.1 Power stage architecture and discontinuous conduction mode inductor currents of (a) two conventional switching power converters and (b) one single-inductor multiple-output converter. (From Zhang, Y., Single-inductor multiple-output power converters: Architectures, control techniques and applications, PhD dissertation, The University of Texas at Dallas, Richardson, TX, 2013.)

I_{L1} drops down until it returns to zero. The inductor current stays at zero during $D_{31}T$. Similar operation scheme applies to the switching power converter 2.

If the two inductor currents can be alternately assigned to occupy different parts of the switching cycle without overlapping, only one single inductor will suffice for their operation, as shown in Figure 3.1b. During the first half switching cycle from 0 to $T/2$, the inductor current is diverted to V_{O1} by properly controlling the power switches S_X and S_1. Similarly, in the second half switching cycle, S_X and S_2 are turned on/off to deliver the inductor current to V_{O2}. With this method, the inductor L is shared by the two subconverters in a time-multiplexing manner. Hence, a SIMO power converter with a time-multiplexing operation scheme is developed.

Obviously, this operation scheme can be readily extended for SIMO power converters with more than two outputs. For SIMO power converters with N subconverters, one switching cycle is divided into N phases, with the inductor current being multiplexed into each output during the corresponding phase. Furthermore, the topology is not only limited to boost converter. It can be easily extended to many existing switching power converter topologies (Ki and Ma 2001), such as buck and noninverting buck–boost.

It should be noted that the time-multiplexing operation is critical for a SIMO power converter. The inductor current is assigned to each subconverter alternately. Each output only occupies its own phase in one switching cycle. These phases are expected to be isolated in order to prevent cross-regulation. When the inductor current is being diverted into one subconverter, the other one is separated from the control loop. In other words, only one switching subconverter is being regulated at a time instant.

3.3 PRIMARY TOPOLOGIES

Similar to the conventional single-output switching power converters, the SIMO power converters can also be categorized into different types based on the topologies. In this section, the three primary topologies for SIMO power converters—boost, buck, and noninverting buck–boost topologies—are introduced.

3.3.1 Boost Topology

The power stage architecture and timing diagram of a SIMO boost switching converter with two outputs are illustrated in Figure 3.2. The two subconverters are regulated by a pair of complementary clocks Φ_1 and Φ_2, with power delivered to each output from the supply V_{IN} in a time-multiplexed manner. The working principle can be described with reference to the timing diagram illustrated in Figure 3.2b. When the first subconverter is operating, $\Phi_1 = 1$ and the switch S_2 is off. No current flows into the output V_{O2}. Meanwhile, S_X is on. The voltage across the inductor is, thus, the supply voltage V_{IN}. The inductor current I_L increases during $D_{11}T$:

$$\frac{di_L}{dt} = \frac{V_{IN}}{L}. \tag{3.1}$$

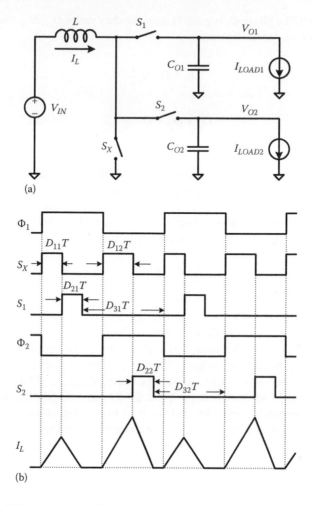

(a)

(b)

FIGURE 3.2 (a) Power stage architecture and (b) timing diagram of single-inductor multiple-output boost converter.

During $D_{21}T$, the switch S_X is off and S_1 is on, causing the inductor to discharge the current into the output V_{O1}. The slope of the inductor current is given by

$$\frac{di_L}{dt} = \frac{V_{IN} - V_{O1}}{L}. \tag{3.2}$$

As the output voltage of a boost converter is greater than the supply voltage, the rate of the inductor current change is negative. As illustrated in Figure 3.2b, the inductor current decreases and delivers the required charge to the output. When it goes to zero, the converter enters $D_{31}T$ and the switch S_1 is turned off. The inductor current stays at zero until the phase Φ_2 begins. Hence, the duty ratios D_{11}, D_{21}, and D_{31} satisfy the requirements shown as follows:

$$D_{11} + D_{21} \leq \frac{1}{2}, \tag{3.3}$$

$$D_{11} + D_{21} + D_{31} = 1. \tag{3.4}$$

During $\Phi_2 = 1$, the inductor current is multiplexed into the output V_{O2} and similar switching actions repeat for the subconverter 2. With this method, the two outputs are regulated in a time-multiplexing manner by sharing the inductor L and the power switch S_X. Compared with two traditional switching boost converters, the number of off-chip magnetic component and power switches is both reduced. Obviously, with the increase of the number of subconverters, this cost reduction effect will become more significant.

3.3.2   BUCK TOPOLOGY

Another primary type of SIMO power converters are constructed with buck topology. Figure 3.3a shows the power stage architecture of a SIMO buck power converter with two outputs. The power stage consists of one inductor L; four power switches S_P, S_N, S_1, and S_2; and two filtering capacitors C_{O1} and C_{O2}. The two subconverters share the inductor L and power switches S_P and S_N. The timing diagram of DCM operation is illustrated in Figure 3.3b. Similar to the boost SIMO converter, the two subconverters are regulated alternately during Φ_1 and Φ_2. The two output voltages are stabilized at lower voltage levels than the input voltage, thereby achieving step-down DC–DC voltage conversions. The inductor current slope during charge and discharge periods is defined by

$$\frac{di_L}{dt} = \frac{V_{IN} - V_{Oi}}{L}, \tag{3.5}$$

$$\frac{di_L}{dt} = -\frac{V_{Oi}}{L}, \tag{3.6}$$

where i equals 1 or 2. Compared with two traditional buck converters, the number of inductors being used is halved. Although the number of power switches is not reduced, the cost effectiveness can still be observed when more subconverters are incorporated.

3.3.3   NONINVERTING BUCK–BOOST TOPOLOGY

The third type of topology for SIMO power converters is noninverting buck–boost topology, which achieves both step-up and step-down voltage conversions. As shown in Figure 3.4a, the two subconverters share the inductor L and power switches

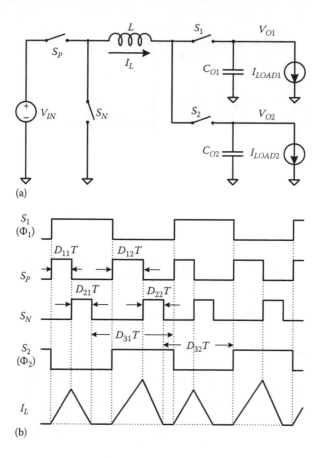

(a)

(b)

FIGURE 3.3 (a) Power stage architecture and (b) timing diagram of single-inductor multiple-output buck converter.

S_P, S_N, and S_X. The corresponding timing diagram is demonstrated in Figure 3.4b. In each phase, during the charge period, S_P and S_X are enabled, thereby charging up the inductor with a slope of V_{IN}/L. Similarly, during the discharge phase, S_N and S_i are turned on. The inductor current ramps down with a slope of $-V_O/L$ ($i = 1$ or 2). Compared with the boost and buck topologies, the noninverting buck–boost topology demonstrates higher flexibility regarding the voltage conversion ratios. The output voltage can be stabilized at higher, lower, or similar voltage levels compared with the input voltage. However, this flexibility is achieved at the cost of more power switches. The corresponding switching power loss and control complexity for non-inverting buck–boost topology are, thus, increased compared with the other two topologies.

In addition to these three topologies, there are also other topologies that can be used in SIMO power converters. For example, inverting buck–boost topology allows the generation of negative output voltages. When combined with the aforementioned topologies, both positive and negative output voltages can be implemented.

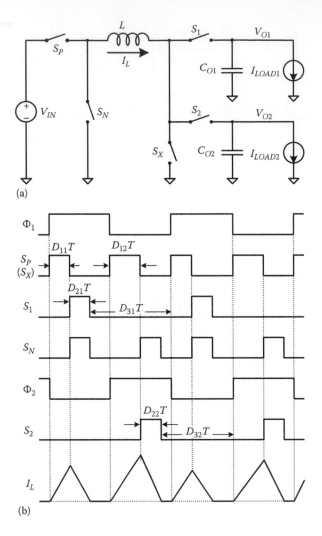

(a)

(b)

FIGURE 3.4   (a) Power stage architecture and (b) timing diagram of single-inductor multiple-output noninverting buck–boost converter.

As a result, it leads to SIMO converters with bipolar power outputs (Ki and Ma 2001, Ma et al. 2001b).

3.4   CONVENTIONAL OPERATION MODES

Historically, based on the inductor current, the operation mode for SIMO power converters can be categorized into two types: DCM (Ki and Ma 2001, Ma et al. 2001a, 2003b, Sze et al. 2008) and CCM (Belloni et al. 2008, Goder and Santo 1997, Li 2000, May et al. 2001). This section addresses both in due course, using a single-inductor dual-output (SIDO) boost topology as an example.

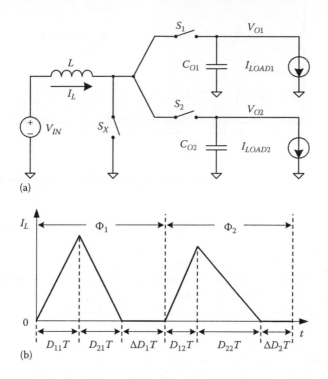

(a)

(b)

FIGURE 3.5 (a) Power stage architecture and (b) inductor current waveform of a single-inductor multiple-output boost converter operating in discontinuous conduction mode.

3.4.1 DISCONTINUOUS CONDUCTION MODE

The typical inductor current waveform in DCM operation is shown in Figure 3.5. At the beginning of each phase, the inductor is charged. The charge process ends when $D_{1i}T$ expires, followed by a discharge period. Instead of occupying the rest of the phase, the discharge period, $D_{2i}T$, ends once the inductor current drops to zero. After it, the inductor current stays at zero until the next phase is enabled. Since each subconverter only operates within its own phase, no two adjacent phases are overlapped, thereby eliminating the potential cross-regulation effect.

3.4.2 CONTINUOUS CONDUCTION MODE

While the DCM operation can effectively suppress cross-regulation, it has certain drawbacks at heavy load conditions. With the same input/output voltages and load current, in order to deliver the same amount of charge with the same switching frequency, the peak inductor current value in DCM operation is usually much higher than in CCM. Hence, under heavy load conditions, DCM operation leads to large voltage/current ripples and switching noise.

Alternatively, another operation mode for SIMO power converters is CCM. As shown in Figure 3.6, the inductor is charged from the beginning of each phase.

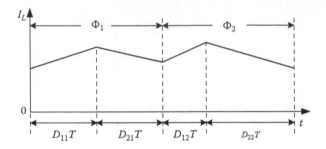

FIGURE 3.6 Inductor current waveform of continuous conduction mode operation.

This charge process continues until the time period $D_{1i}T$ expires, where i equals 1 or 2. Then, the discharge period $D_{2i}T$ is enabled until the discharge phase ends. The inductor current charge and discharge actions are repeated in each phase, thereby delivering the desired power from input to the corresponding output. Since the next phase starts immediately after the current phase expires, the inductor current is kept continuous during the phase transition periods. Such type of operation scheme is thus called CCM.

Due to the continuous conduction property, the inductor current is always maintained above zero. As a result, compared to the DCM operation, the CCM operation facilitates lower inductor current ripples, which in turn reduces output voltage ripples. Moreover, since the inductor current never goes to zero in CCM, the potential negative/reverse inductor current, which may occur in DCM operations, is avoided, thereby improving the power efficiency. From the circuit design perspective, the zero current detector and active diodes can be obviated. The circuit complexity and design challenge are both reduced.

However, CCM operation also incurs some drawbacks. Due to the continuous conduction property, the inductor current at the end of each phase is uncertain. As a result, the initial value of the inductor current to the second subconverter is dependent to the end value of the first one. If a sudden load change occurs in one phase, it will inevitably affect the subsequent phases, causing severe cross-regulation problems (Ma et al. 2003b).

3.5 PSEUDO-CONTINUOUS CONDUCTION MODE

In order to receive the benefits from both CCM and DCM operation schemes, the PCCM was proposed for SIMO power converters (Ma et al. 2002, 2003a). In the PCCM mode, as depicted in Figure 3.7 and similar to a CCM one, the inductor current of a SIMO converter always stays greater than a predetermined DC value I_{dc}, thus reducing the inductor current ripples. This is achieved by shorting the inductor with the aid of a freewheel switch S_{fw} while keeping the other switches off. With reference to Figure 3.7, when Φ_1 is enabled and during the period $D_{21}T$, when the value of the inductor current reaches I_{dc}, S_{fw} is turned on and all the other switches are turned off until Φ_1 expires. This period is known as the freewheel switching period.

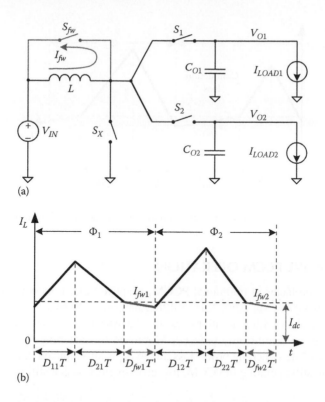

(a)

(b)

FIGURE 3.7 (a) Power stage architecture and (b) inductor current waveform of a single-inductor dual-output boost converter operating in pseudo-continuous conduction mode.

A similar period occurs during Φ_2. As the current continuously stays above I_{dc}, it allows the converter to handle heavy current loads with low ripples. Moreover, since the two subconverters are isolated by the freewheel switching periods, cross-regulation is also avoided.

Despite the aforementioned advantages of PCCM operation, there remain some drawbacks. To avoid cross-regulation, the freewheel switching current level I_{dc} has to be chosen to satisfy the largest load current among all the outputs. If the value of I_{dc} is reduced, then a load change at either output could cause one of the subconverters to enter the CCM mode. For example, assume I_{LOAD1} suddenly increases. The inductor has to be energized to a larger current value in Φ_1 to satisfy the higher power demand. I_L, thus, may not be able to return to I_{dc} before Φ_1 expires. The converter then enters the CCM mode. Meanwhile, if the freewheel switching period $D_{fw}T$ is too long, the turn-on resistance of the freewheel switch and the direct current resistance (DCR) of the inductor can cause significant conduction loss. This is especially pronounced during unbalanced load conditions, as shown in Figure 3.8, in which a long $D_{fw}T$ can be observed in the light-load subconverter. Moreover, the switching power loss of S_{fw} also becomes significant when the number of subconverters is increased.

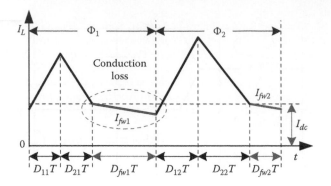

FIGURE 3.8 I_L during pseudo-continuous conduction mode operation in unbalanced load condition.

3.6 ADAPTIVE PCCM OPERATION

To resolve the problems of traditional PCCM, adaptive PCCM operation schemes are developed (Zhang and Ma 2010, 2011, 2012b, Zheng et al. 2010). By adaptively adjusting the freewheel switching durations and the freewheel current level I_{dc}, conduction power loss can be significantly reduced. Moreover, by switching S_{fw} only once per switching cycle, the switching power loss due to S_{fw} is also reduced. Since all the subconverters are still operating in PCCM, low cross-regulation can still be achieved.

3.6.1 POWER LOSS ANALYSIS

Figure 3.9a models a freewheel switching loop. The resistance of the freewheel switch, R_{ON}, and the DCR of the inductor, R_L, both contribute to conduction power loss. To simplify the analysis, the sum of R_{ON} and R_L is named as R_{EQ}. Due to the voltage drop across R_{EQ}, I_L during the freewheel switching period $D_{fwi}T$ ($i = 1$ or 2) decreases gradually as in Figure 3.9b. I_L is, thus, given as

$$I_L(t) = \begin{cases} I_{dc} \cdot e^{-\frac{R_{EQ}}{L}[t-(D_{11}+D_{21})\cdot T]} & (D_{11}+D_{21})\cdot T \leq t \leq \frac{T}{2}, \\[2ex] I_{dc} \cdot e^{-\frac{R_{EQ}}{L}\left[t-\left(\frac{1}{2}+D_{12}+D_{22}\right)\cdot T\right]} & \left(\frac{1}{2}+D_{11}+D_{21}\right)\cdot T \leq t \leq T \end{cases}$$

(3.7)

Consequently, the conduction power loss by S_{fw} is derived as

$$P_C = \sum_{i=1}^{n} \frac{I_{dc}^2 L}{2T}\left(1 - e^{-\frac{2R_{EQ}}{L}\cdot D_{fwi}T}\right),$$

(3.8)

which reveals that the loss is highly related to I_{dc} and the freewheel switching period.

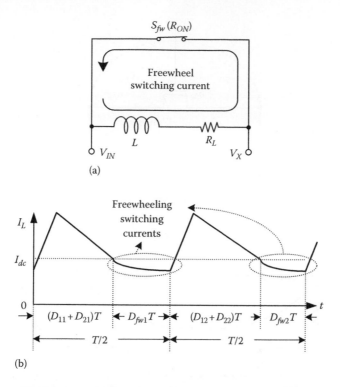

FIGURE 3.9 (a) Circuit model and (b) the inductor current waveform of the freewheel switching. (From Zhang, Y. and Ma, D., *J. Analog Integr. Circuits Signal Process.*, 72(2), 419, August 2012b.)

In addition to the conduction loss, a freewheel switch induces two types of switching losses: the *V–I* overlap power loss and the gate-drive power loss. The cause of *V–I* overlap power loss is illustrated in Figure 3.10b. Due to the parasitic capacitors (C_{GD} and C_{GS} in Figure 3.10a), the gate voltage V_{gate} of S_{fw} cannot step up/down abruptly, leading to the overlapping periods between nonzero voltage V_{SD} and nonzero current I_D. The converter loses power during such *V–I* overlapping transient periods. Meanwhile, as shown in Figure 3.10b, before/after the turn-on/turnoff events, a dead time is usually added to avoid shoot-through current. During such a dead time, all the power transistors are turned off and the inductor current flows through the body diode of S_{fw}. As a floating switch, S_{fw} is usually implemented with a PMOS power switch. The substrate of the PMOS switch S_{fw} is usually tied to the highest voltage in the system (V_{max}), which is one of the output voltages in a SIMO boost converter (Ma 2007, Zhang and Ma 2012a, Zheng and Ma 2011). If the voltage drop across the body diode is neglected, V_{SD} in S_{fw} swings between ($V_{max} - V_{IN}$) and ground. The turn-on process starts at t_1 by discharging the gate capacitance. When the gate voltage drops to $V_{max} - V_T$ at t_2, S_{fw} starts to conduct current. When I_D increases from zero to the full freewheel switching current I_{dc} at t_3, V_{SD} starts to be pulled down. This continues until V_{SD} drops to zero at t_4.

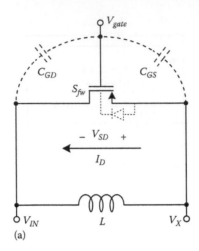

(a)

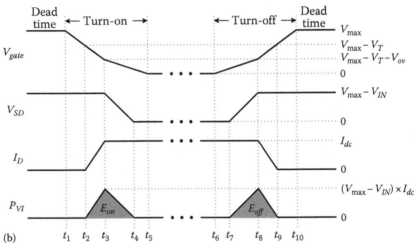

(b)

FIGURE 3.10 (a) Circuit model and (b) key waveforms for switching power loss. (From Zhang, Y. and Ma, D., *J. Analog Integr. Circuits Signal Process.*, 72(2), 419, August 2012b.)

The process ends when V_{gate} becomes zero at t_5. The V–I overlap from t_2 to t_4 results in the corresponding energy loss E_{on}. This is calculated as

$$E_{on} = \frac{1}{2}(V_{max} - V_{IN}) \cdot I_{dc} \cdot (t_{IR} + t_{VF}), \tag{3.9}$$

where t_{IR} and t_{VF} are the current rising time $(t_3 - t_2)$ and voltage falling time $(t_4 - t_3)$, respectively. Similarly, the turn-off transition reverses the sequence of the events and generates the turn-off V–I overlapping energy loss

$$E_{off} = \frac{1}{2}(V_{max} - V_{IN}) \cdot I_{dc} \cdot (t_{VR} + t_{IF}), \tag{3.10}$$

where t_{VR} and t_{IF} are defined as the voltage rising time $(t_8 - t_7)$ and current falling time $(t_9 - t_8)$, respectively. For multiple-charge SIMO converters in PCCM, S_{fw} switches n times per switching cycle. The overall V–I overlap power loss is

$$P_{VI} = \frac{1}{2}(V_{max} - V_{IN}) \cdot I_{dc} \cdot (t_{IR} + t_{VF} + t_{VR} + t_{IF}) \cdot n \cdot f_{sw}, \qquad (3.11)$$

where f_{sw} is the switching frequency of the converter.

Another major switching power loss is the gate-drive power loss. This involves the charge/discharge processes on gate capacitors C_{GD} and C_{GS}. For C_{GD}, the drain terminal is constantly connected to V_{IN} and the gate voltage swings between V_{max} and zero, leading to a traversed voltage of V_{max}. But for C_{GS}, the voltage of source terminal (V_X) is tied to V_{IN} during the conduction period and V_{max} during the dead times. Meanwhile, the gate voltage is equal to zero and V_{max}, respectively. The traversed voltage of C_{GS} is then calculated as V_{IN}. The gate-drive loss of freewheel switch can be derived as

$$P_{DRV} = (C_{GD} \cdot V_{max}^2 + C_{GS} \cdot V_{IN}^2) \cdot n \cdot f_{sw}. \qquad (3.12)$$

The total switching power loss of S_{fw}, thus, can be denoted as

$$P_{SW} = \left[\frac{1}{2}(V_{max} - V_{IN}) \cdot I_{dc} \cdot (t_{IR} + t_{VF} + t_{VR} + t_{IF}) + (C_{GD} \cdot V_{max}^2 + C_{GS} \cdot V_{IN}^2)\right] \cdot n \cdot f_{sw}. \qquad (3.13)$$

In conclusion, to achieve high efficiency in a SIMO converter operating in PCCM, the conduction and switching losses of the freewheel switch should be jointly minimized. These power losses are highly related to switching frequency of S_{fw}, freewheel switching current I_{dc} and freewheel switching period.

3.6.2 Adaptive PCCM Operation Schemes

To identify the optimal operation point, Figure 3.11 illustrates the inductor current I_L in different operation conditions. The proposed adaptive PCCM operation scheme aims to provide a universal method to maintain this optimization in different load conditions. It should be noted that in any case each freewheel switching period should not be too short in order to prevent the converter from entering CCM. Neither should it be too long to cause large conduction loss. This is easy to achieve in balanced load conditions such as in Figure 3.11a. However, in unbalanced load conditions as in Figure 3.11b, due to the fixed phase durations, the freewheel switching period for the light-load output $(D_{fw2}T)$ becomes much longer. The conduction loss of S_{fw}, thus, increases significantly. A straightforward way to reduce it is to lower the freewheel switching I_{dc} level. However, since the phase durations are fixed, this may cause the other subconverter to enter CCM (Figure 3.11c), resulting in cross-regulation. Therefore, to maintain the optimal freewheel switching duration for each subconverter, both the phase durations and I_{dc} level should be adaptively controlled (Figure 3.11d).

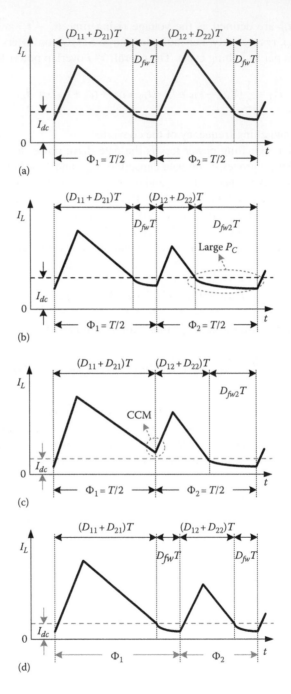

FIGURE 3.11 Inductor current waveform in pseudo-continuous conduction mode with (a) balanced loads, (b) unbalanced loads, (c) lowered I_{dc}, and (d) adaptively controlled phase durations and I_{dc}. (From Zhang, Y. and Ma, D., *J. Analog Integr. Circuits Signal Process.*, 72(2), 419, August 2012b.)

3.6.2.1　Adaptive PCCM with Distributed Freewheel Switching

The adaptive PCCM operation scheme with distributed freewheel switching is illustrated in Figure 3.12. Initially, the two phases of a SIDO converter are set equally as $T/2$, with a preset freewheel switching current at I_{dc}. The duration of $\Sigma D_{ki}T$ is defined as

$$\Sigma D_{ki}T = \sum_{k=1}^{2} D_{ki}T = (D_{1i} + D_{2i})T. \tag{3.14}$$

If the load current I_{LOAD1} of subconverter 1 suddenly increases (case (ii) in Figure 3.12), $\Sigma D_{k1}T$ is extended by ΔDT to accommodate this load power increase, resulting in a decreased freewheel switching duration in Φ_1. The two freewheel switching periods

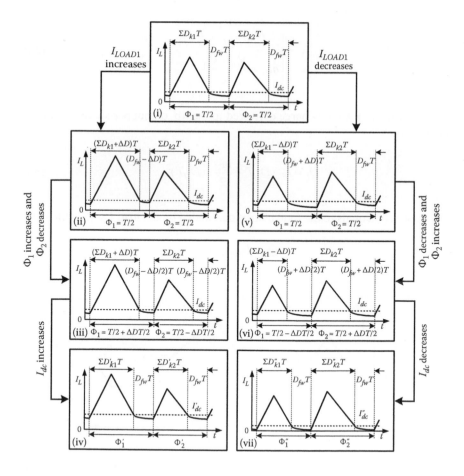

FIGURE 3.12　Adaptive pseudo-continuous conduction mode operation scheme with distributed freewheel switching. (From Zhang, Y. and Ma, D., *J. Analog Integr. Circuits Signal Process.*, 72(2), 419, August 2012b.)

are then equalized by adjusting the phase durations Φ_1 and Φ_2 in case (iii). Since this average freewheel switching duration is smaller than the optimal value, the I_{dc} level is gradually increased to prolong the freewheel switching periods. Thus, the phase adjustment and I_{dc} modulation take place iteratively to maintain the optimal freewheel switching durations. Similar actions apply when I_{LOAD1} suddenly decreases. To quantize the relation between phase duration, I_{dc}, and the instant loads, the inductor current for a SIMO converter in PCCM with distributed freewheel switching is reexamined in Figure 3.13. Here, the ratio of D_{1i} to D_{2i} is defined by

$$\frac{D_{1i}}{D_{2i}} = \frac{m_{2i}}{m_{1i}}, \tag{3.15}$$

where m_{1i} and m_{2i} are the slopes of I_L and can be defined as

$$\begin{cases} m_{1i} = \dfrac{V_{IN}}{L} \\ m_{2i} = \dfrac{V_{Oi} - V_{IN}}{L} \end{cases}. \tag{3.16}$$

On the other hand, the total charge delivered to the ith output per switching cycle is calculated as

$$Q_i = I_{dc} \cdot D_{2i}T + \frac{m_{2i}}{2} \cdot (D_{2i}T)^2. \tag{3.17}$$

In steady state, it equals to the total charge demanded by the load I_{LOADi}, which gives

$$Q_i = I_{LOADi} \cdot T. \tag{3.18}$$

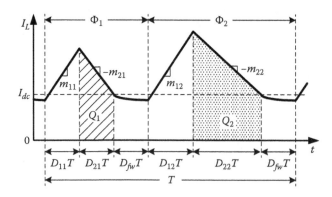

FIGURE 3.13 I_L with distributed freewheel switching. (From Zhang, Y. and Ma, D., *J. Analog Integr. Circuits Signal Process.*, 72(2), 419, August 2012b.)

Substituting Equation 3.18 into Equation 3.17 gives

$$D_{2i}T = \frac{\sqrt{I_{dc}^2 + 2m_{2i}T \cdot I_{LOADi}} - I_{dc}}{m_{2i}}. \tag{3.19}$$

Based on Equations 3.15 and 3.19, the adaptive adjustment on each phase duration follows

$$\Phi_i = \frac{m_{1i} + m_{2i}}{m_{1i} \cdot m_{2i}} \left(\sqrt{I_{dc}^2 + 2m_{2i}T \cdot I_{LOADi}} - I_{dc} \right) + D_{fw}T. \tag{3.20}$$

3.6.2.2 Adaptive PCCM with Unified Freewheel Switching

Alternatively, unified freewheel switching can be implemented, where the freewheel switching only occurs once per switching cycle. According to Equation 3.13, the switching power loss of S_{fw} can be reduced (Zhang and Ma 2011). This saving can be significant when the number of the subconverters in a SIMO converter is large.

Here, a SIDO converter is employed as an example. If I_{LOAD1} suddenly increases from the steady state in case (i) in Figure 3.14, the corresponding phase duration of Φ_1 is extended by $\Delta D_1 T$ to allow more inductor current to be delivered to V_{O1}, thereby stabilizing V_{O1}. Because of using the same start and end levels of the I_{dc}, Φ_2 is time-shifted until I_L in Φ_1 returns to I_{dc}. The duration of Φ_2 remains the same (as in case (ii)). Hence, the total current delivered to V_{O2} in each switching cycle remains unchanged through the adjustment on the duration of the freewheel switching. At the end, the variation $\Delta D_1 T$ is completely absorbed through the freewheel

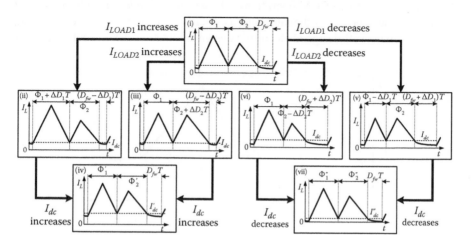

FIGURE 3.14 Adaptive pseudo-continuous conduction mode operation scheme with unified freewheel switching. (From Zhang, Y. and Ma, D., *J. Analog Integr. Circuits Signal Process.*, 72(2), 419, August 2012b.)

switching with a shorter freewheel switching duration $(D_{fw} - \Delta D_1)T$ in case (ii). Similar adaptive operation process can be applied when I_{LOAD2} increases, as shown in case (iii). Conversely, if one load current is suddenly decreased, the corresponding phase duration will be reduced and the freewheel switching period will be prolonged (cases (v) and (vi) of Figure 3.14). It should be noted that, as the regulation period for each subconverter is not isolated from each other, risk of cross-regulation exists in this scheme.

In addition, the freewheel switching current level I_{dc} also needs to be adjusted to maintain the optimal freewheel switching duration. In the proposed operation scheme, if the monitored freewheel switching duration is shorter than the desired value (cases (ii) and (iii) in Figure 3.14), the freewheel switching current will be gradually increased from I_{dc} to a higher level I'_{dc}, thereby recovering the optimal value of $D_{fw}T$ (case (iv) in Figure 3.14). This adjustment prevents the freewheel switching duration from disappearing (CCM) when a sudden load change occurs. Similarly, the freewheel switching current gradually decreases to a lower I''_{dc} (case (vii) in Figure 3.14) when the freewheel switching duration is longer than the desired value (cases (v) and (vi) in Figure 3.14). It should be noted that the duration ratio between Φ'_1 (Φ''_1) and Φ'_2 (Φ''_2) in case (iv) (case (vii)) depends on the previous condition in case (ii) (case (v)) or case (iii) (case (vi)). Similarly, to quantize the relation between phase duration, I_{dc}, and instant loads, the control equation can be obtained from Figure 3.15 such that

$$\Phi_i = \frac{m_{1i} + m_{2i}}{m_{1i} \cdot m_{2i}} \left(\sqrt{I_{dc}^2 + 2m_{2i}T \cdot I_{LOADi}} - I_{dc} \right). \tag{3.21}$$

3.6.3 SYSTEM ARCHITECTURE AND CIRCUIT DESIGN

The proposed adaptive PCCM operations schemes are suitable for both digital and analog implementations. As digital design is usually portable to technology scaling and backed up by well-designed electronic design automation (EDA) tools, it always

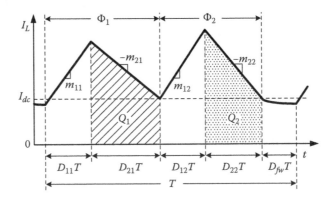

FIGURE 3.15 I_L with unified freewheel switching. (From Zhang, Y. and Ma, D., *J. Analog Integr. Circuits Signal Process.*, 72(2), 419, August 2012b.)

carries a great commercialization potential. Hence, a digital implementation is introduced first using the distributed freewheel switching scheme as major control algorithm. On the other hand, in many cases, analog approach can be very efficient in the aspects of silicon and power consumption. An analog implementation using unified freewheel switching scheme is, thus, discussed as well. In general, the proposed schemes are flexible to either digital or analog processing and can be virtually implemented in any fabrication processes.

3.6.3.1 Implementation of Digital Adaptive PCCM

Figure 3.16 shows the system block diagram of a SIDO boost converter, with the control scheme proposed in Section 3.6.2.1. The operation can be described with reference to Figure 3.17. Consider a SIMO converter has n subconverters. Initially, the phase period for each subconverter and freewheel switching I_{dc} levels are set equal ($\Phi_1 = \Phi_2 = \cdots = \Phi_n$). As the load differs at each output, the freewheel switching duration in each phase differs accordingly.

Such a freewheel switching duration $D_{fwi}T$ ($1 \leq i \leq n$) is measured by a high-frequency digital counter, which also computes the average freewheel switching

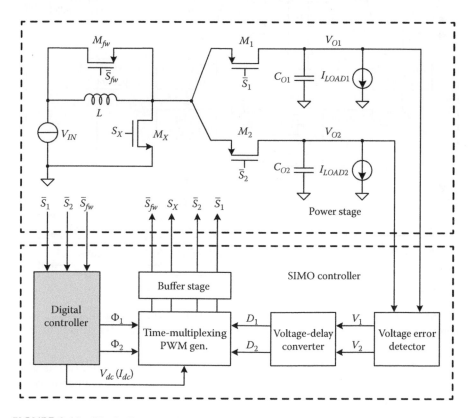

FIGURE 3.16 Block diagram of an adaptive pseudo-continuous conduction mode single-inductor dual-output boost converter with digital distributed freewheel switching. (From Zhang, Y. and Ma, D., *J. Analog Integr. Circuits Signal Process.*, 72(2), 419, August 2012b.)

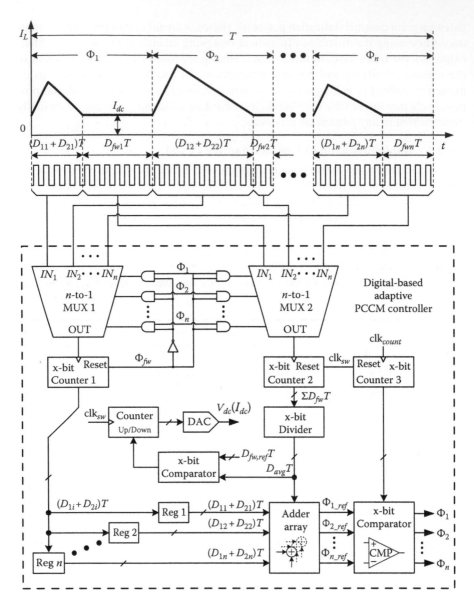

FIGURE 3.17 Implementation of the digital controller with timing diagram. (From Zhang, Y. and Ma, D., *J. Analog Integr. Circuits Signal Process.*, 72(2), 419, August 2012b.)

duration $D_{avg}T$ $(=\Sigma D_{fwi}T/n)$. Each charge and discharge period $(D_{1i} + D_{2i})T$ is also measured. By adding $D_{avg}T$ to $(D_{1i} + D_{2i})T$, the phase duration Φ_i for subconverter i in the next switching cycle is updated as

$$\Phi_i = (D_{1i} + D_{2i} + D_{avg})T. \tag{3.22}$$

In practice, the $D_{avg}T$ should not be too long in order to reduce conduction loss according to Equation 3.8. If it is longer than a predefined limit $D_{fw,max}T$, the adjustment on freewheel switching I_{dc} will be activated. I_{dc} is then decreased by a step of $(I_{dc,max} - I_{dc,min})/2^x$ for each cycle, where x represents the number of bits in the digital-to-analog converter (DAC) of Figure 3.17. As I_{dc} decreases, $(D_{1i} + D_{2i})T$ extends to keep the delivered current to be the same as in the last switching cycle. As a result, the freewheel switching periods $D_{fwi}T$ in all subconverters decrease to reduce $D_{avg}T$ to be shorter than $D_{fw,max}T$. Similarly, reverse adjustment steps will be taken if $D_{avg}T$ is smaller than $D_{fw,min}T$, which is at high risk of cross-regulation. Once the instant $D_{avg}T$ falls between $D_{fw,min}T$ and $D_{fw,max}T$, I_{dc} adjustment is then completed. The iterative averaging on $D_{fwi}T$ leads the optimal operation eventually.

3.6.3.2 Implementation of Analog Adaptive PCCM

The adaptive PCCM operation scheme with unified freewheel switching is demonstrated through an analog implementation. In addition to the error amplifiers EA_i to determine the peak inductor current for each subconverter, the analog controller includes an online analog charge meter to adaptively adjust I_{dc} level, as shown in Figure 3.18. The inductor current is sensed by monitoring the currents flowing through M_X, M_1, and M_2, which are then converted to voltage signals and compared with V_{EAOi} and V_{dc}, thereby defining the duty ratios, phase durations, and freewheel switching durations. To explain the operating mechanism, a key voltage V_{dc} highlighted in Figure 3.18 is illustrated in Figure 3.19.

During the discharge period of $D_{2i}T$ in subconverter i ($i = 1$ or 2), C_{dc} in Figure 3.18 is charged by a constant current I_{ch}. When a freewheel switching action occurs, a constant sink current I_{dch}, which is equal to $m \times I_{ch}$, is activated to discharge C_{dc}. In steady state (solid line in Figure 3.19), it satisfies

$$I_{ch} \cdot (D_{21}T + D_{22}T) = I_{dch} \cdot D_{fw}T. \tag{3.23}$$

V_{dc} keeps constant and an optimal freewheel switching duration can be achieved by setting an appropriate m value. If a load increase occurred in the last phase (Figure 3.19a), the corresponding charge and discharge periods of inductor current ($D'_{12}T$ and $D'_{22}T$) need to be extended in order to deliver more power. The freewheel switching duration ($D'_{fw}T$) is reduced accordingly, resulting in

$$I_{ch} \cdot (D'_{21}T + D'_{22}T) > I_{dch} \cdot D'_{fw}T. \tag{3.24}$$

Consequently, V_{dc} is increased to V'_{dc} and a higher I_{dc} level is thus achieved. Similarly, when a load decrease occurs, V_{dc} decreases due to

$$I_{ch} \cdot (D''_{21}T + D''_{22}T) < I_{dch} \cdot D'''_{fw}T, \tag{3.25}$$

leading to a lower I_{dc} level. On the other hand, when a load change occurs in the foregoing phase (Figure 3.19b), $D_{11}T$ and $D_{21}T$ show similar variations. Since the phase

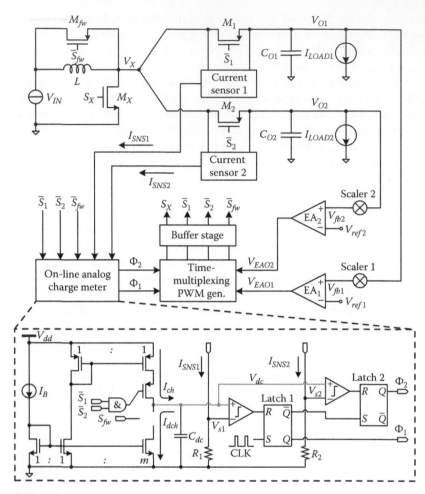

FIGURE 3.18 Circuit detail of single-inductor dual-output converter with proposed analog controller. (From Zhang, Y. and Ma, D., *J. Analog Integr. Circuits Signal Process.*, 72(2), 419, August 2012b.)

duration of subconverter 2 keeps constant (Φ_2, Φ_2', and Φ_2'' are all equal), the variations are added to the freewheel switching period at the end of the switching cycle, thereby adjusting V_{dc} accordingly. With this method, I_{dc} can be adaptively adjusted based on instantaneous load condition, thus improving the conduction loss. Moreover, the frequency of freewheel switching actions is significantly reduced, leading to the reduction of switching power loss.

After the V_{dc} level is stabilized, the phase duration Φ_i of each subconverter is also automatically adjusted by the online charge meter. As shown in Figure 3.18, latch 1 is set by the clock signal *CLK* at the switching frequency of the SIMO converter. The discharge inductor current is sensed and converted to voltage signal V_{s1}. Once V_{s1} becomes lower than V_{dc}, Φ_1 expires and Φ_2 is initiated by latch 2. Similarly, Φ_2 expires when I_L ramps down to I_{dc} level.

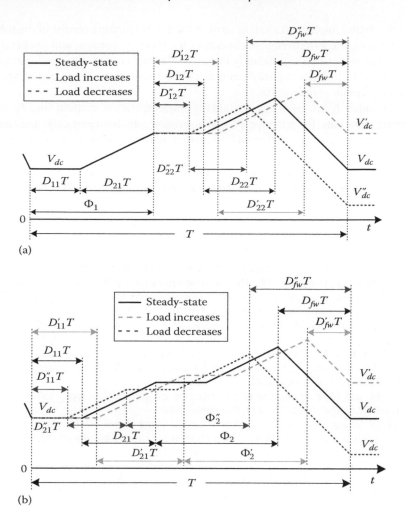

FIGURE 3.19 V_{dc} adjustment with load variation occurring in (a) the last phase and (b) the foregoing phase. (From Zhang, Y. and Ma, D., *J. Analog Integr. Circuits Signal Process.*, 72(2), 419, August 2012b.)

3.7 CONCLUSIONS

In this section, two adaptive PCCM operation schemes with the respective distributed and unified freewheel switching schemes were discussed. Compared with conventional PCCM operations, the phase durations and I_{dc} level can be adaptively modulated, thereby maintaining optimal freewheel switching in various load conditions. As a result, power loss due to freewheel switching can be significantly reduced while retaining low cross-regulation effect. In addition, the two proposed operation schemes were implemented by both digital and analog methods. The advantages for each implementation method are also discussed respectively.

For the digital implementation, as most of the control circuits consist of the robust digital circuits, the system robustness across the process, voltage, and temperature (PVT) variations can be significantly improved. Moreover, with the development of modern VLSI techniques, digital circuits usually consume less silicon area compared with analog circuit counterparts.

On the other hand, by using the analog implementation method, the I_{dc} level and phase durations for adaptive PCCM operation can be accurately fine-tuned. The adjustment resolution is thus well improved, resulting in more accurate control scheme. Based on the specific application background, the digital and analog circuit implementation, as well as the distributed/unified freewheel switching, can also be combined in different ways to define the most appropriate adaptive PCCM operation scheme for SIMO converters.

REFERENCES

Belloni, M., E. Bonizzoni, E. Kiseliovas, P. Malcovati, F. Maloberti, T. Peltola, and T. Teppo, A 4-output single-inductor DC–DC buck converter with self-boosted switch drivers and 1.2A total output current, in *IEEE International Solid-State Circuits Conference Digest of Technical Papers*, San Francisco, CA, February 3–7, 2008, pp. 444–445.

Burd, T. D., T. A. Pering, A. J. Stratakos, and R. W. Brodersen, A dynamic voltage scaled microprocessor system, *IEEE J. Solid State Circuits*, 35(11), 1571–1580, 2000.

Chae, C.-S., H.-P. Le, K.-C. Lee, G.-H. Cho, and G.-H. Cho, A single-inductor step-up DC–DC switching converter with bipolar outputs for active matrix OLED mobile display panels, *IEEE J. Solid State Circuits*, 44(2), 509–524, 2009.

Chang, J.-M. and M. Pedram, Energy minimization using multiple supply voltages, *IEEE Trans. VLSI Syst.*, 5(4), 436–443, December 1997.

Chen, H., Y. Zhang, and D. Ma, A SIMO parallel-string driver IC for dimmable LED backlighting with local bus voltage optimization and single time-shared regulation loop, *IEEE Trans. Power Electron.*, 27(1), 452–462, 2012.

Cheng, W. H. and B. M. Baas, Dynamic voltage and frequency scaling circuits with two supply voltages, in *Proceedings of the IEEE International Symposium on Circuits and Systems (ISCAS)*, Seattle, WA, May 18–21, 2008, pp. 1236–1239.

Goder, D. and H. Santo, Multiple output regulator with time sequencing, US patent 5, 617, 015, April 1, 1997.

Huang, M.-H., Y.-N. Tsai, and K.-H. Chen, Sub-1V input single-inductor dual-output (SIDO) DC–DC converter with adaptive load-tracking control (ALTC) for single-cell-powered systems, *IEEE Trans. Power Electron.*, 25(7), 1713–1724, 2010.

Ki, W.-H. and D. Ma, Single-inductor multiple-output switching converters, in *IEEE Power Electronics Specialists Conference (PESC)*, Vancouver, British Columbia, Canada, June 17–21, 2001, pp. 226–231.

Kim, S. and G. A. Rincon-Mora, Single-inductor dual-input dual-output buck-boost fuel-cell-Li-ion charging DC–DC converter supply, in *IEEE International Solid-State Circuits Conference Digest of Technical Papers*, San Francisco, CA, February 8–12, 2009, pp. 444–445.

Lee, Y.-H., M.-H. Huang, Y.-N. Tsai, M.-Y. Fan, and K.-H. Chen, A single-inductor multiple positive and negative outputs (SIMPNO) converter with a vector current control mode for electronic paper displays (EPDs), in *IEEE European Solid-State Circuits Conference Digest of Technical Papers*, San Francisco, CA, September 2010, pp. 446–449.

Li, T., Single inductor multiple output boost regulator, US patent 6, 075, 295, June 13, 2000.

Luo, J., N. K. Jha, and L.-S. Peh, Simultaneous dynamic voltage scaling of processors and communication links in real-time distributed embedded systems, *IEEE Trans. VLSI Syst.*, 15, 427–437, 2007.

Ma, D., Automatic substrate switching circuit for on-chip adaptive power-supply system, *IEEE Trans. Circuits Syst. II, Express Briefs*, 54(7), 641–645, July 2007.

Ma, D. and R. Bondade, Enabling power-efficient DVFS operations on silicon, *IEEE Circuits Syst. Mag.*, 10(1), 14–30, 2010.

Ma, D., W.-H. Ki, and C.-Y. Tsui, A pseudo-CCM/DCM SIMO switching converter with freewheel switching, in *Proceedings of the IEEE International Solid-State Circuits Conference*, San Francisco, CA, February 7, 2002, pp. 390–391.

Ma, D., W.-H. Ki, and C.-Y. Tsui, A pseudo-CCM/DCM SIMO switching converter with freewheel switching, *IEEE J. Solid State Circuits*, 38(6), 1007–1014, 2003a.

Ma, D., W.-H. Ki, C.-Y. Tsui, and P. K. T. Mok, A 1.8V single-inductor dual-output switching converter for power reduction techniques, in *IEEE Symposium on VLSI Circuits*, Kyoto, Japan, June 14–16, 2001a, pp. 137–140.

Ma, D., W.-H. Ki, C.-Y. Tsui, and P. K. T. Mok, Single-inductor multiple-output switching converters with bipolar outputs, in *Proceedings of the IEEE International Symposium on Circuits and Systems (ISCAS)*, Sydney, New South Wales, Australia, May 6–9, 2001b, pp. 301–304.

Ma, D., W.-H. Ki, C.-Y. Tsui, and P. K. T. Mok, Single-inductor multiple-output switching converters with time-multiplexing control in discontinuous conduction mode, *IEEE J. Solid State Circuits*, 38(1), 89–100, 2003b.

May, M. W., M. R. May, and J. E. Willis, A synchronous dual-output switching dc–dc converter using multibit noise-shaped switch control, in *IEEE International Solid-State Circuits Conference Digest of Technical Papers*, San Francisco, CA, February 5–7, 2001, pp. 358–359.

Sze, N.-M., F. Su, Y.-H. Lam, W.-H. Ki, and C.-Y. Tsui, Integrated single-inductor dual-input dual-output boost converter for energy harvesting applications, in *Proceedings of the IEEE International Symposium on Circuits and Systems (ISCAS)*, Seattle, WA, May 18–21, 2008, pp. 2218–2221.

Usami, K. et al., Automated low-power technique exploiting multiple supply voltages applied to a media processor, *IEEE J. Solid State Circuits*, 33(3), 463–472, 1998.

Zhang, Y., Single-inductor multiple-output power converters: Architectures, control techniques and applications, PhD dissertation, The University of Texas at Dallas, Richardson, TX, 2013.

Zhang, Y., R. Bondade, D. Ma, and S. Abedinpour, An integrated SIDO boost power converter with adaptive freewheel switching technique, in *Energy Conversion Congress and Exposition (ECCE)*, Atlanta, GA, September 12–16, 2010, pp. 3516–3522.

Zhang, Y. and D. Ma, Digitally controlled integrated pseudo-CCM SIMO converter with adaptive freewheel current modulation, in *IEEE Applied Power Electronics Conference and Exposition (APEC)*, Palm Springs, CA, February 21–25, 2010, pp. 284–288.

Zhang, Y. and D. Ma, Integrated SIMO DC-DC converter with on-line charge meter for adaptive PCCM operation, in *Proceedings of the IEEE International Symposium on Circuits and Systems (ISCAS)*, Rio de Janeiro, Brazil, May 15–18, 2011, pp. 245–248.

Zhang, Y. and D. Ma, Input-self-biased transient-enhanced maximum voltage tracker for low-voltage energy-harvesting applications, *IEEE Trans. Power Electron.*, 27(5), 2227–2230, 2012a.

Zhang, Y. and D. Ma, Adaptive pseudo-continuous conduction mode operation schemes and circuit designs for single-inductor multiple-output switching converters, *J. Analog Integr. Circuits Signal Process.*, 72(2), 419–432, August 2012b.

Zheng, C. and D. Ma, Design of a monolithic automatic substrate/supply multiplexer for DVS-enabled adaptive power converters, *IEEE Trans. Circuits Syst. II, Express Briefs*, 58(6), 376–380, 2011.

4 Circuit Techniques for Improving the Power Density of Switched-Capacitor Converters

Yutian Lei and Robert Pilawa-Podgurski

CONTENTS

4.1 INTRODUCTION

The advent of low-voltage digital circuitry has created a need for improved dc–dc converters. Dc–dc converters that can provide a low-voltage output (<2 V) regulated at high bandwidth, while drawing energy from a higher (5–12 V) input voltage, are desirable. In addition, the size, cost, and performance benefits of integration make it advantageous to integrate as much of the dc–dc converter as possible, including control circuits and power switches. Moreover, in some applications it would be desirable—if possible—to integrate the power converter or portions thereof with the load electronics.

One common approach is the use of a switched-mode power converter (e.g., synchronous buck converter, interleaved synchronous buck, three-level buck, and like designs [1,17,38,43–46,48]). For magnetics-based designs operating at low, narrow-range input voltages (e.g., 2 V in and 1 V out), it is possible to achieve extremely high switching frequencies (up to hundreds of MHz [17,38,43]), along with correspondingly high control bandwidths and small passive components (e.g., inductors and capacitors). It also becomes possible to integrate portions of the converter with a microprocessor load in some cases. These opportunities arise from the ability to use fast, low-voltage, process-compatible transistors in the power converter. However, at higher input voltages and wider input voltage ranges, much lower switching frequencies (on the order of a few MHz and below) are the norm due to the need to use slow extended-voltage transistors (on die) or discrete high-voltage transistors. This results in much lower control bandwidth, and large, bulky passive components (especially magnetics) that are not suitable for integration or co-packaging with the devices.

Another conversion approach that has received attention for low-voltage electronics is the use of switched-capacitor (SC)-based dc–dc converters [6,9,26,28,29,31,39]. This family of converters is well suited for integration and/or co-packaging of passive components with semiconductor devices because they do not require any magnetic devices (inductors or transformers). An SC circuit consists of a network of switches and capacitors, where the switches are turned on and off periodically to cycle the network through different topological states. Depending on the topology of the network and the number of switches and capacitors, efficient step-up or step-down power conversion can be achieved at different conversion ratios.

Many techniques have been proposed to improve the efficiency and power density of SC converters. A general methodology is presented in [40] that provides a high-level guideline for designing and optimizing SC converters, and design techniques particular to complementary metal oxide semiconductor (CMOS) process are presented in [22]. Circuit techniques such as resonant gate-driving techniques [6] have been explored to reduce the gate-driving losses. Likewise, the parasitic bottom plate capacitance loss can be reduced through charge recycling techniques, such as [3], where integrated charge recycling transistors are used during the deadtime interval. Moreover, control techniques such as the digital capacitance modulation method introduced in [35] enable improved efficiency with load regulation. Finally, advanced processes such as deep trench capacitors (explored in [4]) and ferroelectric capacitors (explored in [14]) are used for even higher power density and efficiency.

This chapter focuses on the fundamental drawbacks of SC converters, which limit their performance in many applications [37]. Since the capacitors are directly charged/discharged by other capacitors or voltage sources, large transient current spikes can occur, which limit the efficiency and power density of the converter. Moreover, these transient effects increase device stress and can cause undesirable electromagnetic interference problems. To reduce current spikes, either large capacitors or higher switching frequency has to be employed, neither of which is a satisfactory solution. The quasi-SC converter given in [11] manages to reduce the peak of the current transient, but results in the same power loss as conventional SC converters.

Another drawback of SC converters is that high efficiency is only achieved at one or a few conversion ratios. This limits the application of SC converters to mostly low-power applications. In higher-power applications, the solution is usually to cascade a magnetic converter to act as a postregulation stage [2,27,36,44]. However, the overall converter size may increase and the peak conversion efficiency may be reduced.

In Section 4.2, we outline the fundamental challenges associated with reduced power density owing to poor utilization of the capacitors. We also introduce the concept of soft-charging operation, which has recently been proposed to increase the efficiency and power density of SC converters. Sections 4.3 and 4.4 present two practical implementations of soft-charging SC converters using a second-stage converter and an output inductor, respectively. Section 4.5 presents an analytical framework for determining whether *any* SC topology is compatible with soft-charging operation. Finally, in Section 4.6, a recently proposed control technique to achieve soft-charging operation of a Dickson converter is presented.

4.2 CAPACITOR CHARGE SHARING LOSS AND SOFT-CHARGING CONCEPT

A generic SC converter model is shown in Figure 4.1, which consists of an ideal fixed conversion ratio stage with an output-referred impedance [26]. The output impedance directly reflects the efficiency of the converter and incorporates both the conduction loss and the capacitor charging/discharging loss. This impedance is usually plotted against the switching frequency to reveal the characteristics of the SC converters. A typical plot is shown in Figure 4.2, which shows two asymptotic operating regions for SC converters: the fast switching limit (FSL) and the slow switching limit (SSL) [6,18,20,29,40]. The FSL occurs at high switching frequencies, when the dominating

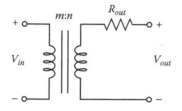

FIGURE 4.1   Generic model of a switched-capacitor converter.

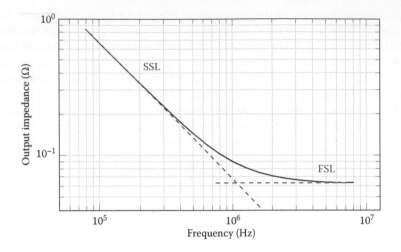

FIGURE 4.2 Output impedance of a typical SC converter.

loss is the conduction loss due to the resistance of the switches as well as the ESR of the capacitors. As can be seen in Figure 4.2, the output impedance in the FSL region is independent of the switching frequency. On the other hand, the SSL occurs at low switching frequencies, when the output impedance is dominated by the charging/ discharging loss of the capacitors during the charge redistribution process at phase transitions. The SSL impedance depends on the switching frequency and capacitor values, and cannot be reduced by lowering the series resistance.

Fundamentally, the SSL power loss is the result of charging/discharging the capacitor with a constant voltage source or another capacitor, as illustrated in Figure 4.3a and b, respectively. We will examine case 2 (Figure 4.3b) more closely since case 1 (Figure 4.3a) can be seen identical to case 2 with C_2 being infinite. Since the capacitor voltages cannot change instantaneously, the mismatch of the initial capacitor voltages after the switch closes will be present across the series resistor, resulting in a large instantaneous current

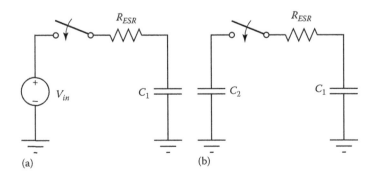

FIGURE 4.3 (a) Capacitor charge by a voltage source. (b) Capacitor charged by another capacitor.

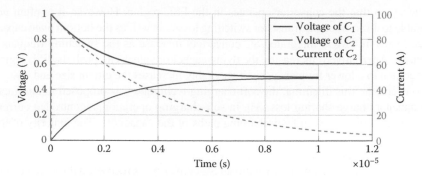

FIGURE 4.4 Capacitor voltages and current waveform in charge redistribution process.

as shown in Figure 4.4. The power loss incurred for complete charge redistribution for the schematic showing in Figure 4.3b can be easily calculated and is given by

$$P_{loss} = \frac{1}{4}C_1\left(V_{C1(t=0)} - V_{C2(t=0)}\right)^2 f_{sw}$$

$$= \frac{1}{4}C_1\Delta V_{(t=0)}^2 f_{sw} \qquad (4.1)$$

assuming $C_1 = C_2$. This equation is valid provided that the duration of each phase is much larger than the time-constant of the circuit, that is, in SSL region of operation. As can be seen, this power loss does not depend on the value of the series resistance. Instead, it depends on the initial voltage difference between the capacitors. Additionally, the initial difference in capacitor voltages is due to the charge transfer in the previous cycle, and thus is proportional to the charge drawn by the load and inversely proportional to the capacitor values. Moreover, the change of charge in the capacitor is proportional to the duration of the charge/discharge, and thus is inversely proportional to the switching frequency. These relations are summarized next:

$$\Delta V \propto \frac{1}{f_{sw}}, \frac{1}{C_{fly}} \qquad (4.2)$$

where C_{fly} represents the overall flying capacitor value, and for the circuit in Figure 4.3b, it is simply C_1. Substituting Equation 4.2 into the power loss equation in (4.1), we have

$$P_{loss} \propto \frac{1}{f_{sw}}, \frac{1}{C_{fly}} \qquad (4.3)$$

For more complex SC converters, more complicated charge sharing scenarios will arise, but the general relationship stays the same, as shown by the analytical results given in [40]. To reduce the charge sharing loss, one may simply increase the switching frequency so that the converter operates in the FSL region. However, it is often not favorable to do so since the transistor switching losses as well as the bottom plate capacitance losses in CMOS integrated SC converters increase as the switching frequency increases. Alternatively, increasing the flying capacitor values can push the FSL region of operation to a lower frequency, but it inevitably increases the circuit size and cost.

To overcome this dilemma, the soft-charging technique was proposed to eliminate the capacitor charge sharing loss [33]. In soft-charging operation, a controlled current load is placed in the charging/discharging paths of the capacitors. The majority of the voltage mismatch between the capacitors and the input/output will be present across the current load instead of across the switch resistance. With this technique, the capacitor charging loss that is present in conventional SC converters is recovered through the controlled current load, which typically is a high-frequency magnetic converter. As a result, smaller capacitance can be used without sacrificing the efficiency despite the resultant larger capacitor voltage ripples. This is the key benefit of soft-charging operation.

There are two requirements for achieving soft-charging operation: The first requirement is that the load must behave as a controlled current source, whose voltage is permitted to change instantaneously to accommodate the voltage mismatch between the flying capacitors and the load during phase transitions. In practice, however, the majority of the loads are voltage-source loads or current-source loads with large decoupling capacitors. Therefore, an interfacing element typically has to be inserted between the SC converters and the voltage-source load. As will be shown in Section 4.3, buck converters can be such an interfacing element, providing controlled charging/discharging of the capacitors while regulating the output voltage [33,34]. In addition, as will be shown in Section 4.4, an inductor at the output of the SC converter can replace the buck converter while still achieving soft-charging operation.

The second requirement for soft-charging operation is that there should be no voltage mismatch within the flying capacitor network during phase transitions. In [33,34], a series-parallel SC stage was chosen as the SC topology for soft-charging operation since the flying capacitors are simultaneously connected to the output either in series or in parallel, and therefore, can always be charged/discharged in the same manner through the load current. Other SC topologies tend to have more complex switching configuration, and it is not immediately apparent whether soft charging of all flying capacitors can be achieved by cascading a second-stage converter (or an inductor) alone. Section 4.5 introduces a formal method for evaluating an arbitrary SC topology and determining whether soft-charging operation is possible. This method replaces the ad hoc approaches that have been employed to date [5,33].

4.3 SOFT-CHARGING OPERATION WITH POWER CONVERTER

Shown in Figure 4.5 is a two-stage architecture that combines a high-efficiency SC transformation stage with a high-frequency regulation stage. The architecture, first introduced in [33], achieves both large voltage step-down and high bandwidth regulation across a wide output and input voltage range. The architecture makes use of the transistors typically available in a CMOS process: slow, moderate blocking voltage devices (e.g., thick gate oxide and/or extended drain transistors) and fast, low-voltage transistors. The SC transformation stage, employing slow-switching moderate

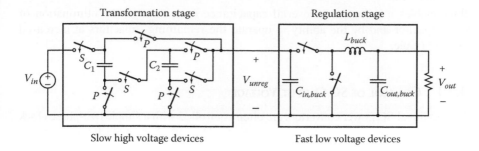

FIGURE 4.5 Block diagram illustrating the two-stage architecture, which enables a single-die power converter providing both large voltage step-down and high-frequency operation.

voltage devices and off-chip capacitors, can be designed for very high-power density and efficiency. The intermediate voltage, V_{unreg}, can be made sufficiently low such that the regulation stage can utilize low-voltage, fast-switching transistors that enable high switching frequency with correspondingly high bandwidth regulation and small passive components.

The separation of the transformation (step-down) and regulation functionality of the converter into two stages provides substantial advantage: the architecture makes use of the inherent advantages of SC power converters (high-voltage step-down, high efficiency), while not tasking it with regulation, which SC converters cannot do efficiently. The regulation functionality is performed by the low-voltage synchronous buck converter, and since that stage operates at low voltage and transformation ratio, it can operate at high frequency with small magnetics size. As shown in [33], substantial advantages in terms of size and efficiency can be realized by employing highly scaled CMOS transistors in switched-mode power converters.

The architecture of Figure 4.5 can provide yet another attractive benefit if designed and operated in a specific manner ("merging" of the two stages). In this case, the regulation stage can provide *soft charging* of the SC stage, a mode of operation that provides increased efficiency and power density of the SC converter as compared to conventional designs. In the circuit of Figure 4.5, the high-frequency regulation stage (a synchronous buck converter) provides soft charging for the series-parallel SC transformation stage. When the SC stage is configured to charge capacitors C_1 and C_2 in series (switches S closed), the capacitors are charged at a rate determined by the power drawn from the regulating stage, ensuring soft-charging operation. When the SC stage is configured to discharge C_1 and C_2 in parallel (switches P closed), the capacitor voltages appear directly across the input terminal of the regulating stage, providing soft discharging of the capacitors. It should be noted that the input capacitor of the buck converter, $C_{in,buck}$, is considerably smaller than C_1 and C_2 as it serves only to filter the high-frequency ripple of the buck converter.

Another advantage of the merged two-stage architecture is that the number of energy transfer capacitors in the SC stage can be reduced from N to $N-1$ for an N-to-1 step-down topology. In the 3-to-1 transformation stage of Figure 4.5, only two capacitors (C_1 and C_2) are used compared to the three capacitors typically required in a conventional 3-to-1 series-parallel SC converter. The proposed architecture thus enables a reduction in overall capacitance both through the elimination of one capacitor and by the ability to operate the remaining capacitors at increased voltage ripple.

4.3.1 CONTROL OF SC OUTPUT VOLTAGE

Conventional SC power converters are often controlled with a simple two-phase clock to alternate between two switch configurations. The soft-charging technique requires a more sophisticated control implementation to ensure that the SC output voltage stays within a suitable range. Figure 4.6 shows a hysteretic control strategy that ensures that the input voltage to the regulation stage (V_{unreg}) is maintained below a maximum value (V_{max}). The value of V_{max} is chosen to be below the maximum operating voltage of the (low-voltage) transistors of the regulation stage. The two reference voltages V_{ref1} and V_{ref2} can be derived from the input voltage V_{IN} and V_{max}, as shown here:

$$V_{ref1} = V_{IN} - 2V_{max} \tag{4.4}$$

$$V_{ref2} = \frac{V_{IN} - V_{max}}{2} \tag{4.5}$$

While the SC stage operates with significant output voltage ripple (as seen in Figure 4.6), the frequency of this ripple is substantially lower than the control bandwidth of the regulation stage (this is possible because the switching frequency of the regulation stage is many times higher than the switching frequency of the SC stage). However, the transitions at the switching intervals nevertheless present a problem to a regular feedback controller, and the sharp edges can appear at the output (audio susceptibility). In order to maintain the output voltage steady despite the sharp voltage transitions at its input terminals, the regulation stage can employ feed-forward control, as shown in the block diagram of Figure 4.7.

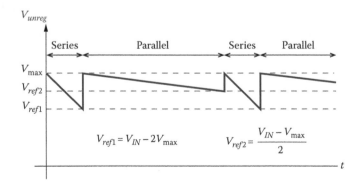

FIGURE 4.6 Idealized waveforms illustrating the switched-capacitor control strategy. This approach maintains the SC stage output voltage below V_{max}, thus enabling the use of low-voltage devices in the regulation stage.

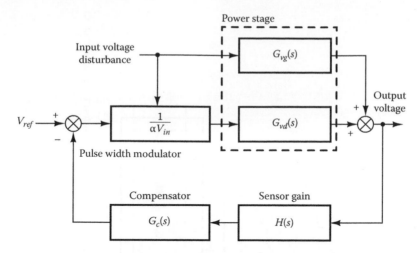

FIGURE 4.7 Block diagram illustrating the feed-forward control used in the system. The gain of the PWM block is inversely proportional to the input voltage, enabling cycle-by-cycle feed-forward control with fast response.

In this implementation, the pulse width modulator gain is inversely proportional to the input voltage, rather than fixed, as is done in conventional feedback-only control. The effect of this change is that any input voltage disturbance that is propagated from the input to the output through the line-to-output transfer function $G_{vs}(s)$ can be cancelled by an immediate change in the input to the control-to-output transfer function $G_{vd}(s)$ [15].

A schematic drawing illustrating an implementation of the feed-forward control is shown in Figure 4.8. The feed-forward control is accomplished by making the amplitude of the triangle wave reference of the comparator in the comparator in the pulse-width modulation (PWM) block proportional to the buck converter input voltage (V_{unreg}). With this method, any sharp edges in the input voltage will immediately appear as an increased triangle amplitude at the input of the comparator, which controls the PWM signal to the gate. The overall feedback loop (controlled by the compensation network of the error amplifier) can still be kept slow enough to ensure stability, while the response of the feed-forward control can be made very fast and is only limited by the speed of the transconductance amplifier that controls the triangle waveform amplitude. In the work of [34], the feed-forward circuitry of Figure 4.8 was developed in a 180 nm CMOS process, as shown in the die photograph of Figure 4.9. Shown in Figure 4.10 are experimental waveforms, capturing a load-step event, as well as illustrating the significant attenuation of the input voltage ripple at the input of the regulating converter.

Detailed descriptions of the design and implementation of the circuit blocks of Figure 4.8 are provided in [32].

4.4 SOFT-CHARGING OPERATION WITH AN OUTPUT INDUCTOR

Since an inductor allows instantaneous change of its terminal voltage, it can also act as a controlled current load [25,41]. In fact, the buck converter is able to facilitate soft-charging operation precisely because of the inductor it contains. In this section,

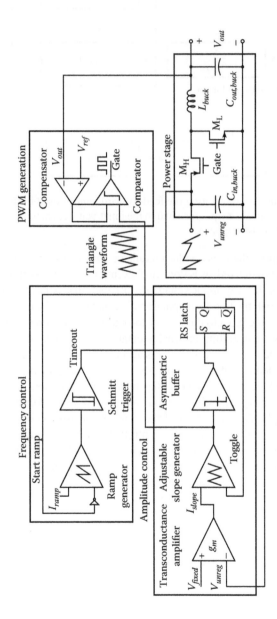

FIGURE 4.8 Schematic block diagram illustrating the feed-forward control implementation. The height of the triangle waveform is proportional to the regulation stage input voltage (V_{unreg}).

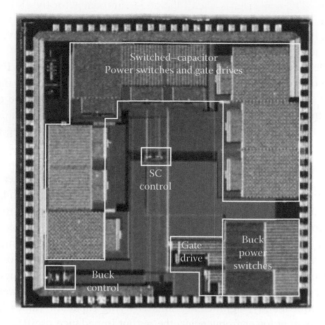

FIGURE 4.9 Die photograph of the converter implemented in 180 nm CMOS technology. The total die area is 3.6 × 3.6 mm, although much of that area is not used.

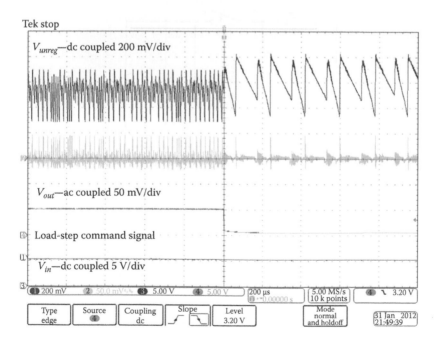

FIGURE 4.10 Experimental waveforms showing the converter response to a load-step between 90% and 10% of full load. Note that the input voltage is 5 V, and the output voltage is steady at 1 V, despite the large voltage swings at the input of the buck converter (V_{unreg}).

the technique of adding an inductor alone to achieve soft charging is presented. Furthermore, it will be shown that resonant operation can also be achieved using the same technique.

To illustrate the technique, a simple 2-to-1 SC converter is shown in Figure 4.11a, and a modified structure is shown in Figure 4.11b. As can be seen in Figure 4.11b, the technique to achieve resonant and soft-charging operation is to add an inductor at the output of the SC converter, immediately before the output capacitor. The simple circuit structure in Figure 4.11b allows direct circuit analysis using differential equations [41]. Figure 4.12 plots the simulated output impedance as a function of frequency for both the original SC converter (Figure 4.11a) and the modified converter (Figure 4.11b). It can be seen that for the original SC converter the output impedance reduces as the frequency increases, while leveling off at high frequencies, marking the transition from SSL to FSL. The output impedance curve is more complicated for the modified converter, but a few key observations can be made. First, with the additional inductor, the modified converter is able to reach the same minimum impedance at a much lower switching frequency due to the elimination of the current transient and associated loss. Therefore, the proposed converter can achieve the same efficiency as conventional SC converters while using significantly lower switching frequency, or equivalently, significantly smaller flying capacitor values. The second observation is that, at lower frequencies, the output impedance oscillates around the SSL impedance of the conventional SC converter.

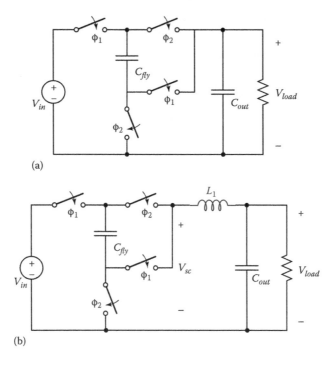

FIGURE 4.11 (a) Example 2-to-1 SC converter. (b) Example 2-to-1 SC converter with an inductor.

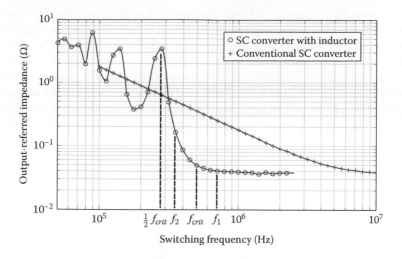

FIGURE 4.12 Simulated output impedance versus frequency.

The minimum frequency at which the converter is able to stay in FSL operation can be defined as f_{crit}, and for the modified converter in Figure 4.11b, it is given by

$$f_{crit} = \frac{1}{2\pi\sqrt{LC}} \qquad (4.6)$$

where
  L is the added inductance
  C is the collective capacitance in series with the inductor

In the case of the example SC converter in Figure 4.11b, the capacitance is simply C_{fly}. For more complex SC converter topologies, this equivalent capacitance can be obtained by calculating the equivalent series and parallel connected capacitance at the inductor input node for each phase. For converter topologies that have different equivalent capacitance in each phase (such as the Fibonacci and series-parallel), there is a critical frequency for each phase, and the overall critical frequency is the weighted average (by duty ratio) of the individual frequencies. Note that the critical frequency corresponds to the resonant frequency of the circuit. To understand the frequency-dependent behavior of the modified SC converter, the terminal voltage before the inductor (V_{sc} in Figure 4.11b) as well as the inductor current are shown in Figure 4.13 at three different frequencies—the resonant frequency (f_{crit}) as well as below (f_2) and above (f_1) the resonant frequency. It can be seen that, above the resonant frequency, the current waveform (Figure 4.13a) is smooth and has small ripple due to the filtering effect of the inductor. Moreover, since the flying capacitor is always in the same current path as the inductor, the conventional current spikes of the capacitor are eliminated, and the capacitors transfer charges in soft-charging mode, with no charge transfer loss. The effect of this can be seen directly from Figure 4.12, where for switching frequencies larger than the critical frequency the SC converter has the minimum FSL output impedance. As the switching frequency is reduced, the current waveform has larger ripple, while having the same average value, since the load current is kept constant. At the resonant frequency, f_{crit}, the inductor current takes the shape of a rectified sinusoid, and the current reaches zero at moments of phase transitions, as shown in Figure 4.13b. Thus, zero current switching can be achieved at the resonant frequency. As can be seen in Figure 4.12, the impedance of the converter at resonance is slightly larger than the FSL impedance. This is because the sinusoidal current has larger RMS value than the near constant current in FSL operation. As the switching frequency is reduced further (Figure 4.13c), the inductor current drops negative during each cycle, resulting in a much larger RMS current for the same average

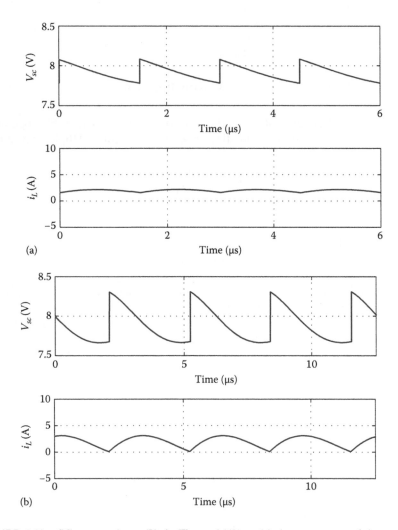

(a)

(b)

FIGURE 4.13 SC stage voltage (V_{sc} in Figure 4.11b) and inductor current of the modified 2-to-1 converter. (a) $f_{sw} = f_1 > f_{crit}$. (b) $f_{sw} = f_{crit}$.   (*Continued*)

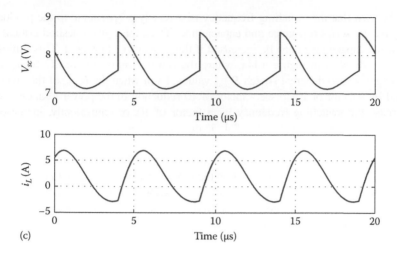

FIGURE 4.13 (*Continued*)   SC stage voltage (V_{sc} in Figure 4.11b) and inductor current of the modified 2-to-1 converter. (c) $f_{sw} = f_2 < f_{crit}$.

power delivered. This is why the impedance increases sharply for $f_{sw} < f_{crit}$. At one half of the resonant frequency defined by Equation 4.6, the current becomes nearly a full-wave sinusoid, giving a peak impedance in Figure 4.12. This peak repeats itself at lower frequencies when the current waveform has multiple periods of the full-wave sinusoid, at $f_{sw} = 1/nf_{crit}$, where n is an integer and $n \geq 2$. Therefore, f_{crit} given in Equation 4.6 sets the lower bound on the switching frequency for which near FSL impedance in soft-charging operation can be achieved.

4.4.1   TRADE-OFF ANALYSIS

While the soft-charging technique enables a reduction in switching frequency and/or total required capacitance, this comes at the expense of adding an inductor. While the trade-off between the capacitor values and inductor values should be evaluated on a case-by-case basis, in general, adding an inductor results in better utilization of passive components than simply using larger capacitance. This energy utilization must then be carefully weighted against the relative ease at which inductors can be integrated on-die in the case of a fully integrated CMOS power converter. Alternatively, off-chip or bond-wire-based inductors may be used. For a traditional SC converter circuit, the value of the charge redistribution current transient is determined by the time constant of the circuit, $R_{ESR}C$, where R_{ESR} is the series resistance in each conducting branch (including both switch resistance and capacitor parasitic resistance) and C is the capacitance in each branch. Thus, the critical frequency at which the conventional SC converter enters FSL operation is

$$f_{crit} = \frac{1}{2\pi R_{ESR}C} \tag{4.7}$$

It can be seen that the switching frequency is inversely proportional to the product of the equivalent series resistance and capacitance. Thus, for a given desired critical frequency, the capacitance must be increased if the resistance is lowered. This limitation can be clearly seen in Figure 4.14a, where the power loss of a pure SC converter is plotted against two different switch $R_{ds,on}$ values. Even when the $R_{ds,on}$ of the switch is reduced by a factor of 10, to see a factor of 10 reduction in the power loss, one needs to increase the switching frequency by a factor of 10, or equivalently, increase the capacitor values by a factor of 10. This is due to the inversely proportional relationship among the resistance, capacitance, and switching frequency. On the other hand, with the additional inductor presented here, the critical frequency is decoupled from the series resistance and only depends on the inductance and the capacitance, as shown in Equation 4.6. The effect can be seen in Figure 4.14b, where a reduction in the series resistance instantly brings a nearly equal reduction in power loss, without the need to increase the capacitance or the frequency.

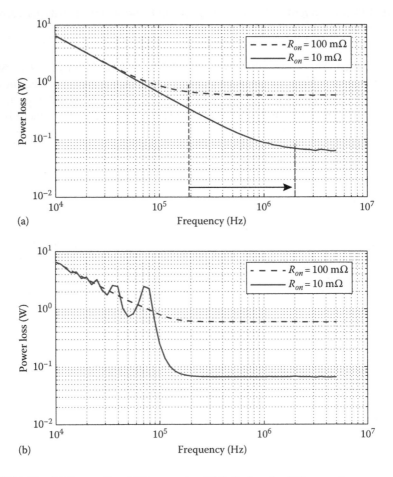

FIGURE 4.14   (a) Conventional SC converter. (b) Soft-charging SC converter.

Therefore, the addition of the inductor gives the designer the choice of using smaller on-state-resistance switches and introduces a new design dimension in which the converter can be optimized. In discrete implementations, the addition of the inductors often results in overall improvement in energy utilization of the passive components. While the inductor is more difficult to integrate than the capacitors given the current IC technology, the energy density and quality of integrated inductors are improving as more advanced processes are adopted [12,37,42], and the proposed converter is able to take advantage of the progress and advancement of technologies in inductors, capacitors, and switches simultaneously. Furthermore, the additional inductor enables lossless regulation of output voltage, which is not possible with conventional SC converters. A technique is given in [25], which integrates regulation function using existing switches from the SC converter, and demonstrated higher power density and efficiency over a wide input voltage range.

4.5 ANALYSIS OF SC CONVERTERS FOR SOFT-CHARGING/RESONANT OPERATION

It was shown in the previous section that resonant and soft-charging operation are closely related and both modes of operation can be achieved with a single additional inductor. The example used was a 2-to-1 SC converter, which easily satisfies the second condition given in Section 4.2, which states that to achieve full soft-charging operation there can be no voltage mismatch among the flying capacitors during phase transitions. The example 2-to-1 SC converter satisfies this condition easily since it only has a single flying capacitance. More complicated SC converters have multiple flying capacitors connected in a number of different configurations. Thus, it is of great interest to determine whether this proposed technique can be broadly applied to other SC converter topologies. To answer this question, a general method is derived in this section to determine whether an arbitrary SC converter topology can operate in resonance or soft-charging operation with the addition of an output inductor [24]. Since resonance with the inductor at the output can be viewed as soft-charging operation at a special switching frequency, only the term, soft-charging, is used in this section for convenience.

In essence, the proposed method examines the charge flow characteristics of an SC converter topology and observes the change in capacitor voltage subject to Kirchhoff's voltage law (KVL) constraints. In each phase of the SC converter, the voltage across a capacitor changes according to the charge flow in the given phase. When the converter switches to the next phase, KVL poses new constraints on each component. Complete soft-charging is achieved if and only if the ideal capacitor network satisfies KVL at all times, including at phase transitions. If during any period the KVL constraint is not satisfied, the voltage discrepancy will appear across the series resistances, resulting in a charge transfer impulse and attendant losses. The KVL constraint is present whether soft-charging or resonant operation is of interest, and thus the analysis presented in this section applies to both modes of operation.

4.5.1 GENERAL ANALYSIS USING DICKSON CONVERTER AS AN EXAMPLE

The analysis method in this work is illustrated with a 4-to-1 SC converter in Dickson configuration [13,16,47] shown in Figure 4.15. To simplify the analysis, a constant current source is used as the load for this and all following examples, while we note that a practical implementation would use a magnetic-based converter or an inductor. The two phases of the Dickson topology are shown in Figure 4.16a and b, respectively. In each phase of Figure 4.16, the circuit consists of a number of closed loops and a KVL equation can be written for each loop. For example, the following two independent KVL equations can be written for phase 1 of the Dickson converter (Figure 4.16a):

$$\begin{cases} V_{in} - V_{C3} - V_{out} = 0 \\ V_{C2} - V_{C1} - V_{out} = 0 \end{cases} \tag{4.8}$$

and for phase 2 (Figure 4.16b):

$$\begin{cases} V_3 - V_{C2} - V_{out} = 0 \\ V_{C1} - V_{out} = 0 \end{cases} \tag{4.9}$$

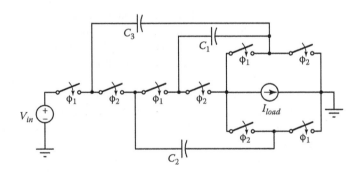

FIGURE 4.15 4-to-1 Dickson topology.

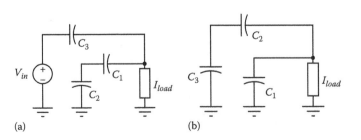

FIGURE 4.16 (a) Phase 1. (b) Phase 2.

These KVL equations can be written in a matrix–vector–product form as

$$\mathbf{A_i v^i} = 0 \tag{4.10}$$

where $\mathbf{A_i}$ is called the reduced loop matrix of the ith phase [10] and the voltage vector $\mathbf{v}$ is defined as

$$\mathbf{v} = \begin{bmatrix} v_{in} & v_{c1} & v_{c2} & v_{c3} & v_{out} \end{bmatrix}^T \tag{4.11}$$

$$= \begin{bmatrix} v_{in} & \mathbf{v_c^T} & v_{out} \end{bmatrix}^T \tag{4.12}$$

In this analysis, the entries of the loop matrices are positive if the circuit element is traversed from the negative terminal to the positive terminal and vice versa. Combining the definitions in Equations 4.10 and 4.11, and the KVL equations in Equations 4.8 and 4.9, the loop matrices are found to be

$$\mathbf{A_1} = \begin{bmatrix} 1 & 0 & 0 & -1 & -1 \\ 0 & -1 & 1 & 0 & -1 \end{bmatrix} \quad \text{and} \quad \mathbf{A_2} = \begin{bmatrix} 0 & 0 & -1 & 1 & -1 \\ 0 & 1 & 0 & 0 & -1 \end{bmatrix}$$

Denoting the voltage vector at the start of phase 1 as $\mathbf{v^1}$, KVL analysis yields

$$\mathbf{A_1 v^1} = 0 \tag{4.13}$$

which captures the KVL constraints given by Equation 4.8 at the moment when the converter has begun phase 1 operation. At the end of phase 1, the voltage vector becomes $\mathbf{v^1 + Av^1}$, creating a second KVL constraint:

$$A_1(v^1 + \Delta v^1) = 0 \tag{4.14}$$

where $\mathbf{\Delta v}$ represents the change in voltage due to charge being delivered to the load, and is in the form of $[\Delta v_{in} \ \Delta v_c^T \ \Delta v_{out}]^T$, similar to the voltage vector in Equation 4.11. Since *both* Equations 4.13 and 4.14 must be satisfied, a resulting constraint is that the vector $\mathbf{\Delta v}$ must satisfy

$$A_1 \Delta v^1 = 0 \tag{4.15}$$

From a circuit intuition point of view, Equation 4.15 describes the fact while the individual node voltages can (and will) change as a result of charge transfers, the sum of the changes in a KVL loop must be zero. Similarly, for phase 2, we have

$$A_2 \Delta v^2 = 0 \tag{4.16}$$

Note that the Δv_{in} component of $\Delta \mathbf{v^i}$ is typically zero since the input voltage is considered constant. This information can be included in the loop matrices by adding a row of $[1\,0\,0\,0\,0]$ to both $\mathbf{A_1}$ and $\mathbf{A_2}$, resulting in $\mathbf{A_{1m}}$ and $\mathbf{A_{2m}}$, respectively, where the subscript m indicates a modified reduced loop matrix. Correspondingly, Equations 4.15 and 4.16 become

$$A_{1m}\Delta v^1 = 0 \tag{4.17}$$

$$A_{2m}\Delta v^2 = 0 \tag{4.18}$$

The solution to Equations 4.17 and 4.18 represents the set of permissible voltage changes that satisfy KVL and $\Delta v_{in} = 0$. This solution is the nullspace of $\mathbf{A_{1m}}$ and $\mathbf{A_{2m}}$, by definition. Let $\mathbf{w}$ and $\mathbf{u}$ be the collective basis for nullspaces of $\mathbf{A_{1m}}$ and $\mathbf{A_{2m}}$, respectively. It follows that any solution to Equations 4.17 and 4.18 can be represented by a linear combination of the basis:

$$\Delta v^1 = a_1 w_1 + a_2 w_2 \tag{4.19}$$

$$\Delta v^2 = b_1 u_1 + b_2 u_2 \tag{4.20}$$

In the case of the 4-to-1 Dickson converter, such basis can be found* as

$$\mathbf{w} = \left\{ \begin{bmatrix} 0 \\ 0.607 \\ 0.763 \\ -0.157 \\ 0.157 \end{bmatrix} \begin{bmatrix} 0 \\ -0.482 \\ 0.131 \\ -0.613 \\ 0.613 \end{bmatrix} \right\} \quad \text{and} \quad \mathbf{u} = \left\{ \begin{bmatrix} 0 \\ 0.362 \\ 0.398 \\ 0.761 \\ 0.362 \end{bmatrix} \begin{bmatrix} 0 \\ 0.518 \\ -0.664 \\ -0.146 \\ 0.518 \end{bmatrix} \right\}$$

For conventional SC converters, we have the additional constraint that

$$\Delta v^1 = -\Delta v^2 \tag{4.21}$$

from the condition of periodic steady-state operation. This is because in a capacitive network the voltage changes must sum up to zero in a full switching cycle. Combining Equations 4.19 through 4.21, we have

$$a_1 \mathbf{w_1} + a_2 \mathbf{w_2} + b_2 \mathbf{u_1} + b_2 \mathbf{u_2} = 0 \tag{4.22}$$

---

* For example, by using the command "null" in MATLAB®, which will yield a set of orthonormal basis.

Note that Equation 4.22 can be written in a matrix form as

$$
\begin{bmatrix} \mathbf{w_1} & \mathbf{w_2} & \mathbf{u_1} & \mathbf{u_2} \end{bmatrix}
\begin{bmatrix} a_1 \\ a_2 \\ b_1 \\ b_2 \end{bmatrix} = 0
\tag{4.23}
$$

For the conventional Dickson SC converter, no solution for Equation 4.23 can be found, except for the trivial case of zero. This means that no voltage change exists for the circuit that satisfies KVL at all times. This result is reassuring and consistent with the behavior of conventional SC converters, where it is well known that this instantaneous voltage mismatch at phase transitions is what gives rise to the power loss from charge redistribution [40]. Hence, conventional SC converters have to rely on high switching frequency or larger capacitor values to minimize the voltage mismatch and the associated power loss.

However, with soft-charging operation, the SC stage output node is connected to an inductor, and the inductor voltage is allowed to change instantaneously during phase transitions, as opposed to the capacitor voltages, which must be continuous. Stated in another way, the parameter Δv_{out} defined previously is no longer a state variable in a switch-linear circuit, and can be discontinuous. As a result, the change in output voltage in phase 1 due to the current load, Δv_{out}^1, does not necessarily equal $-\Delta v_{out}^2$. Therefore, the inductor introduces one more degree of freedom to the system. To mathematically express this additional degree of freedom, the basis $\mathbf{w}$ and $\mathbf{u}$ can be modified by removing the last element in each column (the entry that represents Δv_{out}), resulting in the new basis $\bar{\mathbf{w}}$ and $\bar{\mathbf{u}}$. Now, replacing the basis in Equation 4.23 with the newly formed $\bar{\mathbf{w}}$ and $\bar{\mathbf{u}}$, we obtain for the soft-charging converter:

$$
\begin{bmatrix} \bar{\mathbf{w}}_1 & \bar{\mathbf{w}}_2 & \bar{\mathbf{u}}_1 & \bar{\mathbf{u}}_2 \end{bmatrix}
\begin{bmatrix} a_1 \\ a_2 \\ b_1 \\ b_2 \end{bmatrix} = 0
\tag{4.24}
$$

Mathematically, the matrix in Equation 4.24 has a reduced rank compared to the one in Equation 4.23, and thus a nonzero solution can be found. Solving Equation 4.24 for the Dickson converter, we obtain

$$
\begin{bmatrix} a_1 \\ a_2 \\ b_1 \\ b_2 \end{bmatrix} =
\begin{bmatrix} 0.120 \\ -0.697 \\ -0.607 \\ -0.364 \end{bmatrix}
\tag{4.25}
$$

The voltage change vectors in each phase can then be found using Equations 4.19 and 4.20, yielding

$$\Delta v^1 = \begin{bmatrix} 0 \\ 0.408 \\ 0 \\ 0.4088 \\ -0.408 \end{bmatrix} \quad \text{and} \quad \Delta v^2 = \begin{bmatrix} 0 \\ -0.408 \\ 0 \\ -0.408 \\ -0.408 \end{bmatrix} \tag{4.26}$$

Note that the original basis are used to obtain the voltage change at each node. From Equation 4.26, it can be seen that the net change in the capacitor voltages is zero (i.e., each column adds to zero), except for the last entry. This entry represents Δv_{out}, which is the node that can be discontinuous owing to the soft-charging operation. Having obtained the change in capacitor voltage required to satisfy KVL in each phase, we can calculate the required capacitance for soft-charging operation as

$$C_j = \frac{q_j}{\Delta v_{cj}} \tag{4.27}$$

for each capacitor j. Equation 4.27 requires the charge that flows into each flying capacitor to be found for each phase. For any well-posed SC topology, a charge flow vector can be obtained for each phase either by inspection [40] or Kirchhoff's current law [29]. In this work, the charge flow vector is defined as the vector of charge that flows into the positive terminal of each element in the circuit and is given in the form of

$$\mathbf{q} = [q_{in}\ q_{c1}\ q_{c2}\ q_{c3}\ q_{out}]$$

The charge flow vectors for the Dickson converter of this example are found to be

$$\mathbf{q}^1 = \begin{bmatrix} -1 & 1 & -1 & 1 & 2 \end{bmatrix} \quad \text{and} \quad \mathbf{q}^2 = \begin{bmatrix} 0 & -1 & 1 & -1 & 2 \end{bmatrix}$$

Together with the voltage change vector found in Equation 4.26, the capacitor values are obtained using Equation 4.27 and are simplified as follows:

$$\begin{bmatrix} C_1 \\ C_2 \\ C_3 \end{bmatrix} = \begin{bmatrix} 1 \\ \infty \\ 1 \end{bmatrix}$$

It can be seen that the Dickson topology can achieve complete soft charging only when $C_2 = \infty$ and $C_1 = C_3$. In practice, this means that it can *approach* soft charging with a C_2 large enough compared to C_1 and C_3. As can be seen from Equation 4.26, the output voltage ripple has the same magnitude as the voltage ripple of C_1 and C_3 under soft-charging operation. Thus, for soft-charging operation with the Dickson converter, a designer would want to minimize the charging/discharging loss by maintaining a relatively large C_2/C_1 ratio while keeping the output ripple of the SC stage tolerable with a second-stage converter or an inductor.

To summarize, the following steps are used to determine whether any given SC topology is compatible with soft-charging or resonant operation using a single inductor connected to the output of the SC stage:

1. Obtain the reduced loop matrix for each phase (A_1 and A_2) using KVL analysis.
2. Add a row of [1 0 0 ... 0] to A_1 and A_2, obtaining A_{1m} and A_{2m}.
3. Find the collective nullspace basis of A_{1m} and A_{2m} (**w** and **u**, respectively).
4. Remove the last row of **w** and **u** to obtain $\bar{\mathbf{w}}$ and $\bar{\mathbf{u}}$.
5. Use Equations 4.19, 4.20, and 4.24 to find the change in capacitor voltages.
6. Find the charge transfer vector for each capacitor [29,40].
7. Use Equation 4.27 to find the capacitance values required for soft charging.

As demonstrated in this section, for a two-phase SC dc–dc converter, if a capacitor voltage change vector, $\Delta\mathbf{v}_c$, can be found to satisfy KVL at all times, and the resultant capacitor values required are practical (finite and positive), the given topology is able to perform soft-charging and resonant operation and will exhibit no charging/discharging loss. Otherwise, at least one loop of the circuit will not be able to perform soft charging, and the benefit will be limited.

4.5.2 SIMULATION VERIFICATION

To verify that the analytical method is correct, the Dickson converter shown in Figure 4.15 is simulated using LTSpice with simulation parameters given in Tables 4.1 and 4.2. A total capacitance of 30 µF is used for the flying capacitors. In hard-charging (conventional) operation, an additional 100 µF output capacitor is added in parallel to the current load while there is no output capacitance in the soft-charging simulation. The converters are operated at a fixed duty ratio of 0.5. The simulated power losses are plotted in Figure 4.17. It can be seen that the hard-charging power loss decreases linearly as switching frequency increases, while leveling off at high

TABLE 4.1

Simulation Parameters

V_{in}	5 V
I_{load}	2 A
R_{on}	10 mΩ
R_{ESR}	1 mΩ
$C_{o,hard\text{-}charging}$	100 µF
$C_{o,soft\text{-}charging}$	0.1 µF

TABLE 4.2

Flying Capacitor Values

Configuration	C_1 (µF)	C_2 (µF)	C_3 (µF)
Hard-charging	10	10	10
Soft-charging 1	10	10	10
Soft-charging 2	5	20	5

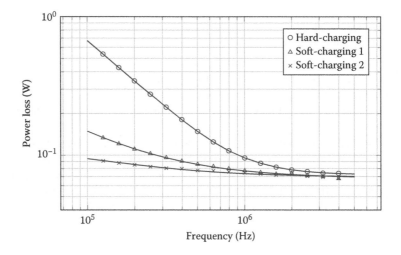

FIGURE 4.17   Power loss of Dickson converter at different frequencies.

frequency, showing the transition from SSL to FSL. In soft-charging operation, a significant reduction in power loss at lower frequencies is seen when the capacitors are such that $C_1 = C_2 = C_3$, as in the hard-charging case. However, a more prominent reduction is seen when the capacitor values are chosen such that $C_2/C_1 = 4$, plotted in Figure 4.17 as "Soft-charging 2." This confirms that the Dickson converter can approach full soft charging by maintaining a high C_2/C_1 ratio, as predicted by the analysis in this work.

The currents through the capacitor C_2 of the Dickson SC converter (Figure 4.15) in hard-charging and soft-charging operations with the converter switching at 250 kHz are shown in Figure 4.18. The different waveforms are shifted apart in the time axis for clearer observation. It can be seen that under hard-charging condition the capacitor current resembles the exponential discharge, as expected. With soft-charging 1, both the magnitude and the width of the impulse are reduced, while the tail of the exponential decay is raised. In the soft-charging 2 case, with capacitor values selected according to the analysis result, the height and the width of the impulse is further reduced and current waveform resembles more of a square wave. The transient effect is not completely eliminated due to the fact that perfect soft charging cannot

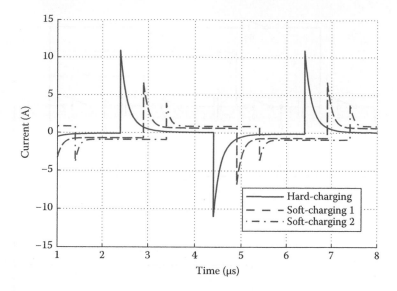

FIGURE 4.18   Current waveform of capacitor C_2 of the Dickson SC converter.

TABLE 4.3
RMS and Average Current of Capacitor C_2 in a Single Phase

Configuration	RMS Current (A)	Average Current (A)
Hard-charging	2.19	1.00
Soft-charging 1	1.40	1.00
Soft-charging 2	1.11	1.00

be achieved. To quantify the change in the current waveform, the RMS and the average currents through the capacitor for one phase duration are calculated and tabulated in Table 4.3. It can be seen that while in all cases the capacitor supplies the same average current over a phase duration the RMS value of the current decreases from hard-charging operation to soft-charging operation. Again soft-charging 2 is an improvement over soft-charging 1. Thus, the soft-charging operation reduces the impedance in the SSL region due to the improvement in the charging and discharging current waveform.

4.5.3   APPLICATION TO OTHER TOPOLOGIES

The general analysis method proposed is applied to four additional commonly used two-phase SC converter topologies: series-parallel, ladder, Fibonacci, and doubler. The schematics are shown in Figure 4.19a through d, respectively. The same analysis is repeated for each of them, and the results are shown in Figure 4.20.

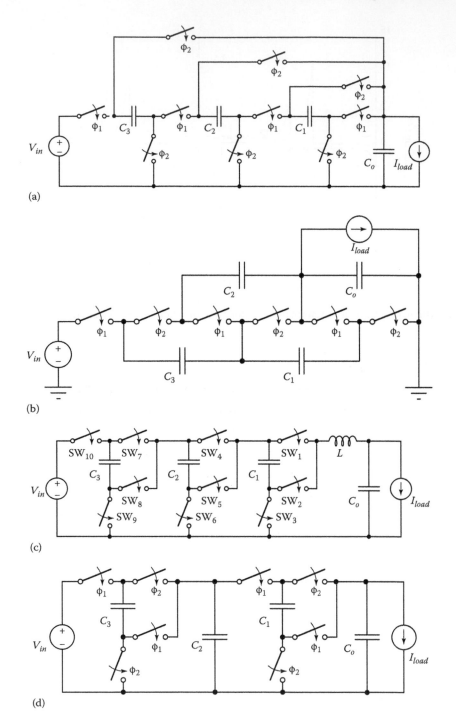

FIGURE 4.19 Common switched-capacitor converter topologies. (a) 4-to-1 series-parallel. (b) 3-to-1 ladder. (c) 5-to-1 Fibonacci. (d) 4-to-1 doubler.

$$\Delta v = \begin{bmatrix} \Delta v_{in} \\ \Delta v_{c1} \\ \Delta v_{c2} \\ \Delta v_{c3} \\ \Delta v_{out} \end{bmatrix}, \quad C = \begin{bmatrix} C_1 \\ C_2 \\ C_3 \end{bmatrix};$$

(a)

$$\Delta v^1 = \begin{bmatrix} 0 \\ 1 \\ 1 \\ 1 \\ -3 \end{bmatrix}, \quad \Delta v^2 = \begin{bmatrix} 0 \\ -1 \\ -1 \\ -1 \\ -1 \end{bmatrix}, \quad C = \begin{bmatrix} 1 \\ 1 \\ 1 \end{bmatrix};$$

(b)

$$\Delta v^1 = \begin{bmatrix} 0 \\ 1 \\ 1 \\ 1 \\ -2 \end{bmatrix}, \quad \Delta v^2 = \begin{bmatrix} 0 \\ -1 \\ -1 \\ -1 \\ -1 \end{bmatrix}, \quad C = \begin{bmatrix} 1 \\ -2 \\ 1 \end{bmatrix};$$

(c)

$$\Delta v^1 = \begin{bmatrix} 0 \\ 2 \\ -1 \\ 1 \\ -3 \end{bmatrix}, \quad \Delta v^2 = \begin{bmatrix} 0 \\ -2 \\ 1 \\ -1 \\ -2 \end{bmatrix}, \quad C = \begin{bmatrix} 1 \\ 1 \\ 1 \end{bmatrix};$$

(d)

$$\Delta v^1 = \begin{bmatrix} 0 \\ 1 \\ 0 \\ 0 \\ -1 \end{bmatrix}, \quad \Delta v^2 = \begin{bmatrix} 0 \\ -1 \\ 0 \\ 0 \\ -1 \end{bmatrix}, \quad C = \begin{bmatrix} 1 \\ \infty \\ \infty \end{bmatrix};$$

(e)

FIGURE 4.20 Voltage change vectors and relative capacitor values for soft-charging operation. (a) General, (b) series-parallel, (c) ladder, (d) Fibonacci, and (e) doubler.

It can be seen that for the series-parallel converter a simple requirement for soft
charging is that all the flying capacitors have the same value. Under soft-charging
condition, the output voltage ripple is shown to be equal to $N-1$ times the change
in any of the capacitor voltages, where N is the conversion ratio. These observations
agree with the experimental work in [33]. In addition, the Fibonacci converter is also
found capable of soft-charging operation with equal capacitors. On the other hand,
for the ladder configuration, one can see that a negative capacitance is needed on C_2
for soft-charging operation, which is not achievable. This means that the change in
capacitor voltage of C_2 is in the opposite direction of what is required to satisfy KVL.
Thus, the single-output ladder topology is not compatible with soft charging without
modification, and a limited improvement is expected. As for the doubler converter,
both C_1 and C_2 have to be infinite for complete soft charging, indicating a partial
soft-charging capability similar to that of the Dickson converter.

To verify the analysis results, the circuits shown in Figure 4.19 are simulated with
the same parameters as those of the Dickson converter. Equal flying capacitors are
used in all cases. Again, a constant current source is used as the load instead of an
inductor to simplify the simulation and remove the effect of resonance, since in practice operation below the critical frequency is to be avoided as shown in Section 4.3.
The corresponding power loss curves are plotted in Figure 4.21. It should be noted
that the power loss values are not intended for cross-comparison between different
SC topologies. Rather, it is the reduction of the power loss by changing from hard-
charging operation to soft-charging operation that is of key interest here. For both the

series-parallel converter and the Fibonacci converter, soft-charging operation results in a significantly lower power loss in the SSL region than in the hard-charging case, and the loss is almost independent of the frequency. The ladder configuration only receives very limited benefit from soft charging and a strong frequency dependency is still seen on the power loss plotted in Figure 4.21b. The doubler converter shows moderate improvement with soft charging. These simulation results agree with the prediction of the analytical technique presented earlier.

Since the Fibonacci converter is shown to be able to achieve full soft-charging operation, it is useful to examine the current waveform, as shown in Figure 4.22. It can be seen that the waveform in soft-charging operation is a square wave, confirming that the current transient associated with capacitor charge redistribution has been eliminated. Table 4.4 shows the RMS and average of the absolute values of the current through capacitor C_3 of the Fibonacci converter. It can be seen that now the RMS current is almost equal to the average current in the soft-charging case, ensuring the lowest power loss.

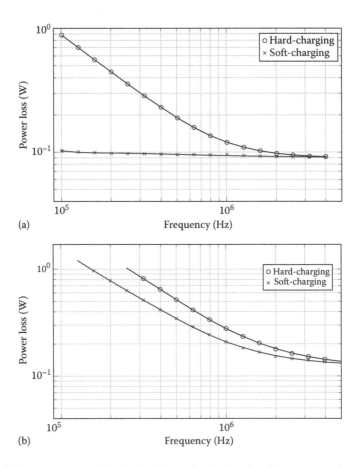

FIGURE 4.21 Power loss of hard-charging and soft-charging SC converters from LTSpice simulation. (a) Series-parallel. (b) Ladder.   (*Continued*)

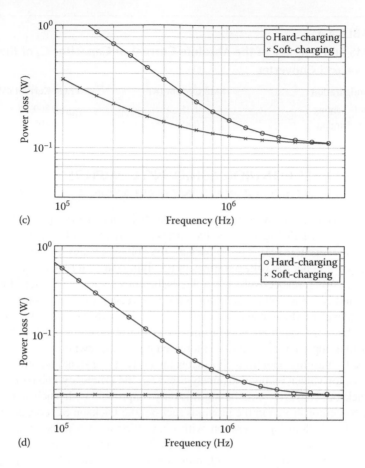

(c)

(d)

FIGURE 4.21 (*Continued*) Power loss of hard-charging and soft-charging SC converters from LTSpice simulation. (c) Doubler. (d) Fibonacci.

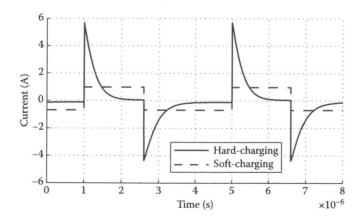

FIGURE 4.22 Current waveform of capacitor C_3 of the Fibonacci converter.

RMS and Average of the Absolute Current of Capacitor C_2 of the Fibonacci Converter

Configuration	RMS Current (A)	Average Current (A)
Hard-charging	1.45	0.800
Soft-charging	0.816	0.800

4.6 SPLIT-PHASE CONTROL: ACHIEVING COMPLETE SOFT-CHARGING OPERATION OF DICKSON SC CONVERTERS

Among the various SC converter topologies, the Dickson converter [13,16,47], whose schematic is shown in Figure 4.23, has efficient utilization of switches but poor utilization of capacitors [41]. This is because the voltage ratings on the switches are either $2V_{out}$ or V_{out}, but the voltage ratings of the flying capacitors (C_1, C_2, ..., C_N) are V_{out}, $2V_{out}$, ..., NV_{out}, respectively, where V_{out} is the output voltage of a step-down configuration and N is the voltage step-down ratio. A variant of the Dickson converter is the multilevel modular capacitor clamped converter (MMCCC) [19]. MMCCC converters have the same equivalent circuits as the Dickson converters but take a more modular approach at the cost of increased number of switches.

Since soft-charging operation eliminates the current transient and allows smaller flying capacitor values without adversely affecting the efficiency, the power density of the Dickson converter can be significantly improved with such a technique. However, as demonstrated in the previous section, the Dickson SC converter cannot achieve full soft-charging operation with conventional, two-phase control and a single inductor [24]. Likewise, zero current switching operation for MMCCC converters are only achieved so far with more than one inductors in the circuit [7,8]. Here, we present a new control technique that achieves soft-charging operation with a single inductor at the output of a Dickson converter. A technique to achieve full

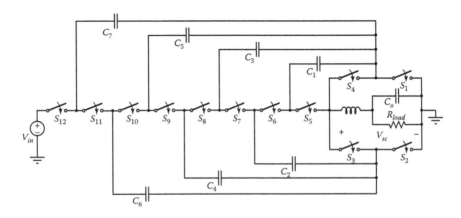

FIGURE 4.23   4-to-1 Dickson topology.

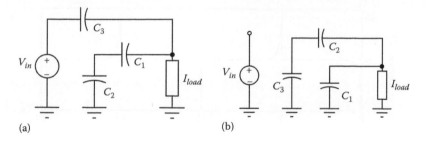

FIGURE 4.24 Two-phase operation of a 4-to-1 Dickson converter. (a) Phase 1 and (b) phase 2.

soft-charging operation is proposed by splitting the original two switching phases into four phases [30]. The proposed technique does not introduce any additional components to the two-phase soft-charging converter and can be realized with a small additional control effort.

4.6.1 CONVENTIONAL DICKSON SC CONVERTER

A conventional 4-to-1 step-down Dickson SC converter is shown in Figure 4.23, and the two conventional switching phases are shown in Figure 4.24. It should be clarified that "phase" here refers to the state of the switching circuit and should not be confused with the use of "multiphase," which is sometimes used to mean interleaved SC designs [21]. For conventional (hard-charging) operation, the flying capacitor network is directly connected to the output, with a large output capacitance C_o acting as a voltage-source load. Thus, a large transient current occurs during the phase switching instances due to the capacitor voltages mismatch and the resultant charge redistribution process. This is the characteristic of the SSL of SC converters [40]. The current through one of the capacitors (C_2) is shown in Figure 4.25. It can be seen that there is a large impulse current at phase transitions in the conventional, hard-charging case (top plot). To minimize the voltage mismatch among the capacitors and the resultant current impulse, large capacitors or high switching frequency has to be employed such that the converter operates in the FSL [40].

4.6.2 SOFT-CHARGING OPERATION WITH CONVENTIONAL TWO-PHASE CONTROL

In soft-charging operation, the output capacitance is removed so that the output voltage of the SC stage can change instantaneously to compensate for the difference in capacitor voltages. By eliminating the voltage mismatch and the resultant current impulse, soft-charging SC converters exhibit the same behavior as an SC converter in FSL, while operating at a switching frequency corresponding to the SSL region of a conventional design. In order to shield the load from the large voltage ripple at the output port of the soft-charging SC converter, an interfacing element must be placed between the SC stage and the load. As discussed in the preceding sections, practical implementations of the interfacing element can be either a magnetic converter [33] or an LC filter [25,41], but for the purpose of clear illustration, an ideal constant current load is used here to represent the interfacing element and the actual load.

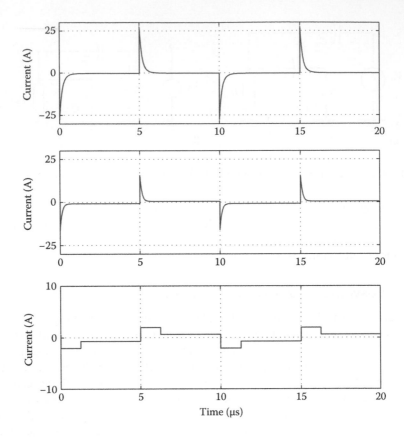

FIGURE 4.25 Current waveform of capacitor C_2 of the Dickson SC converter. Simulation parameters: $C_1 = C_2 = C_3 = 10$ μF, $f_{sw} = 100$ kHz, $I_{load} = 2$ A. Note the different scales on the y-axis.

While the current load ensures that there is no transient current drawn from the SC converter due to the voltage mismatch between the capacitor network and the final output node, complete soft-charging operation also requires that there is no voltage mismatch among the internal capacitor connections of the SC stage, so that there is no circulating current within the flying capacitor network. For example, for the two-phase control, by applying KVL to the circuits in Figure 4.24, the following requirements can be found for soft-charging operation:

$$\text{Phase 1} : V_{in} - V_{C_3} = V_{C_2} - V_{C_1} \tag{4.28}$$

$$\text{Phase 2} : V_{C_3} - V_{C_2} = V_{C_1} \tag{4.29}$$

However, constraints (4.28) and (4.29) *cannot* be satisfied during the transition between phases when the Dickson converter is operated with a conventional, two-phase control scheme.

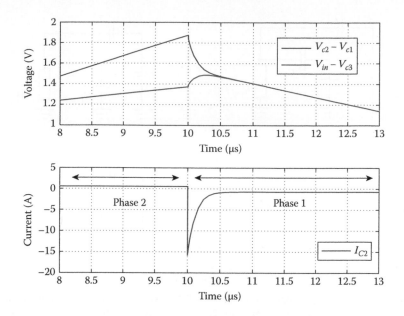

FIGURE 4.26 Voltage and current waveforms for two-phase soft-charging operation of the Dickson converter.

Figure 4.26 shows the voltage and current waveforms of interest during the transition from phase 2 to phase 1. It can be seen that, due to the charging and discharging operation in phase 2, the two voltage values, $(V_{C_2} - V_{C_1})$ and $(V_{C_3} - V_{C_2})$, are different and diverging. Thus, when the converter transitions to phase 1 and forces the two nodes to have the same voltage, charge redistribution occurs, which results in a the large current impulse as seen in Figure 4.26, which shows the current through C_2 of Figure 4.23. A similar scenario happens at the start of phase 2 when $(V_{C_3} - V_{C_2})$ is always greater than V_{C_1}, making it also a hard-charging transition from phase 1 to phase 2.

From a circuit intuition point of view, the voltage mismatch is due to the asymmetry in the capacitor connection, particularly for the outer most (C_3) and inner most capacitor (C_1). As can be seen in Figure 4.24, these two capacitors are in series with another capacitor in one phase but not in the other phase. As a result, not all current paths have the same equivalent capacitance, giving rise to voltage mismatch when transitioning to the other phase of operation. Therefore, unlike topologies such as series-parallel and Fibonacci, the Dickson SC converter cannot achieve full soft-charging operation despite using a current load. As can be seen in the middle plot of Figure 4.25, while the magnitude and width of the current impulse are reduced with two-phase soft-charging operation, there is still significant transient effect and associated losses, owing to the internal capacitor voltage mismatch.

4.6.3 SOFT-CHARGING OPERATION WITH SPLIT-PHASE CONTROL

To ensure that the capacitor network results in the same voltage at the output node, we propose the split-phase control of the Dickson converter, with two secondary phases

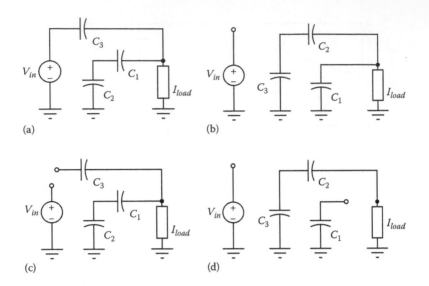

FIGURE 4.27 Split-phase operation of a 4-to-1 Dickson converter. Switching sequence: 1b → 1a → 2b → 2a. (a) Phase 1a, (b) phase 2a, (c) phase 1b, and (d) phase 2b.

introduced [23], as shown in Figure 4.27. Phase 1a and 2a are the same as phase 1 and 2 in the original operation, while the phase 1b configuration is a subset of phase 1 and the phase 2b configuration is a subset of phase 2. The switching sequence is phase 1b → phase 1a → phase 2b → phase 2a. As can be seen from the schematic in Figure 4.27c, in phase 1b, C_2 discharges and C_1 charges, and thus $(V_{C_2} - V_{C_1})$ decreases while $(V_{in} - V_{C_3})$ remains constant. The circuit can transition from phase 1b to phase 1a when $(V_{C_2} - V_{C_1})$ equals $(V_{in} - V_{C_1})$, that is, when Equation 4.28 is satisfied. This process is illustrated in the voltage and current waveforms of Figure 4.28. As can be seen, with the introduction of the additional "buffer" phase, 1b, KVL is satisfied during phase transitions and the current transient can be eliminated. Similarly, the circuit transitions from phase 2b to phase 2a when Equation 4.29 is satisfied. The effect on the overall current waveform can be seen in the bottom plot of Figure 4.25, which shows that the current has no transient component and is a constant value in each phase.

To quantify the improvement in the power transfer, the average, RMS, and peak values of capacitor current for a half-period duration are calculated and tabulated in Table 4.5. For SC converters, the average capacitor current represents the delivered power and is hence held fixed in this comparison. The RMS current reflects the conduction loss of the switches and capacitors, and should be as low as possible for high-efficiency operation. It can be seen that two-phase soft-charging operation reduces the RMS and peak values of the capacitor current, but to a limited extent. On the other hand, the proposed split-phase control achieves both the lowest RMS values and the lowest peak values. By eliminating the current transient, the converter efficiency can be improved and the current stress of the devices can be reduced.

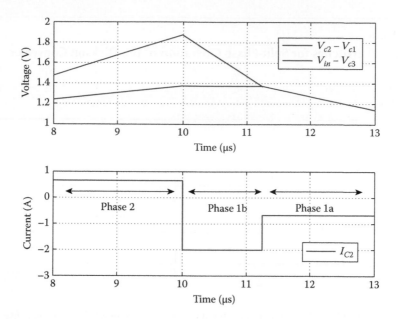

FIGURE 4.28 Voltage and current waveforms for split-phase soft-charging operation of the Dickson converter. The circuit changes from phase 2 to phase 1b at the 10 μs mark and changes to phase 1a at 11.25 μs mark.

TABLE 4.5

Average, RMS, and Peak Values of Current of Capacitor C_2 in a Single Half Period

Configuration	Average (A)	RMS (A)	Peak (A)
Hard-charging	1.00	3.52	27.3
Soft-charging two-phase	1.00	1.91	15.8
Soft-charging split-phase	1.00	1.15	2.00

Note: Simulation parameters are as in Figure 4.25.

The control signals for both the original two-phase and the proposed split-phase operations are shown in Table 4.6 and Figure 4.29. It can be seen that compared to the original two-phase control the proposed switching sequence only delays the turn-on of two switches (S_5 and S_8). Thus, generating the extra phases in the split-phase operation does not increase the switching frequency of the switches, ensuring no added switching loss. Another advantage of the proposed split-phase control is the scalability of the technique. Even though the technique is illustrated with a 4-to-1 Dickson converter with only three flying capacitors, it is applicable to Dickson converters with larger conversion ratios without introducing more

TABLE 4.6

Control of Switches for Split-Phase Operation of the Dickson Converter

Switches	S_8	S_7	S_6	S_5	S_4	S_3	S_2	S_1
Two-phase	q_2	q_1	q_2	q_1	q_2	q_1	q_2	q_1
Split-phase	q_4	q_1	q_2	q_3	q_2	q_1	q_2	q_1

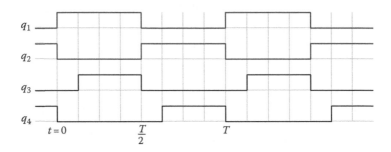

FIGURE 4.29   Control signal diagrams for split-phase operation of the Dickson converter.

secondary phases [30]. This is because only the switch that connects to the input voltage and the switch that connects to the innermost capacitor (C_1) needs to be delayed. Extra switches for higher conversion ratios follow the original gate signals (q_2 and q_1) as do switches S_6 and S_7. Therefore, there is no increase in control complexity as the conversion ratio increases.

4.6.4   SPLIT-PHASE ANALYSIS CONTROL IMPLEMENTATION

While the preceding section presents an intuitive understanding why the split-phase control eliminates the current transient in the operation of the Dickson converter, it is beneficial to formulate a general analysis. An analytical method was presented in Section 4.5 [24] that determines whether an arbitrary SC topology is able to achieve full soft-charging operation. However, the method was developed for an SC converter with two phases, and for the proposed split-phase control, a total of four different circuit states are present. Hence, the method in [24] was extended in [23] to a higher number of phases. With a higher number of phases, the duty ratio of each phase is not known. Therefore, unlike in [24], where the capacitor values required by soft-charging operation are found given the fixed 50% ratio, the objective in the split-phase analysis of [23] was to find the corresponding duty ratios for full soft-charging operation, given a set of capacitor values. As explained in the previous section, complete soft-charging can be achieved if and only if the internal capacitor network satisfies KVL at all times, including during phase transitions. The aim of the analysis is thus to find the set of charge flow vectors for the capacitors, such that the corresponding capacitor voltage changes result in node voltages that satisfy KVL during all phase transitions.

In [23], it was shown that for complete soft-charging operation of the Dickson converter with equal flying capacitance the duty ratio of each phase is

$$T^{1a} = \frac{3}{8}, T^{2a} = \frac{3}{8}, T^{1b} = \frac{1}{8}, T^{2b} = \frac{1}{8} \qquad (4.30)$$

Another useful result that can be obtained from the analysis is that soft-charging operation can be achieved regardless of the order of the switching phases since the derivation does not depend on the sequence of the phases. With the proposed split-phase control, there are four phases. Therefore, six possible periodic sequences exist in total and three representative ones are shown later. While sequence 1 is the same sequence obtained from the voltage balance intuition in Section 4.2, sequence 2 is the reverse of sequence 1; and in sequence 3, the two original phases (1a and 2a) are adjacent instead of being separated by the secondary phases. The duration of each phase is still given by the constraint of (4.30).
Switching sequences:

1. Phase 1b → phase 1a → phase 2b → phase 2a
2. Phase 2a → phase 2b → phase 1a → phase 1b
3. Phase 1a → phase 2a → phase 1b → phase 2b

Figure 4.30 shows the simulated current waveforms for these switching sequences. It can be seen that all three of the switching sequences result in a nonimpulse current, showing that complete soft-charging operation can be achieved for each switching sequence. However, while all six sequences are equivalent from the point of achieving soft-charging operation, they have some different implications in practical implementations. In particular, depending on the direction of current flow, the voltage blocking requirements of the switches for the different sequences vary. In most discrete implementations using power MOSFETS with native body diodes, it is desirable to employ switches that do not require reverse blocking capabilities. While CMOS transistors generally are less stringent in this regard, often time the gate driving circuitry can be greatly simplified if it is known that the transistor source voltage is lower than the drain voltage at all times.

4.6.5 QUANTIFIED PERFORMANCE IMPROVEMENT

To illustrate the benefit of the split-phase soft-charging operation, the 4-to-1 step-down Dickson converter shown in Figure 4.23 is simulated using LTSpice with simulation parameters given in Table 4.7. In the hard-charging operation, the duty ratio is fixed to 0.5 (as is convention) while the duty ratio of the split-phase operation is as found analytically in Section 4.3.

The output referred impedance of a SC converter encapsulates both the capacitor charge transfer loss and the conduction loss of the converter and is widely used to characterize the performance of such converters [6,26,29]. The output impedance for the Dickson converter is plotted in Figure 4.31. It can be seen that the conventional hard-charging Dickson converter shows two regions of asymptotic behaviors as found

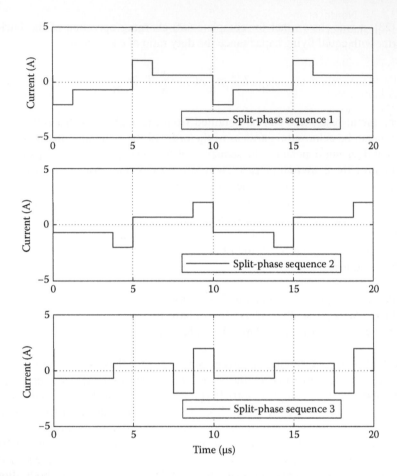

FIGURE 4.30 Current waveform of capacitor C_2 of the Dickson SC converter under different switching sequences. Simulation parameters are as in Figure 4.25.

TABLE 4.7
Simulation Parameters

V_{in}	40 V
I_{load}	2 A
$R_{ds,on}$	10 mΩ
R_{ESR}	1 mΩ
C_1, C_2, C_2	10 μF
$C_{o,hard\text{-}charging}$	100 μF
$C_{o,soft\text{-}charging}$	None

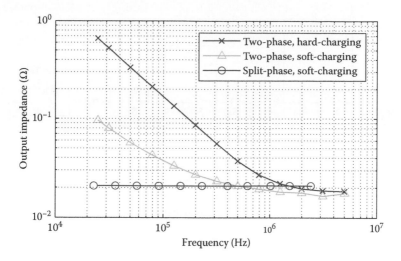

FIGURE 4.31 Simulated output impedance of the Dickson converter.

in previous literature [40]. At low frequencies (SSL), when the power loss due to the current transient dominates, the impedance decreases as the switching frequency increases. The impedance reaches a constant at high frequencies (FSL), when the conduction loss dominates. As can be seen in Figure 4.31, with two-phase soft-charging operation, the impedance in the SSL region is reduced significantly owing to the fact that the output voltage is allowed to swing. However, there is still non-negligible frequency-dependent behavior since complete soft-charging operation cannot be achieved with a conventional Dickson converter. With the proposed split-phase control, however, it can be seen that now the output impedance is independent of the switching frequency due to the complete elimination of the charge transfer losses. It should be noted that the impedance at high frequencies is slightly higher than the FSL impedance of the conventional two-phase operation. This is due to the fact that in the added phases (1b and 2b), there is one path less that delivers current to the load, resulting in a slightly increased effective switch resistance. However, this increase in conduction loss will diminish as the converter conversion ratio increases. Therefore, using split-phase control, soft-charging Dickson converters can achieve significant efficiency and power density improvement.

In [23], the performance of the split-phase control technique was verified on a hardware prototype, with a voltage step-down ratio of 8 to 1. The schematic of the converter is shown in Figure 4.32. The prototype, implemented using discrete transistors and passive components, used 12 GaN switches and 7 flying capacitors.

In the experimental verification, the same converter prototype was used for both hard-charging and soft-charging operation. The difference was that in soft-charging operation an extra inductor was added to make an LC filter. The additional inductor incurred an approximately 15% increase in the components volume of the power stage, but the penalty in volume was much less significant when the total enclosed box volume of the power stage was considered.

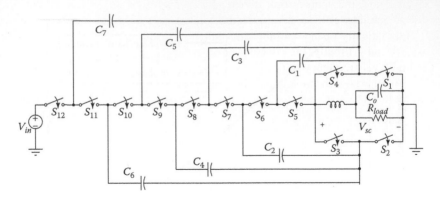

FIGURE 4.32 The schematic of the proposed experimental prototype 8-to-1 Dickson converter.

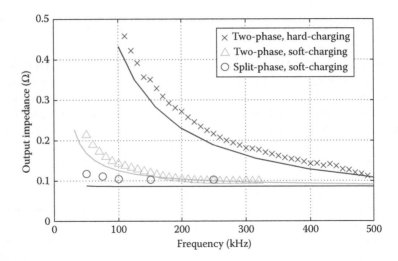

FIGURE 4.33 Output impedance calculated from measured data. Simulated values are shown as solid lines.

The output impedance is plotted against the switching frequency in Figure 4.33 and is calculated from the measured data using

$$R_{out} = \frac{(V_{in}/N) - V_{out}}{I_{out}}$$

where $N = 8$ is the conversion ratio. It can be seen that similar to the simulation results the output impedance in hard-charging operation increases as frequency decreases. Two-phase soft-charging operation reduces the impedance at low switching frequencies while the proposed split-phase operation results in the lowest output impedance. For example, to achieve the same output impedance as the split-phase operation at 100 kHz, the two-phase soft-charging operation requires a switching frequency of approximately 200 kHz while the hard-charging converter has to switch at over 500 kHz. This means that the split-phase converter can reduce the capacitor values by a factor of 5 compared to a hard-charging converter if both are switching at 500 kHz. Consequently, the reduced capacitor requirement more than compensates for the additional volume of the added inductor. Moreover, it can be seen in Figure 4.33 that the measured data closely match the simulated values, especially at higher switching frequencies. In the lower switching frequencies, the experimental values are larger than expected. This can be attributed to the tolerance of the flying capacitor used (up to 20%). Since the split-phase control assumed equal capacitor values for soft-charging operation, different capacitor values will result in more capacitor charging/discharging loss and is not modeled in the simulation.

In addition, the efficiencies of the converter in the SSL region are plotted in Figure 4.34. It can be seen that soft-charging operation brings significant efficiency improvement while the proposed split-phase control has the highest efficiency. The split-phase soft-charging operation also has the smallest drop in efficiency as the load increases due to its smallest output impedance. At a load current of 2 A, the split phase reduces power loss by 30% compared to two-phase soft-charging operation and by 75% compared to the projected power loss with the hard-charging operation. It should be noted that both the output impedance measurements and the efficiency measurements are obtained using reduced input voltage and output current rather than the rated values. This is to prevent the converter from breaking from the excessive heat when the converter is operating with high output impedance in the SSL region, especially for the conventional hard-charging operation. The hardware results also show that indeed the split-phase control is effective for a Dickson converter with a conversion ratio that is higher than the analyzed 4-to-1.

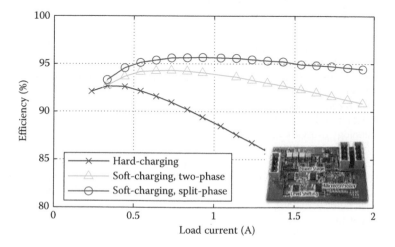

FIGURE 4.34 Measured efficiency of the Dickson converter in deep SSL region. $V_{in} = 40$ V, $f_{sw} = 100$ kHz.

4.7 CONCLUSIONS

In this chapter, we demonstrated the concept of soft-charging operation for SC converters, which enables improved energy utilization of capacitors. Example architectures included a merged two-stage converter, where a secondary, high-frequency power converter provides the soft-charging path, and a topology where a single inductor enables soft-charging operation. An analytical framework was developed that enabled the evaluation of *any* SC converter topology to determine whether soft-charging or resonant operation is achievable with a single inductor. This method was applied to a number of SC topologies, where it was observed that some topologies are inherently soft-charging compatible (series-parallel and Fibonacci), whereas others are not able to utilize this technique for improved power density and efficiency.

Finally, a multiphase control method was introduced, which enabled soft-charging operation of a Dickson converter, which is not possible with a conventional two-phase design. In addition to an intuitive derivation of the control parameters, results from a formal method were presented, as well as a number of possible control sequences. Simulations and experimental results demonstrated the increased power density and efficiency possible through this new control method.

REFERENCES

1. K. Abe, K. Nishijima, K. Harada, T. Nakano, T. Nabeshima, and T. Sato. A novel three-phase buck converter with bootstrap driver circuit. In *Power Electronics Specialists Conference*, Orlando, FL, June 17–21, 2007, pp. 1864–1871.
2. O. Abutbul, A. Gherlitz, Y. Berkovich, and A. Ioinovici. Step-up switching-mode converter with high voltage gain using a switched-capacitor circuit. *IEEE Transactions on Circuits and Systems I*, 50(8): 1098–1102, August 2003.
3. T.M. Andersen, F. Krismer, J.W. Kolar, T. Toifl, C. Menolfi, L. Kull, T. Morf, M. Kossel, M. Brandli, P. Buchmann, and P.A. Francese. A 4.6 w/mm² power density 86 dc-dc converter in 32 nm SOI CMOS. In *2013 28th Annual IEEE Applied Power Electronics Conference and Exposition (APEC)*, Long Beach, CA, March 17–21, 2013, pp. 692–699.
4. T.M. Andersen, F. Krismer, J.W. Kolar, T. Toifl, C. Menolfi, L. Kull, T. Morf, M. Kossel, M. Brandli, P. Buchmann, and P.A. Francese. 4.7 A sub-ns response on-chip switched-capacitor DC-DC voltage regulator delivering 3.7 W/mm² at 90% efficiency using deep-trench capacitors in 32nm SOI CMOS. In *2014 IEEE International Solid-State Circuits Conference Digest of Technical Papers (ISSCC)*, San Francisco, CA, February 9–13, 2014, pp. 90–91.
5. M. Araghchini, J. Chen, V. Doan-Nguyen, D.V. Harburg, D. Jin, J. Kim, M.S. Kim et al. A technology overview of the powerchip development program. *IEEE Transactions on Power Electronics*, 28(9): 4182–4201, 2013.
6. B. Arntzen and D. Maksimovic. Switched-capacitor dc/dc converters with resonant gate drive. *IEEE Transactions on Power Electronics*, 13(5): 892–902, September 1998.
7. D. Cao, S. Jiang, and F.Z. Peng. Optimal design of a multilevel modular capacitor-clamped dc-dc converter. *IEEE Transactions on Power Electronics*, 28(8): 3816–3826, August 2013.
8. D. Cao, X. Lu, X. Yu, and F.Z. Peng. Zero voltage switching double-wing multilevel modular switched-capacitor dc-dc converter with voltage regulation. In *2013 28th Annual IEEE Applied Power Electronics Conference and Exposition (APEC)*, Long Beach, CA, March 17–21, 2013, pp. 2029–2036.

9. S.V. Cheong, S.H. Chung, and A. Ioinovici. Development of power electronics converters based on switched-capacitor circuits. In *Proceedings of 1992 IEEE International Symposium on Circuits and Systems, 1992 (ISCAS'92)*, Vol. 4, San Diego, CA, May 10–13, 1992, pp. 1907–1910.

10. L.O. Chua, C.A. Desoer, and E.S. Kuh. *Linear and Nonlinear Circuits*. McGraw-Hill, New York, 1987.

11. H. Chung, B. O, and A. Ioinovici. Switched-capacitor-based dc-to-dc converter with improved input current waveform. In *1996 IEEE International Symposium on Circuits and Systems, 1996 (ISCAS'96), Connecting the World*, Vol. 1, Atlanta, GA, May 12–15, 1996, pp. 541–544.

12. J.T. Dibene II, P.R. Marrow, C.-M. Park, H.W. Koertzen, P. Zou, F. Thenus, X. Li, S.W. Montgomery, E. Stanford, R. Fite, and P. Fischer. A 400 amp fully integrated silicon voltage regulator with in-die magnetic coupled embedded inductors. In *Proceedings of Applied Power Electronic Conference, Special Session on On-Die Voltage Regulators*, Palm Springs, CA, 2010.

13. J.F. Dickson. On-chip high-voltage generation in MNOS integrated circuits using an improved voltage multiplier technique. *IEEE Journal of Solid-State Circuits*, 11(3): 374–378, June 1976.

14. D. El-Damak, S. Bandyopadhyay, and A.P. Chandrakasan. A 93 using on-chip ferro-electric capacitors. In *2013 IEEE International Solid-State Circuits Conference Digest of Technical Papers (ISSCC)*, February 2013, pp. 374–375.

15. R.W. Erickson and D. Maksimovic. *Fundamentals of Power Electronics*. Kluwer Academic, Dordrecht, the Netherlands, 2000.

16. H. Greinacher. ber eine methode, wechselstrom mittels elektrischer ventile und konden-satoren in hochgespannten gleichstrom umzuwandeln. *Zeitschrift fr Physik A Hadrons and Nuclei*, 4: 195–205, 1921.

17. P. Hazucha, G. Schrom, J. Hahn, B.A. Bloechel, P. Hack, G.E. Dermer, S. Narendra et al. A 233-MHz 80%–87% efficient four-phase DC-Dc converter utilizing air-core inductors on package. *IEEE Journal of Solid-State Circuits*, 40(4): 838–845, April 2005.

18. J.M. Henry and J.W. Kimball. Practical performance analysis of complex switched-capacitor converters. *IEEE Transactions on Power Electronics*, 26(1): 127–136, 2011.

19. F.H. Khan and L.M. Tolbert. A multilevel modular capacitor clamped dc-dc converter. In *41st IAS Annual Meeting Conference Record of the 2006 IEEE Industry Applications Conference, 2006*, Vol. 2, Tampa, FL, October 8–12, 2006, pp. 966–973.

20. J.W. Kimball, P.T. Krein, and K.R. Cahill. Modeling of capacitor impedance in switch-ing converters. *IEEE Power Electronics Letters*, 3(4): 136–140, December 2005.

21. S. Kiratipongvoot, S.-C. Tan, and A. Ioinovici. Switched-capacitor converters with multiphase interleaving control. In *2011 IEEE Energy Conversion Congress and Exposition (ECCE)*, Phoenix, AZ, September 17–22, 2011, pp. 1156–1161.

22. H.-P. Le, S.R. Sanders, and E. Alon. Design techniques for fully integrated switched-capacitor dc-dc converters. *IEEE Journal of Solid-State Circuits*, 46(9): 2120–2131, September 2011.

23. Y. Lei, R. May, and R.C.N. Pilawa-Podgurski. Split-phase control: Achieving com-plete soft-charging operation of a Dickson switched-capacitor converter. In *2014 IEEE 15th Workshop on Control and Modeling for Power Electronics (COMPEL)*, Cantabria, Spain, June 22–25, 2014, pp. 1–7.

24. Y. Lei and R.C.N. Pilawa-Podgurski. Analysis of switched-capacitor dc-dc converters in soft-charging operation. In *2013 IEEE 14th Workshop on Control and Modeling for Power Electronics (COMPEL)*, Salt Lake City, UT, June 23–26, 2013, pp. 1–7.

25. Y. Lei and R.C.N. Pilawa-Podgurski. Soft-charging operation of switched-capacitor dc-dc converters with an inductive load. In *2014 IEEE the Applied Power Electronics Conference and Exposition (APEC)*, Fort Worth, TX, March 16–20, 2014, pp. 2112–2119.
26. P. Lin and L. Chua. Topological generation and analysis of voltage multiplier circuits. *IEEE Transactions on Circuits and Systems*, 24(10): 517–530, October 1977.
27. F.-L. Luo. Switched-capacitorized dc/dc converters. In *Fourth IEEE Conference on Industrial Electronics and Applications, 2009 (ICIEA 2009)*, Xi'an, China, May 25–27, 2009, pp. 1074–1079.
28. M.S. Makowski. Realizability conditions and bounds on synthesis of switched-capacitor DC-DC voltage multiplier circuits. *IEEE Transactions on Circuits and Systems-I*, 44(8): 684–691, August 1997.
29. M.S. Makowski and D. Maksimovic. Performance limits of switched-capacitor dc-dc converters. In *26th Annual IEEE Power Electronics Specialists Conference, 1995. PESC'95 Record*, Vol. 2, Atlanta, GA, June 18–22, 1995, pp. 1215–1221.
30. R. May. Analysis of soft charging switched capacitor power converters. Master's thesis, ECE Department, University of Illinois at Urbana Champaign, Urbana, IL, August 2013.
31. K.D.T. Ngo and R. Webster. Steady-state analysis and design of a switched-capacitor DC-DC converter. *IEEE Transactions on Aerospace and Electronic Systems*, 30(1): 92–101, January 1994.
32. R.C.N. Pilawa-Podgurski. Architectures and circuits for low-voltage energy conversion and applications in renewable energy and power management. PhD thesis, Department of Electrical Engineering and Computer Science, Massachusetts Institute of Technology, Cambridge, MA, 2012. http://dspace.mit.edu/handle/172L1/71485 (accessed on June 12, 2015).
33. R.C.N. Pilawa-Podgurski, D.M. Giuliano, and D.J. Perreault. Merged two-stage power converter architecture with soft-charging switched-capacitor energy transfer. In *IEEE Power Electronics Specialists Conference, 2008 (PESC 2008)*, Rhodes, Greece, June 15–19, 2008, pp. 4008–4015.
34. R.C.N. Pilawa-Podgurski and D.J. Perreault. Merged two-stage power converter with soft charging switched-capacitor stage in 180 nm CMOS. *IEEE Journal of Solid-State Circuits*, 47(7): 1557–1567, July 2012.
35. Y.K. Ramadass, A.A. Fayed, and A.P. Chandrakasan. A fully-integrated switched-capacitor step-down dc-dc converter with digital capacitance modulation in 45 nm CMOS. *IEEE Journal of Solid-State Circuits*, 45(12): 2557–2565, December 2010.
36. M. Saad, A. Shawky, and M. Orabi. Design of integrated POL dc-dc converters based on two-stage architectures. In *2013 28th Annual IEEE Applied Power Electronics Conference and Exposition (APEC)*, Long Beach, CA, March 17–21, 2013, pp. 2098–2105.
37. S.R. Sanders, E. Alon, H.-P. Le, M.D. Seeman, M. John, and V.W. Ng. The road to fully integrated dc-dc conversion via the switched-capacitor approach. *IEEE Transactions on Power Electronics*, 28(9): 4146–4155, 2013.
38. G. Schrom, P. Hazucha, F. Paillet, D.J. Rennie, S.T. Moon, D.S. Gardner, T. Kamik et al. A 100 MHz eight-phase buck converter delivering 12 A in 25 mm^2 using air-core inductors. In *Applied Power Electronics Conference*, Anaheim, CA, February 25–March 1, 2007, pp. 727–730.
39. M.D. Seeman and S.R. Sanders. Analysis and optimization of switched-capacitor DC-DC converters. In *IEEE Computers in Power Electronics Workshop*, Troy, NY, July 16–19, 2006, pp. 216–224.
40. M.D. Seeman and S.R. Sanders. Analysis and optimization of switched-capacitor dc-dc converters. *IEEE Transactions on Power Electronics*, 23(2): 841–851, March 2008.

41. M.D. Seeman. A design methodology for switched-capacitor DC–DC converters. PhD thesis, EECS Department, University of California, Berkeley, CA, May 2009.
42. N. Sturcken, E.J. O'Sullivan, N. Wang, P. Herget, B.C. Webb, L.T. Romankiw, M. Petracca et al. A 2.5 d integrated voltage regulator using coupled-magnetic-core inductors on silicon interposer. *IEEE Journal of Solid-State Circuits*, 48(1): 244–254, January 2013.
43. J. Sun, J.-Q. Lu, D. Giuliano, T.P. Chow, and R.J. Gutmann. 3D power delivery for microprocessors and high-performance ASICs. In *Applied Power Electronics Conference*, Anaheim, CA, February 25–March 1, 2007, pp. 127–133.
44. J. Sun, M. Xu, Y. Ying, and F.C. Lee. High power density, high efficiency system two-stage power architecture for laptop computers. In *37th IEEE Power Electronics Specialists Conference, 2006 (PESC'06)*, Jeju, South Korea, June 18–22, 2006, pp. 1–7.
45. N.X. Sun, X. Duan, X. Zhang, A. Huang, and F.C. Lee. Design of a 4 MHz, 5V to 1V monolithic voltage regulator chip. In *Proceedings of the 14th International Symposium on Power Semiconductor Devices and ICs*, Santa Fe, NM, June 2002, pp. 217–220.
46. J. Wei and F.C. Lee. Two novel soft-switched, high frequency, high-efficiency, non-isolated voltage regulators—The phase-shift buck converter and the matrix-transformer phase-buck converter. *IEEE Transactions on Power Electronics*, 20(2): 292–299, March 2005.
47. J.-T. Wu and K.-L. Chang. MOS charge pumps for low-voltage operation. *IEEE Journal of Solid-State Circuits*, 33(4): 592–597, April 1998.
48. V. Yousefzadeh, E. Alarcon, and D. Maksimovic. Three-level buck converter for envelope tracking applications. *IEEE Transactions on Power Electronics*, 21(2): 549–552, March 2006.

5 Resonant Switched Capacitor Power Converters and Architectures

Jason Stauth

CONTENTS

5.1 INTRODUCTION

The invention and rapid scaling of the transistor have had a profound impact on the performance, cost, and applicability of power electronics. However, the same scaling trends have also enabled a new range of applications that require efficient, accurate, and fast voltage regulation. For example, modern computing, communications, and mobile devices are increasingly limited by trade-offs among power consumption, size, and performance. The maximum clock rate of performance microprocessors (e.g., PCs and servers) has been constrained by thermal limits and power loss density for more than a decade [1–3]. On the other hand, mobile computing and "Internet of Things" devices are increasingly limited by both battery life and form factor, which drives the need for efficient (ultralow-power) computation [4]. In response to these trends, microprocessor designs have shifted toward increasing parallelism to continue to scale throughput at constant power density [5]. This has driven an increase in the scope and prevalence of multicore microprocessor architectures and a corresponding need for deeper integration of power management blocks and subcomponents [2]. In order to "catch up" with the capabilities of computational blocks in scaled deep-submicron CMOS, new directions are needed in power management technology to address wide input and output voltage ranges while maintaining high efficiency and power density.

In recent years, switched capacitor (SC) converters have gained academic and commercial interest for high-density and monolithic power management. The advantages of the SC approach compared to more traditional "inductor-based" buck and boost converters is severalfold. SC converters can be configured and hierarchical and cascaded structures that can interface across wide conversion ratios with low-voltage transistors and passive components. Compared to traditional buck or boost topologies, SC converters can achieve theoretically lower conduction loss for a given voltage–current (V–A) rating of the power devices [6]. The SC approach relies on capacitors that can have multiple-order-of-magnitude higher-energy density than inductors at small form factors [7]. Finally, SC converters are amenable to monolithic integration in scaled CMOS processes as the approach is increasingly favorable with stacked arrays of high f_t transistors [8].

However, the SC approach also has limitations. First, SC converters are not well suited for variable conversion ratio designs as they are most efficient with integer conversion ratios (e.g., 2:1, 3:2, 1:3). In order to regulate the output voltage, they typically use a mixture of linear (loadline) regulation and architectural reconfiguration [9,10]. The former method is accomplished by tuning the switching frequency to regulate the output voltage via a resistive drop, which is akin to using a linear regulator at the output [8]. The latter method requires a reconfigurable SC architecture but necessitates a large number of switching devices that can both increase conduction loss and parasitic common mode capacitance [9]. SC designs are also subject to fundamental limitations on efficiency. These limitations are set by finite capacitance density, resistance and capacitance of active devices, as well as metal interconnect, and bottom-plate parasitics of flying capacitors [11]. For example, "bottom-plate" capacitance, the parasitic common mode capacitance between flying capacitors and the substrate or other fixed nodes in the circuit, can represent a significant source of power loss. While the SC approach shows significant promise for a variety of applications and can favorably leverage emerging technologies such as "deep-trench" capacitors [11], it may not be suitable for all applications, power levels, and voltage ranges [12].

Resonant switched capacitor (ReSC) converters can be considered an extension of the SC approach in that similar hierarchical and cascaded topologies are used, which provide comparable device utilization and scaling advantages [13–16]. However, the use of a small amount of magnetic energy storage to resonate with flying capacitor components can facilitate soft switching with zero-current transitions. By eliminating hard-switching losses and using resonant operation, the ReSC approach can improve trade-offs between switching and conduction loss, which, in principle, can provide higher efficiency at a given power level.

The fundamental disadvantage of the ReSC approach is the need for magnetic components, which are a potential source of cost, size, and power loss. Conventional planar-spiral integrated circuit (IC) inductors tend to be lossy and require significant die area to achieve even modest inductance [17,18]. However, more advanced silicon-integrated inductors that leverage high-frequency magnetic materials such as laminated nanocomposites continue to improve in terms of power density, efficiency, and operating frequency [19–21]. Additional directions such as the use of PCB-embedded trace inductors and small-footprint air-core topologies are also promising for the ReSC application [22,23].

Another important advantage of the ReSC approach is the ability to implement variable step-up and step-down conversion ratios while maintaining high efficiency. As discussed further in this chapter, variable regulation can be achieved by expanding the operating states of the converter to include specific nonresonant modes. This is enabled by the use of magnetic components that prevent hard switching of the flying capacitor component(s) and address one of the fundamental limitations of the SC approach.

In this chapter, we focus on the power electronics and architectures that are applicable for a range of low-voltage systems spanning renewable energy, energy storage, performance computing, and mobile communications. In particular, we will explore the ReSC architecture, which is emerging as a promising candidate for monolithic DC–DC conversion, as well as high-density discrete DC–DC converters. ReSC converters build on a wealth of literature in the SC space and utilize techniques and analytical treatments common to other resonant converter families and RF power amplifiers.

Section 5.2 will provide an overview of basic SC and ReSC architectures, providing a discussion of fundamental trade-offs, advantages, and disadvantages of each approach. Section 5.3 will discuss the implementation of variable conversion ratios with the nominally 2:1 ReSC topology. Section 5.4 will provide an overview of other hierarchical and cascaded ReSC architectures, including ladder, series-parallel, and Dickson approaches. This section will also provide an example of a hierarchical ReSC ladder converter used to improve the performance of photovoltaic (PV) power management systems. Section 5.5 will present a conclusion and future opportunities for high-density resonant power management.

5.2   SWITCH CAPACITOR AND RESONANT SWITCHED CAPACITOR CONVERTERS: OVERVIEW AND COMPARISON

The field of power electronics can trace its origins to the development of municipal power generation and distribution in the nineteenth century [24]. Throughout this long history, power electronics have relied heavily on magnetic energy storage in the form of transformers and inductors. Even in our current era, the majority of power electronics, from the utility-scale to low-power portable electronics, rely on architectures that leverage inductors and transformers. However, the growth of the semiconductor industry, combined with a range of recent advances in materials, processing techniques, and manufacturing, is starting to open the possibility of new power converter architectures that rely increasingly on electrostatic energy storage in the form of capacitors.

5.2.1 Switched Capacitor Overview

SC circuits have been known for several decades in the academic literature and have also been deployed commercially in filters, data conversion, and power management applications [25]. In the form of "charge pumps" or voltage multipliers, SC converters have achieved commercial success for a range of moderate- to low-power applications [26–28]. However, recent experimental and analytical work has demonstrated the promise of SC architectures to extend to a wide range of power and voltage levels [6,7,9–11,29,30]. For example, high-power modular-multilevel converters utilize SC stages to reduce the voltage stress on active and passive devices, while also shaping the harmonic content of AC waveforms; these examples have shown that the SC concept can extend even to the 100 MW or even GW power level, operating at many hundreds of kV [31,32].

Figure 5.1 shows a representative circuit schematic for a nominally 2:1 SC converter. In general, SC converters work by transferring energy among series-connected voltage taps. This process is achieved by alternately connecting flying capacitors (e.g., C_X in Figure 5.1) in parallel with different voltage domains in the system. The first operating state, shown in Figure 5.1b, consists of C_X in parallel with the high-side bypass capacitor, $C_{BP,1}$. In a second operating state, shown in Figure 5.1c, C_X is in parallel with the low-side bypass capacitor, $C_{BP,2}$. If there is a voltage difference between the two bypass capacitors (or more generally between the two voltage taps), charge is transferred in the direction of high voltage to low voltage via the switching process of C_X to drive the series voltage domains toward equalization with a nominal 2:1 conversion ratio.

To implement the switching process, semiconductor devices M_1–M_4 (here shown as complementary mosfets) are configured between the flying capacitance

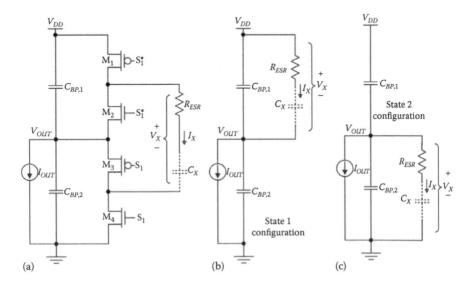

FIGURE 5.1 Nominally 2:1 SC converter: (a) circuit schematic, (b) operating state with S_1 low, and (c) operating state with S_1 high.

and the series voltage taps. In state one, switches M_1 and M_3 are closed, while in state 2, switches M_2 and M_4 are closed. A single control signal, S_1, is shown as controlling the complementary switching stage (S_1^* is in phase with S_1, but is referenced to the common mode of the source of M_2). In practice, there is a deadtime period between the M_2–M_4 on time and M_1–M_3 on time to prevent shoot-through current, as is common in most DC–DC converters. Resistance, R_{ESR} captures the resistance in the loop containing the switching devices, flying capacitance, and bypass capacitance.

Figure 5.2 shows the transient waveform for 2:1 SC operation. It is implicit that there is some load current, which drives a small voltage difference in V_X in state 1 and state 2. The current flowing into C_X follows a decaying exponential in each operating state with a time constant set by the $R_{ESR} * C_X$ product. The DC behavioral operation of the SC converter can be captured by the transformer model in Figure 5.3. The turns ratio $N{:}1$ represents the nominal, unloaded conversion ratio of the SC stage, while the output impedance, R_{EFF}, captures the conduction loss and slope of the loadline for the SC converter. Similar to SC filters and other applications, when the switching frequency is low relative to the ESR-defined pole in the SC stage, R_{EFF} scales inversely with the product of switching frequency, f_{SW}, and flying capacitance, C_X. This mode of operation is known as the *slow switching*

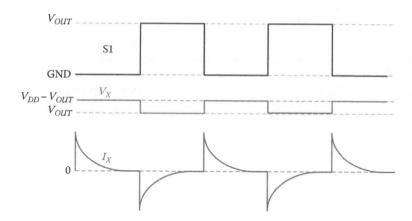

FIGURE 5.2   Transient waveforms in nominal 2:1 conversion mode.

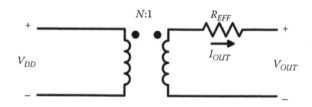

FIGURE 5.3   Transformer model for the switched capacitor and resonant switched capacitor converter.

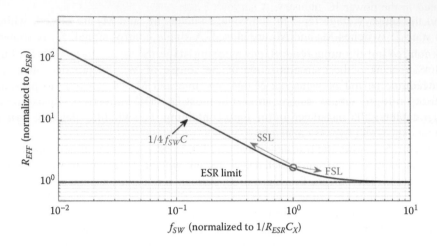

FIGURE 5.4 Effective resistance versus frequency for 2:1 SC converter. (Adapted from Sangwan, R. et al., High-density power converters for sub-module photovoltaic power management, *2014 IEEE Energy Conversion Congress and Exposition (ECCE)*, September 14–18, 2014, Pittsburgh, PA, pp. 3279–3286.)

limit (SSL) because the output impedance is inversely proportional to switching frequency. At high switching frequencies, resistance in the circuit will limit the energy transfer per cycle, resulting in *fast switching limit (FSL)* operation. In this mode, the transient current, I_X, begins to approach a square wave and R_{EFF} saturates at an ESR-defined limit.

As shown in Figure 5.4, the SSL and FSL operation map out the minimum achievable R_{EFF} as a function of switching frequency. Effective resistance, R_{EFF}, is an important parameter for converter design as it relates the voltage drop at the output of the converter as a function of load current. It also captures conduction losses in the circuit, and therefore, is a criterion for both power density and efficiency. A reasonably accurate expression for R_{EFF} in the 2:1 SC circuit was derived in [16,20]:

$$R_{EFF} = \frac{1}{4 f_{SW} C_X} \left[\tanh\left(\frac{1}{4 f_{SW} C_X R_{ESR}} \right) \right]^{-1}, \tag{5.1}$$

where
   f_{SW} is the switching frequency
   C_X is the flying capacitance
   R_{ESR} is the parasitic resistance in the energy-transfer loop

The tanh (hyperbolic tangent) expression in Equation 5.1 captures the transition from SSL to FSL operation. It can be shown that in the SSL this expression converges to $R_{EFF} = (4 f_{SW} C_X)^{-1}$, where the factor of 4 results from the nominally 2:1 impedance transformation. In the FSL, R_{EFF} saturates at R_{ESR} as the ESR limits the current transferred by the flying capacitor each cycle.

Several points should be noted about the configuration of the 2:1 SC converter in Figure 5.1. First, many SC architectures are known in the literature. Popular topologies include the Dickson, series-parallel, ladder, and Fibonacci architectures, among others [6,28,34,35]. However, in the 2:1 configuration these architectures converge to the same circuit topology, for example, Figure 5.1. A secondary consideration is the use of bypass capacitors. For example, in multiphase interleaved designs, bypass capacitance may be redundant. Figure 5.5 shows an example of a 2:1 multiphase interleaved SC converter, for example, [8]. Interleaving presents a fundamental advantage to the SC topology as there is usually no disadvantage to splitting the flying capacitor into subunits in an IC implementation [10]. Through the use of interleaving, output voltage ripple can be reduced to negligible values [30]. However, because roughly ½ of the flying capacitance is configured across the high-side voltage tap and the other ½ across the low-side voltage tap at any given time, it is important to model the converter correctly by including charge-sharing losses as if there were appreciable bypass capacitance in the circuit [16].

SC converters have several fundamental loss metrics. A primary source of power loss is due to charge sharing in the flying capacitor component. Charge sharing loss results from hard-switching capacitor components that have different initial conditions for their DC voltages, and the corresponding energy loss of CV^2 per cycle. Fortunately, this is captured in the output impedance model in Equation 5.1, where the loss is equivalently computed as the value of $I_{OUT}^2 * R_{EFF}$. Another important loss mechanism is due to bottom-plate switching loss in flying capacitor components. Bottom-plate capacitance is a parasitic quantity that can be viewed as the common

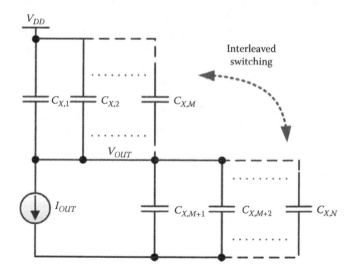

FIGURE 5.5 Multiphase interleaved 2:1 switched capacitor converter where flying capacitance is split into "N" separate interleaved units.

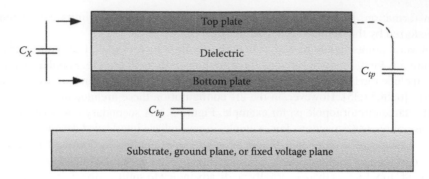

FIGURE 5.6 Representation of bottom plate capacitance in a switched (flying) capacitor component.

mode capacitance of a flying (differential) capacitor with respect to a fixed voltage node in the circuit. As shown in Figure 5.6, the predominant parasitic of a flying capacitor component is typically the common mode capacitance of the "bottom-plate" as this is closest to the substrate, which is commonly connected to ground. Bottom-plate capacitance limits the maximum efficiency of SC converters. For example, it can be shown that if the bottom-plate ratio (or the ratio of common mode to differential capacitance) is a modest 1%, then the maximum efficiency of an SC converter is 90.9% [8].

To fully characterize and optimize a given SC converter, it is also necessary to consider frequency-dependent losses in the converter. These losses generally comprise gate-drive losses for the power semiconductor devices and bottom-plate switching loss. SC circuits are, therefore, optimized at a given output current, output power, or power density level by considering both conduction losses and switching losses, as is common for more traditional buck or boost converters. The optimization is slightly more complex as both switching frequency and power semiconductor device sizes need to be optimized for maximum efficiency. For example in [8], this optimization is completed analytically through an approximate expression for R_{EFF}, where optimum device sizes and frequency can be computed using partial derivatives.

Alternatively, the expression in Equation 5.1 can be optimized for device width by taking a partial derivative, resulting in an expression for the optimum device width for the 2:1 SC converter, the result being

$$W_{opt-SC} = 8C_X R_{on,sp} f_{SW} \tanh^{-1}\left(\frac{I_{OUT}}{\sqrt{I_{OUT}^2 + 128 C_{in,sp} V_G^2 f_{SW}^2 C_X^2 R_{on,sp}}} \right), \quad (5.2)$$

where
$R_{on,sp}$ is the specific "on" resistance
$C_{in,sp}$ is the specific gate capacitance of the device technology

The value of Equation 5.2 is that it provides an optimum switch width as a function of fundamental device and circuit parameters, switching frequency and load current. Switching frequency ends up being an independent design variable with a transcendental solution, so a numerical optimization procedure is needed to complete the design. Therefore, as discussed in [15], a given 2:1 SC converter can be designed for maximum efficiency at a given load current by co-optimizing the power devices and switching frequency for minimum power loss.

5.2.2 Resonant Switched Capacitor Overview

While SC circuits are gaining increased traction due to favorable characteristics of high power density and high efficiency without the use of inductors, the appreciable power loss due to charge sharing and bottom plate loss results in a fundamental limit on conversion efficiency. While the advantages of the SC approach, which eliminates the need for magnetic components, are substantial, the use of a modest amount of magnetic energy storage can significantly improve conversion efficiency by enabling soft switching and lower frequency operation that can mitigate the fundamental loss mechanisms in SC converters [13,36–38].

ReSC converters have many similarities to SC converters at the architectural level, but differ in the use of magnetic components to facilitate resonant energy transfer [13,14,39]. At the millimeter scale, ReSC converters benefit also from high capacitor density available in MIM, MOS, and deep-trench technologies, but leverage a small amount of magnetic energy storage to shape the transient current waveform, tune out the reactive impedance of flying capacitors, and maximally utilize the available energy density of on-chip passive components [15,16].

Figure 5.7a shows a circuit schematic for a nominal 2:1 ReSC converter that can be compared to the 2:1 SC topology in Figure 5.1. The circuit works in an identical fashion to the SC converter: here, a complementary power train is also shown that is controlled by a single clock reference, S_1. As in Figure 5.1, M_1–M_4 are power semiconductor devices used to configure a flying impedance between two series-connected voltage taps. However, a small inductor is placed in series with the flying capacitor, which is used to facilitate resonant energy transfer. In Figure 5.8, it can be seen that the timing and voltage waveforms across the series resonant impedance are similar to the SC case. If the switching frequency is the same as the fundamental resonant frequency of the L–C network, defined by $\omega_0 = (L_X C_X)^{-\frac{1}{2}}$, the current, I_X, that flows in the flying impedance is approximately sinusoidal.

The current waveform in Figure 5.8 can be compared with the SC waveform in Figure 5.2, resulting in several notable differences. First, the SC waveform has inherent hard switching transitions and relatively high peak currents. The transient current follows an exponential decay profile that is related to the RC time constant and may or may not decay to zero depending on whether the SC circuit is in the SSL or FSL. The resonant waveform has the benefit of providing zero current switching on both

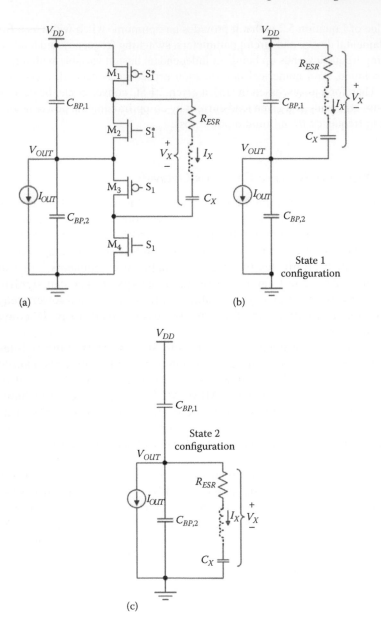

FIGURE 5.7 Nominally 2:1 ReSC converter: (a) circuit schematic, (b) operating state with S_1 low, and (c) operating state with S_1 high.

transitions, reducing switch stress and peak current levels. First-order advantages of the ReSC topology can be discussed using the transformer model abstraction in Figure 5.3. Specifically, the value for output impedance, R_{EFF}, can be computed by quantifying the time-averaged output current as a function of the output voltage, V_{OUT}. This process has been detailed in [15] for R_{EFF} over all frequency, but the

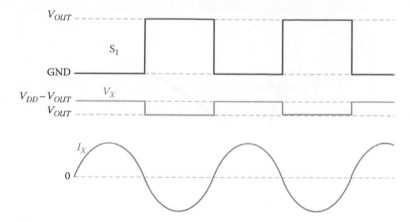

FIGURE 5.8   Transient waveforms in nominal 2:1 conversion mode.

expression can be simplified for operation exclusively at fundamental and subharmonic resonant minima as

$$R_{EFF} = \frac{1}{4 f_{SW} C_X} \tanh\left(\frac{R_{ESR}}{8 f_{SW} L_X} \right),$$ (5.3)

where
R_{ESR} is the series resistance in the resonant loop containing switches, capacitors, and the inductor
f_{SW} assumes the value of the fundamental or odd-subharmonic resonant frequency

Further, assuming a modest quality factor, typically $Q > 2$, Equation 5.3 further simplifies to

$$R_{EFF} \approx \frac{\pi^2}{8} R_{ESR},$$ (5.4)

Figure 5.9 shows a plot comparing R_{EFF} for SC and ReSC converters. Here, R_{EFF} is normalized to R_{ESR} and switching frequency is normalized to the SSL–FSL boundary: $\omega_{fsl} = (R_{ESR} C_X)^{-1}$. The plot assumes the same total flying capacitance (e.g., C_X is the same for both converters), but in the ReSC case, a modest inductor is included to provide resonance. The plot also compares the performance of the ReSC converter for a range of R_{ESR} values and associated quality factors (here it should be noted that the resulting quality factor is dominated by the inductor resistance; e.g., the $R_{ESR} = 1 \times$ case assumes the same ESR in both converters, the $R_{ESR} = 8 \times$ assumes eight times higher ESR for the ReSC converter).

The benefit of resonant operation, as expected, is highly dependent on the quality factor of the resonant circuit,

$$Q_{ReSC} = \frac{\omega_0 L_X}{R_{ESR}} = \frac{1}{R_{ESR}} \sqrt{\frac{L_X}{C_X}},$$

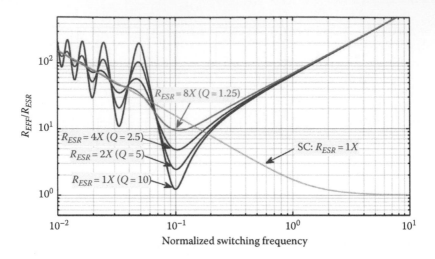

FIGURE 5.9 Effective resistance (R_{EFF}) versus frequency for 2:1 SC and ReSC converters.

where $\omega_0 \approx 1/\sqrt{L_X C_X}$. For example, using expressions (5.1) and (5.3), it can be shown that for the SC and ReSC circuits operating both at the same fundamental resonant switching frequency, f_0, the ReSC approach reduces R_{EFF} by approximately the quality factor of the circuit:

$$\frac{R_{EFF\_SC}}{R_{EFF\_ReSC}} @ f_0 = \frac{4}{\pi} Q_{ReSC} \approx Q_{ReSC}. \tag{5.5}$$

Similarly, for the SC circuit to achieve the same R_{EFF} as the equivalent ReSC circuit, and assuming slow-switching operation, the SC circuit needs to switch approximately Q times faster:

$$\frac{f_{SW\_SC}}{f_{SW\_ReSC}} @ same\ R_{EFF} = \frac{4}{\pi} Q_{ReSC} \approx Q_{ReSC}. \tag{5.6}$$

The perspective highlighted in Equations 5.5 and 5.6 can be extended to a fundamental figure of merit (FOM) for DC–DC converters, which is the product of *equivalent output impedance* times *switching frequency* (e.g., $f_{SW} \cdot R_{EFF}$). This FOM is related to total power loss in the circuit that includes both switching loss (proportional to f_{SW}) and conduction loss (proportional to R_{EFF}). This FOM, computed for SC and ReSC topologies and only considering the SSL, is as follows:

$$f_{SW} R_{EFF\_SC} \approx \frac{1}{4C_X} \tag{5.7}$$

and

$$f_{SW}R_{EFF\_ReSC} \approx \frac{1}{4C_X}\frac{\pi}{4Q_{ReSC}}. \tag{5.8}$$

In Equations 5.7 and 5.8, it can be again seen that the benefit of resonant operation is again related to the achievable quality factor. In the SC case, Equation 5.7, the $f_{SW} \cdot R_{EFF}$ product is reduced by increasing the total flying capacitance; this underlies relationships between flying capacitance value and efficiency as well as capacitance density and power density.

In the ReSC case, the flying capacitance value is important, but also the need to achieve low ESR, or high Q to improve efficiency and power density. It can also be seen in Equation 5.8 that the ReSC converter is improved by the quantity $\pi/4Q_{ReSC}$. As seen by comparing Equations 5.7 and 5.8, *in theory*, this quantity should be less than one for the resonant architecture to have a benefit. Therefore, by using the analysis described earlier, a constraint can be placed on the resonant quality factor: $Q_{ReSC} > \pi/4 \approx 0.785$, which sets a reasonable lower limit for the ReSC converter.

However, to amend the aforementioned discussion, it should be noted that while the ReSC circuit analysis considered the ESR present in the resonant loop, the SC circuit was treated in the SSL (e.g., ESR for the SC circuit was ignored). Figure 5.10 provides an improved comparison of the FOMs described in Equations 5.7 and 5.8. Here, the FOM of $f_{SW}R_{EFF}$ is plotted versus quality factor for 2:1 SC and ReSC circuits with the same flying capacitance, C_X. The expression is also normalized to the $4C_X$ quantity that is common to both converters. However, it can be seen that for $Q < 1$ the SC converter starts to enter the FSL (in fact, the frequency where $Q = 1$ is identical to the FSL–SSL boundary). Also, while the ReSC topology is clearly favorable in the high-Q regime, at low values of Q, it gradually approaches SC operation. In other words, a ReSC converter with low Q approximates an SC converter. In this case,

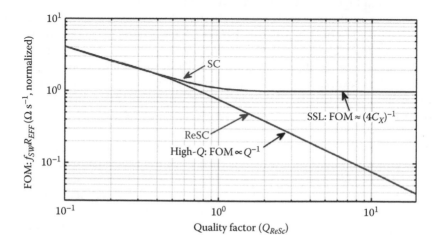

FIGURE 5.10   Figure of merit (FOM) comparison: $f_{SW}R_{EFF}$ versus quality factor (lower is better).

where we have assumed the same ESR, it may be appreciated that the ReSC converter is therefore always favorable compared to the SC converter, but the FOM is asymptotically identical as Q approaches zero.

While the aforementioned analysis is favorable for the ReSC topology, it should be noted that the critical assumption was made that the ESR was the same for both converters. In practice, the introduction of an inductor into the flying capacitor loop is likely to increase ESR. In Figure 5.10, this would result in a shift in the $f_{SW}R_{EFF}$ curve up and to the right. However, a realistic design is complicated by the fact that higher inductance typically results in higher resistance assuming the total volume of the inductor is fixed [21]. It should be further noted that Q is dependent on a number of factors including switch resistance (therefore, the size of the switches), parasitic interconnect and capacitor ESR, the AC resistance of the inductor, and the operating frequency. Therefore, the ReSC design requires careful analysis of passive components and co-optimization of the resonant (switching) frequency and switch sizing. A full analysis of these factors is provided in [15,16], which also compares the operating regimes, process specifications, and geometries of the magnetic components that are needed for the ReSC topology to be favorable from the perspectives of efficiency and power density.

5.3   VARIABLE CONVERSION RATIOS WITH ReSC CONVERTERS

Another potential advantage of the ReSC architecture is the ability to implement variable voltage regulation while maintaining high efficiency. Variable conversion ratios are possible in the ReSC architecture because the inductor can prevent hard switching of the flying capacitor (CV^2 loss) in a variety of operating states that extend beyond the normal resonant operating mode.

Figure 5.11 shows an extension of the operating states in Figure 5.7, including two additional states that can be used to provide higher or lower conversion ratios than the nominal 2:1 mode. It can be seen that states "B" and "D" are the normal resonant states shown in Figure 5.7. In state "A," M_1 and M_4 are on, connecting the resonant impedance across the full supply voltage. In state "C," M_2 and M_3 are on, effectively shorting out the resonant impedance. In "boost" mode, which can provide voltage higher than the nominal 2:1 level, the state sequence and transient waveforms for the resonant impedance are shown in Figure 5.12. The resonant tank is first exposed to the full stack voltage, state A, for a time d_1T, then connected in state B for a time d_2T until the zero-current crossing, and finally, connected in state D to complete a resonant half-cycle. Because the average voltage across the tank in state A–B is higher than $V_{DD}/2$, the output voltage is nominally regulated to a higher value.

In "buck" mode, which can provide voltage lower than the nominal 2:1 level, the state sequence and transient waveforms for the resonant impedance are shown in Figure 5.12. The resonant tank is first configured across the high-side voltage tap, state A, for a time d_1T, then shorted in state C for a time d_2T until the zero-current crossing, and finally, connected in state D to complete a resonant half-cycle. Because the average voltage across the tank in state A–B is lower than $V_{DD}/2$, the output voltage is nominally regulated to a lower value.

As shown in Figures 5.12 and 5.13, variable conversion ratio operation requires at least one switching transition that is not at a zero-current state. Therefore, it is important

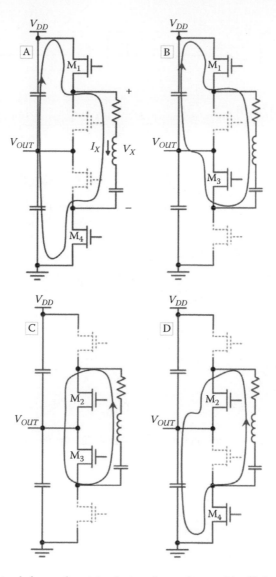

FIGURE 5.11 Extended operating states that can be used to provide efficient variable voltage regulation.

that the switching transition is fast to avoid diode conduction and higher power loss. However, the remaining transitions occur at the zero-current crossing. Unlike the SC topology, there is no charge-sharing loss as the inductor limits the current flow through the flying capacitor in all states. Therefore, to achieve high efficiency the converter is designed in the normal sense by minimizing conduction loss in the resonant loop and optimizing the switch sizes to balance conduction and switching loss.

Figure 5.14 shows simulation results for a representative ReSC converter providing variable output voltages, normalized to V_{DD}. The simulation captures

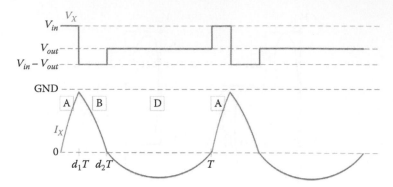

FIGURE 5.12 State sequence and transient waveforms for the resonant impedance in "boost" mode.

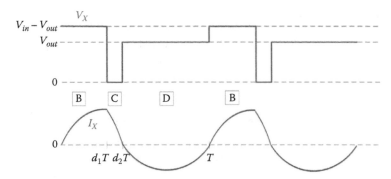

FIGURE 5.13 State sequence and transient waveforms for the resonant impedance in "buck" mode.

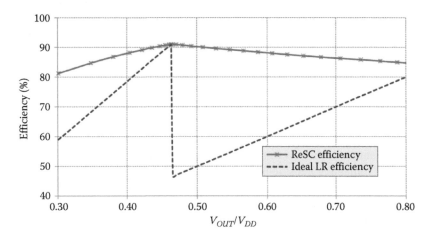

FIGURE 5.14 Efficiency versus normalized output voltage (representative ReSC converter simulated in SPICE).

realistic loss mechanisms such as conduction loss from switch resistance, inductor ESR, and interconnect. Switching losses include both gate drive losses and bottom plate switching losses. The highest efficiency is achieved in the nominal 2:1 resonant mode. It can be seen that the output voltage in this mode is less than $V_{DD}/2$ due to the loadline of the converter: $V_{OUT}(2:1) = V_{DD}/2 - I_{DC} * R_{EFF}$. However, by using "boost" mode, higher output voltages are possible. It can be seen that as output voltage increases from $V_{DD}/2$, efficiency degrades due to the higher RMS current flowing in the active and passive components. However, when compared to an ideal linear regulator that is connected to V_{DD}, the efficiency is higher over a wide range.

"Boost" mode is valuable in many applications where the converter needs to compensate its own loadline. For example, at higher load current, the converter can maintain a fixed output voltage at or above $V_{DD}/2$. This is challenging for typical SC circuits, which need to reconfigure the architecture (e.g., switch from a 2:1 mode to a 3:2 mode) in order to maintain a fixed output voltage.

The "buck" or step-down mode efficiency is also shown in Figure 5.14. Again, it can be seen that while the efficiency degrades as the voltage moves away from the nominal 2:1 mode, "buck" mode provides higher efficiency than the case where an ideal linear regulator steps the voltage down from the 2:1 output. Therefore, a combination of "buck" and "boost" modes can provide relatively high efficiency over a wide range without reconfiguring the circuit architecture.

It should be noted that the curve in Figure 5.14 is primarily representative of the potential of the ReSC topology and is based on behavioral simulation. In a realistic converter, the efficiency profile may be better or worse than shown in Figure 5.14. For example, in the case that the inductor is small (e.g., <10 nH), the timing of the switching transitions is important. For example, in [22], the nonresonant transition required a timing interval with single-digit nanosecond resolution. It was also difficult to prevent some diode conduction during the dead-times, which further increases power loss at conversion ratios further from the nominal 2:1 operating point.

5.4 HIERARCHICAL AND SCALABLE ReSC ARCHITECTURES

In addition to the nominal 2:1 ReSC architecture, many other configurations and topologies are possible. For example, it can be shown that a number of classic SC architectures are suitable for resonant operation. The primary requirement for ReSC suitability is that there is no inherent charge sharing within the topology. For example, the method discussed in [40,41] can be used to identify suitable topologies. There must also be a suitable location for an inductor (or inductors) to be placed in resonance with flying capacitors.

Figure 5.15 shows an example of the ReSC topology extended to the well-known SC ladder converter. The ladder topology can be considered a cascaded network of 1:1 conversion stages each with switching structure and operating modes similar to the 2:1 ReSC converter discussed in the previous section. The ladder converter interfaces across "N" series-stacked voltage domains with "$N - 1$" converter stages, and therefore requires "$N - 1$" inductors to resonate with the same number

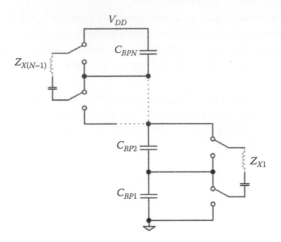

FIGURE 5.15 Resonant "ladder" converter architecture.

of flying capacitors. Ladder topologies are advantageous in that active and passive components need to be rated only to the voltage of the individual voltage taps, but have the disadvantage that the current may need to flow through many series-connected stages.

The ladder architecture has been successfully deployed for PV applications using the "partial" or "differential" power processing architecture [13,36,42]. As shown in Figure 5.16, partial power processing is used in order to balance power flow in series-connected energy sources or loads by connecting a power converter array in parallel with the series stack. In the example in [13], the ReSC ladder converter was used to provide voltage equalization for series-connected PV substrings. Partial or differential power processing converters are typically configured in a hierarchical or cascaded architecture and provide a path for current to flow in parallel with series-connected PV units [43–47]. The extra current flow path provides a mechanism for PV units to operate independently from adjacent PV units [13]. It is primarily important to allow subregions of the PV string to operate at different current levels due to the fact that short-circuit or maximum power current is correlated with the insolation (or number of photons incident) on a given subunit in a given time period.

It has been shown experimentally and through the Sandia Array Performance Model Database [14,22] that maximum power voltage (V_{MPP}) does not vary significantly even under large variations in insolation. In contrast with variable conversion ratio architectures, voltage equalization is relatively simple and inherently stable [47]. In practice, equalization can be operated open-loop as long as R_{EFF} is small relative to the inherent output impedance of the PV subunits. As discussed in [13,47], voltage equalization can be described as an effective parallelization of strings, providing the energy capture of a parallel-configured system, but still maintaining the higher voltage of a series-configured system.

The ladder architecture in Figure 5.16 enforces voltage equalization through the equivalent transformer model shown in Figure 5.3. It can be shown that in the 1:1

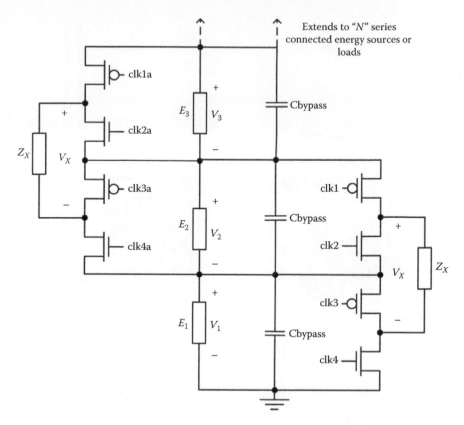

FIGURE 5.16 ReSC ladder converter with "$N - 1$" stages managing power flow in "N" series-connected energy sources or loads (e.g., batteries or photovoltaic elements).

configuration the output impedance, R_{EFF}, is modified from Equation 5.4 in the high-Q assumption to

$$R_{EFF} \approx \frac{\pi^2}{2} R_{ESR} \approx 5 * R_{ESR}. \qquad (5.9)$$

In short, R_{EFF} is higher in the 1:1 configuration than in the 2:1 configuration due to the nominal unity conversion ratio.

Figure 5.17 shows two examples of ReSC ladder converter implementations for the PV partial power processing application. The first converter, detailed in [13], used a discrete powertrain, implemented on a 4.6 in. × 2.6 in. printed circuit board. This converter was suitable to manage submodule PV strings with open-circuit voltage between 10 and 20 V and open circuit current up to 10 A. The discrete solution included a ~5 mm^2 IC controller. The IC provided local power management (a DC–DC converter and linear regulator to provide local power supplies from the PV module), a crystal oscillator providing an accurate timing reference, gate drivers

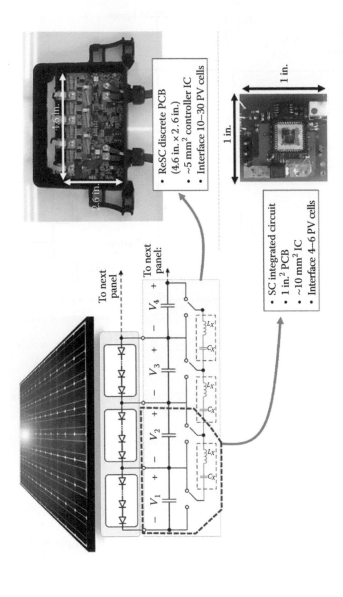

FIGURE 5.17 Example ReSC ladder converter implementations for PV applications. (Adapted from Stauth, J.T. et al., *IEEE Trans. Power Electron.*, 28(3), 1189, 2013; Sangwan, R. et al., High-density power converters for sub-module photovoltaic power management, *2014 IEEE Energy Conversion Congress and Exposition (ECCE)*, September 14–18, 2014, pp. 3279–3286.)

for the ReSC converter stage, and a powerline communication system to transmit performance data up and down the stack.

The second solution, detailed in [33], was a high-density ReSC converter with all bypass and flying capacitors implemented on-chip. This prototype was designed in 180 nm bulk CMOS with MIM capacitors that had 6.6 fF/μm^2 density and also used a thick top metal with sheet resistance ~10 mΩ/sq. The converter was designed with a maximum input voltage of 6.6 V and was suitable to manage PV substrings with four to six cells per string.

The IC solution was able to achieve 90%–95% conversion efficiency in the 2:1 mode for load currents between 200 mA to 1.0 A. The converter was tested with real PV cells in a series array. In a gradient mismatch scenario, the converter was able to achieve higher power than the conventional solution for mismatch levels above ~3%.

5.4.1 OTHER ReSC TOPOLOGIES

ReSC converters are configurable in a variety of other topologies that achieve various power management functions and conversion ratios. For example, consider Figure 5.18 shows a series-parallel ReSC converter. Importantly, for this topology, the converter can be designed with a single inductor that resonates with two sequential capacitor configurations. In an N:1 converter, in state A, $N - 1$ flying capacitors are in series with the resonant inductor, L_X, between V_{DD} and V_{OUT}. In state B, the same flying capacitors are in parallel with their bottom plate referenced to ground.

The corresponding inductor current waveform for the series-parallel topology is shown in Figure 5.19. In state A, the current through the inductor is positive and acts to deliver power to the load and simultaneously charge the flying capacitors. In state B, the current is also positive, but results from energy being discharged through the

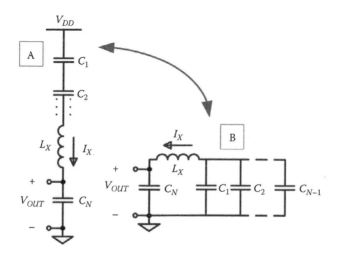

FIGURE 5.18 Series-parallel ReSC topology.

FIGURE 5.19 Current waveforms (I_X) in series-parallel topology.

flying capacitors that are now in the parallel arrangement. The current waveform is not symmetrical in the two operating states because of the different capacitor configuration. For example, in state A, with flying capacitors in series, the effective resonant capacitance is $C_X/(N-1)$. In state B, the effective resonant capacitance is $(N-1) * C_X$. Therefore, the ratio of the two resonant frequencies is related to the conversion ratio and is approximately $N-1$.

Figure 5.20 shows an example of a Dickson ReSC converter. In this case, the converter shown has a nominally 4:1 conversion ratio. The Dickson topology is well known to be a favored SC architecture due to its efficient utilization of switching devices [6,40]. This topology can be implemented as an ReSC converter by including a single inductor coupled to the output terminal. As seen in Figure 5.20, the resonant frequency is set by the series and parallel configuration of capacitors C_1–C_3. The additional degree of freedom in choosing the specific capacitor values allows the resonant frequency to be the same or different in each state depending on the design requirements.

As outlined in [40,41], a limitation to the Dickson approach is that there is inherent charge sharing in the capacitor network. This charge sharing occurs in the transition from state A to state B. In state A, the switching node voltage, V_{SW}, starts at some value higher than V_{OUT}, and ends at a value that is ΔV lower. By the end of state A, the voltage on C_3 increases by ΔV. Because C_1 is configured in a negative polarity, and assuming the values of C_1 and C_2 are equal, the voltage on C_1 increases by $\Delta V/2$, and the value of C_2 decreases by $\Delta V/2$.

With these new values, the capacitor network is reconfigured in state B. Here, the polarities of all capacitors are flipped (where they were charging in state A, they are now discharging in state B and vice versa). It can be seen that, considering only the ΔV quantities, C_1 will try to set V_{SW} to a value nominally $\Delta V/2$ higher than its

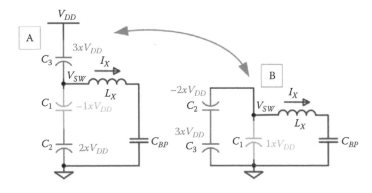

FIGURE 5.20 Dickson ReSC topology.

previous state. However, the series combination of C_2 and C_3 will try to set VSW to a value of $\Delta V + \Delta V/2$ higher than in the previous state. The discrepancy between the capacitor voltages in these state transitions leads to charge sharing losses (e.g., charge will flow to equalize the voltage at across the capacitor networks, resulting in CV^2 loss).

The notion of inherent charge sharing is an important limitation to the ReSC architecture with arbitrary SC configurations. However, the work of Lei et al. [40,41] points to a possible solution in the "split-phase control" scheme. For example, charge sharing occurs due to the hard connection between the two branches of the capacitor network. Lei et al. propose to split state A and B into four separate phases where the inductor component is connected to each capacitor network branch for an intermediate phase, which can, in principle, equalize the voltage before they are hard-switched to the common V_{SW} node.

The notion of using split-phase control and the possibility of other future improvements related to the integration of magnetic energy storage with SC converters is promising and may open up a variety of efficient, high conversion ratio solutions. In general, the ReSC topology can be considered as an intermediate between more common magnetics-based DC–DC converters and promising, but still emerging SC converters.

5.5   CONCLUSION

In a variety of applications, there is a growing need for high-density, low-cost, and high-efficiency power management architectures. To meet this need requires exploring new DC–DC converter topologies that can maximize the energy and power density utilization of both active and passive components. ReSC converters show promise because they leverage the favorable active device utilization of switched-capacitor converter and the relatively high-energy density capacitors in many current and voltage ranges. However, the ReSC approach can improve utilization of high-density capacitors by eliminating hard switching losses and tuning out the reactive impedance, which increases energy stored per switching cycle.

Ultimately, the benefit of the ReSC approach can be quantified through the quality factor of the converter. This can be expanded into a FOM as it captures the product of switching frequency times output impedance as it is reduced relative to a comparable SC converter. This corresponds to a reduction of total loss that includes both switching and conduction loss.

This chapter also discussed important peripheral benefits of using the resonant approach such as the ability to implement efficient step-up and step-down conversion ratios around a nominal integer level. The extension of the ReSC approach to hierarchical and scalable architectures was also discussed, in particular, the ladder, series-parallel, and Dickson converter architectures.

In the future, there is growing promise for both SC and ReSC converters, and a variety of applications that can benefit. The PV application was presented as a specific example where a scalable, nominally 1:1 converter scheme can dramatically improve energy capture and reliability. However, other applications spanning low-voltage power delivery, battery management, and monolithic DC–DC conversion for mobile computing and communication devices are also promising and exciting.

REFERENCES

1. S. Borkar, Design challenges of technology scaling, *IEEE Micro.*, 19(4), 23–29, 1999.
2. E. A. Burton, G. Schrom, F. Paillet, J. Douglas, W. J. Lambert, K. Radhakrishnan, and M. J. Hill, FIVR—Fully integrated voltage regulators on 4th generation Intel Core SoCs, in *IEEE Applied Power Electronics Conference*, Fort Worth, TX, March 16–20, 2014, pp. 432–439.
3. A. Ramachandran, J. Vienne, R. Van Der Wijngaart, L. Koesterke, and I. Sharapov, Performance evaluation of NAS Parallel Benchmarks on Intel Xeon Phi, in *2013 42nd International Conference on Parallel Processing*, Lyon, France, October 1–4, 2013, pp. 736–743.
4. Y. K. Ramadass, A. A. Fayed, and A. P. Chandrakasan, A fully-integrated switched-capacitor step-down DC-DC converter with digital capacitance modulation in 45 nm CMOS, *IEEE J. Solid-State Circuits*, 45(12), 2557–2565, 2010.
5. A. P. Chandrakasan and R. W. Brodersen, Minimizing power consumption in digital CMOS circuits, *Proc. IEEE*, 83(4), 498–523, April 1995.
6. M. D. Seeman and S. R. Sanders, Analysis and optimization of switched-capacitor DC-DC converters, *IEEE Trans. Power Electron.*, 23(2), 841–851, 2008.
7. S. R. Sanders, E. Alon, H. Le, M. D. Seeman, M. John, and V. W. Ng, The road to fully integrated DC-DC conversion via the switched-capacitor approach, *IEEE Trans. Power Electron.*, 28(9), 1–14, 2013.
8. H.-P. Le, S. R. Sanders, and E. Alon, Design techniques for fully integrated switched-capacitor DC-DC converters, *IEEE J. Solid-State Circuits*, 46(9), 2120–2131, September 2011.
9. L. G. Salem and P. P. Mercier, 4.6 An 85%-efficiency fully integrated 15-ratio recursive switched-capacitor DC-DC converter with 0.1-to-2.2V output voltage range, in *2014 IEEE International Solid-State Circuits Conference, Digest of Technical Papers (ISSCC)*, San Francisco, CA, February 9–13, 2014, pp. 88–89.
10. H.-P. Le, M. Seeman, S. R. Sanders, V. Sathe, S. Naffziger, and E. Alon, A 32 nm fully integrated reconfigurable switched-capacitor DC-DC converter delivering $0.55W/mm^2$ at 81% efficiency, in *2010 IEEE International Solid-State Circuits Conference (ISSCC)*, San Francisco, CA, February 7–11, 2010, pp. 210–211.
11. T. M. Andersen, K. Florian, J. W. Kolar, T. Toifl, C. Menolfi, L. Kull, T. Morf, M. Kossel, M. Brandli, P. Buchmann, and P. A. Francese, A sub-ns response on-chip switched-capacitor DC-DC voltage regulator delivering $3.7W/mm^2$ at 90% efficiency using deep-trench capacitors in 32 nm SOI CMOS, in *International Solid State Circuits Conference (ISSCC)*, San Francisco, CA, February 9–13, 2014, pp. 90–92.
12. G. V. Pique, H. J. Bergveld, and E. Alarcon, Survey and benchmark of fully integrated switching power converters: Switched-capacitor versus inductive approach, *IEEE Trans. Power Electron.*, 28(9), 4156–4167, 2013.
13. J. T. Stauth, M. D. Seeman, and K. Kesarwani, Resonant switched-capacitor converters for sub-module distributed photovoltaic power management, *IEEE Trans. Power Electron.*, 28(3), 1189–1198, March 2013.
14. A. Cervera, M. Evzelman, M. M. Peretz, and S. S. Ben-Yaakov, A high efficiency resonant switched capacitor converter with continuous conversion ratio, in *IEEE Energy Conversion Congress and Exposition (ECCE)*, Denver, CO, September 15–19, 2013, pp. 4969–4976.
15. K. Kesarwani, R. Sangwan, and J. T. Stauth, Resonant switched-capacitor converters for chip-scale power delivery: Modeling and design, in *IEEE Workshop on Control and Modeling for Power Electronics*, Salt Lake City, UT, June 23–26, 2013, pp. 1–7.
16. K. Kesarwani, R. Sangwan, and J. T. Stauth, Resonant switched-capacitor converters for chip-scale power delivery: Design & implementation, *IEEE Trans. Power Electron.*, 30(12), 6966–6977, 2014.

17. J. Wibben and R. Harjani, A high-efficiency DC–DC converter using 2 nH integrated inductors, *IEEE J. Solid-State Circuits*, 43(4), 844–854, April 2008.
18. W. Kim, D. M. Brooks, and G.-Y. Wei, A fully-integrated 3-level DC/DC converter for nanosecond-scale DVS with fast shunt regulation, in *International Solid-State Circuits Conference (ISSCC)*, San Francisco, CA, February 20–24, 2011, pp. 268–270.
19. G. J. Mehas, K. D. Coonley, and C. R. Sullivan, Design of microfabricated inductors for microprocessor power delivery, in *IEEE Applied Power Electronics Conference (APEC)*, Dallas, TX, March 14–18, 1999, pp. 1181–1187.
20. C. R. Sullivan, D. V. Harburg, J. Qiu, C. G. Levey, and D. Yao, Integrating magnetics for on-chip power: A perspective, *IEEE Trans. Power Electron.*, 28(9), 4342–4353, September 2013.
21. D. J. Perreault, J. Hu, J. M. Rivas, Y. Han, O. Leitermann, R. C. N. Pilawa-podgurski, A. D. Sagneri, and C. R. Sullivan, Opportunities and challenges in very high frequency power conversion, in *IEEE Applied Power Electronics Conference (APEC)*, Washington, DC, February 15–19, 2009, pp. 1–14.
22. C. Schaef and J. T. Stauth, A variable-conversion-ratio 3-phase resonant switched capacitor converter with 85% efficiency at 0.91 W/mm² using 1.1 nH PCB-trace inductors, in *IEEE International Solid State Circuits Conference (ISSCC), Digest of Technical Papers*, San Francisco, CA, February 22–26, 2015, pp. 1–3.
23. K. Kesarwani, R. Sangwan, and J. T. Stauth, 4.5 A 2-phase resonant switched-capacitor converter delivering 4.3W at 0.6W/mm² with 85% efficiency, in *2014 IEEE International Solid-State Circuits Conference, Digest of Technical Papers (ISSCC)*, San Francisco, CA, February 9–13, 2014, pp. 86–87.
24. F. Bedell, History of A-C wave form, its determination and standardization, *Trans. Am. Instrum. Electr. Eng.*, 61(12), 864–868, December 1942.
25. D. J. Allstot, R. W. Brodersen, and P. R. Gray, MOS switched capacitor ladder filters, *IEEE J. Solid-State Circuits*, 13(6), 806–814, December 1978.
26. P. Lin and L. Chua, Topological generation and analysis of voltage multiplier circuits, *IEEE Trans. Circuits Syst.*, 24(10), 517–530, October 1977.
27. J. S. Brugler, Theoretical performance of voltage multiplier circuits, *IEEE J. Solid-State Circuits*, 6(3), 132–135, June 1971.
28. M. S. Makowski and D. Maksimovic, Performance limits of switched-capacitor DC-DC converters, in *IEEE Power Electronics Specialists Conference (PESC)*, Atlanta, GA, June 18–22, 1995, pp. 1215–1221.
29. H. Meyvaert, T. Van Breussegem, and M. Steyaert, A monolithic 0.77 W/mm² power dense capacitive DC-DC step-down converter in 90 nm bulk CMOS, in *2011 Proceedings of ESSCIRC (ESSCIRC)*, Helsinki, Finland, September 12–16, 2011, pp. 483–486.
30. G. V Pique, A 41-phase switched-capacitor power converter with 3.8 mW output ripple and 81% efficiency in baseline 90 nm CMOS, in *International Solid State Circuits Conference (ISSCC)*, San Francisco, CA, February 19–23, 2012, Vol. 23(2), pp. 98–100.
31. A. Lesnicar and R. Marquardt, An innovative modular multilevel converter topology suitable for a wide power range, in *2003 IEEE Bologna PowerTech Conference Proceedings*, Bologna, Italy, June 23–26, 2003, Vol. 3, pp. 272–277.
32. R. Marquardt, Modular multilevel converter: An universal concept for HVDC-networks and extended DC-bus-applications, in *The 2010 International Power Electronics Conference (ECCE ASIA)*, Sapporo, Japan, June 21–24, 2010, pp. 502–507.
33. R. Sangwan, K. Kesarwani, and J. T. Stauth, High-density power converters for sub-module photovoltaic power management, in *2014 IEEE Energy Conversion Congress and Exposition (ECCE)*, Pittsburgh, PA, September 14–18, 2014, pp. 3279–3286.
34. D. Wolaver, Basic constraints from graph theory for dc-to-dc conversion networks, *IEEE Trans. Circuit Theory*, 19(6), 640–648, 1972.

35. J. F. Dickson, On-chip high-voltage generation in MNOS integrated circuits using an improved voltage multiplier technique, *IEEE J. Solid-State Circuits*, 11(3), 374–378, June 1976.

36. J. Stauth, M. D. Seeman, and K. Kesarwani, A high-voltage CMOS IC and embedded system for distributed photovoltaic energy optimization with over 99% effective conversion efficiency and insertion loss below 0.1%, in *International Solid-State Circuits Conference (ISSCC)*, San Francisco, CA, February 19–23, 2012, 23(2), pp. 100–102.

37. R. C. N. Pilawa-podgurski, D. M. Giuliano, and D. J. Perreault, Merged two-stage power converter architecture with soft charging switched-capacitor energy transfer, in *IEEE Power Electronics Specialists Conference (PESC)*, Rhodes, Greece, June 15–19, 2008, pp. 4008–4015.

38. R. C. N. Pilawa-podgurski and D. J. Perreault, Merged two-stage power converter with soft charging switched-capacitor stage in 180 nm CMOS, *IEEE J. Solid-State Circuits*, 47(7), 1557–1567, 2012.

39. S. S. Ben-yaakov, A. Blumenfeld, A. Cervera, and M. Evzelman, Design and evaluation of a modular resonant switched capacitors equalizer for PV panels, in *IEEE Energy Conversion Congress and Exposition (ECCE)*, Raleigh, NC, September 15–20, 2012, pp. 4129–4136.

40. Y. Lei and R. C. N. Pilawa-Podgurski, Soft-charging operation of switched-capacitor DC-DC converters with an inductive load, in *IEEE Applied Power Electronics Conference (APEC)*, Fort Worth, TX, March 16–20, 2014, pp. 2112–2119.

41. Y. Lei, R. May, and R. C. N. Pilawa-podgurski, Split-phase control: Achieving complete soft-charging operation of a Dickson switched-capacitor converter, in *IEEE Control and Modeling for Power Electronics (COMPEL)*, Santander, Spain, June 22–25, 2014, pp. 736–743.

42. J. Stauth, M. Seeman, and K. Kesarwani, A high-voltage CMOS IC and embedded system for distributed photovoltaic energy optimization with over 99% effective conversion efficiency and insertion loss below 0.1%, in *2012 IEEE International Solid-State Circuits Conference*, San Francisco, CA, February 19–23, 2012, pp. 100–102.

43. M. S. Agamy, M. Harfman-todorovic, A. Elasser, S. Chi, R. L. Steigerwald, J. A. Sabate, A. J. Mccann, L. Zhang, and F. J. Mueller, An efficient partial power processing DC/DC converter for distributed PV architectures, *IEEE Trans. Ind. Electron.*, 29(2), 674–686, 2014.

44. P. S. Shenoy and K. A. Kim, Differential power processing for increased energy production and reliability of photovoltaic systems, *IEEE Trans. Power Electron.*, 28(6), 2968–2979, 2013.

45. C. Olalla, C. Deline, D. Clement, Y. Levron, M. Rodriguez, and D. Maksimovic, Performance of power limited differential power processing architectures in mismatched PV systems, *IEEE Trans. Power Electron.*, 30(2), 1, 2014.

46. P. S. Shenoy and P. T. Krein, Differential power processing for DC systems, *IEEE Trans. Power Electron.*, 28(4), 1795–1806, April 2013.

47. S. Poshtkouhi, A. Biswas, and O. Trescases, DC-DC converter for high granularity, substring MPPT in photovoltaic applications using a virtual-parallel connection, in *2012 Twenty-Seventh Annual IEEE Applied Power Electronics Conference and Exposition*, Orlando, FL, February 5–9, 2012, pp. 86–92.

6 Design of Recursive Switched-Capacitor DC–DC Converters

Loai G. Salem and Patrick P. Mercier

CONTENTS

6.1 INTRODUCTION

Modern digital integrated circuits achieve a balance between performance and energy efficiency through dynamic voltage scaling (DVS) of individual processing cores in accordance with performance needs. As the number of voltage domains increases in state-of-the-art system-on-chips, generation of each supply voltage must occur not only efficiently but within a small area. While linear regulators are compact and achieve fast response times [17,23], their efficiencies are determined by the ratio of output to input voltage, potentially limiting system-level energy efficiency [7,21,22]. On the other hand, switched-inductor DC–DC converters can achieve high efficiencies, yet typically require large off-chip inductors [3] or increased packaging complexity [12,13,18,26,31], limiting their ability to power many independent voltage domains in a small volume. To simultaneously address the efficiency/size trade-off, fully integrated switched-capacitor (SC) DC–DC converters utilize high-Q capacitors available in typical CMOS processes to convert and regulate power in an energy- and area-efficient manner [1,2,8,14–16,22,30,32].

Unlike switched-inductor DC–DC converters, however, SC converters are only efficient at discrete ratios of input-to-output voltages, constricting efficient DVS operation to small supply voltage ranges. Increasing the number of reconfigurable ratios can solve this; however, doing so introduces two main challenges: capacitance utilization and relative sizing. In a fully integrated SC converter, the achievable efficiency is limited by the amount of committed capacitance; disabling even a small fraction of such capacitance significantly lowers the efficiency. Additionally, ensuring optimal relative sizing among the constituent capacitors can improve efficiency considerably [28]. Unfortunately, the complexity of conventional topologies, including the number of necessary capacitors and reconfigurations switches, increases significantly with the number of ratios, making simultaneous 100% capacitance utilization and optimal relative sizing extremely challenging. Thus, most SC converter designs employ only a small number of ratios [5,8,10,16,20], often resulting in large efficiency drops in-between the available ratios.

This chapter presents the first demonstration of an SC converter that is reconfigurable amongst $2^N - 1$ ratios without disconnecting a single capacitor while ensuring optimal relative sizing and high efficiency across a large output voltage range. The proposed *recursive* SC DC–DC converter topology (RSC) [24], shown in Figure 6.1, recursively divides the delivered output charge across N 2:1 cells connected in cascade to generate N-bit ratios. By maximizing the number of input voltage and ground connections, charge-sharing losses are minimized, and in fact become a convergent geometric series with minimal additional losses incurred beyond 4-bit ratios. Given the inherently modular nature of the converter, 100% capacitance utilization is ensured by reconfiguring cell connections either in cascade (for high resolutions) or parallel (for lower resolutions), with binary slicing of the largest cascaded cell in order to enable reconfiguration among odd and even resolutions, all the while ensuring optimal relative sizing.

This chapter is organized as follows: Section 6.2 introduces the RSC topology and discusses its theoretical performance compared with prior topologies. Section 6.3 presents architectural implementation details of a 4-bit RSC converter, while Section 6.4 presents detailed circuit design. Experimental results of the test chip that verify the predicted performance are provided in Section 6.5.

6.2   RECURSIVE SWITCHED-CAPACITOR TOPOLOGY

The most basic RSC building block is a 2:1 SC converter. As shown in Figure 6.1a, a 2:1 SC can be considered as a three-port circuit that includes two input ports IN_{top} and IN_{bottom} to receive a high and low input voltages, respectively, and an output port *MID* that provides the average of the voltages at the input ports, that is, $(IN_{top} + IN_{bottom})/2$. The 2:1 SC cell equally loads its output port current on the two input ports (IN_{top}, IN_{bottom}). The following sections discuss how 2:1 SC building cells can be connected to realize $2^N - 1$ conversion ratios while minimizing losses.

6.2.1 Topology Definition and Steady-State Loss Analysis

Figure 6.1b shows the *recursive* SC topology pseudo code. Starting with a single 2:1 SC that divides the converter input voltage, V_{in}, into two intervals (0-to-$V_{in}/2$, $V_{in}/2$-to-V_{in}), the topology inserts a 2:1 SC cell in series between the previous cell output *MID* and the converter ground 0, or stacked between V_{in} and the output *MID* of the

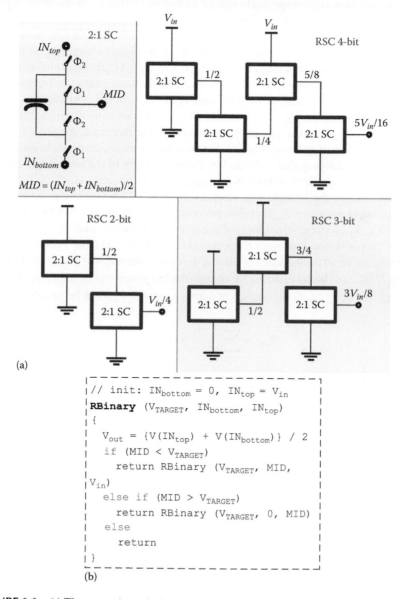

(a)

```
// init: INbottom = 0, INtop = Vin
RBinary (VTARGET, INbottom, INtop)
{
    Vout = {V(INtop) + V(INbottom)} / 2
    if (MID < VTARGET)
        return RBinary (VTARGET, MID,
Vin)
    else if (MID > VTARGET)
        return RBinary (VTARGET, 0, MID)
    else
        return
}
```

(b)

FIGURE 6.1 (a) The *recursive* switched-capacitor realization of the ratios 1/4, 3/8, and 5/16. (b) Toplogy pseudo code. Each SC cell comprises two out-of-phase 2:1 SC for a well-posed SC network.

previous 2:1 cell, repeatedly, until the desired binary conversion ratio $m/2^N$ is realized, where $m < 2^N$. Figure 6.1a demonstrates examples of the ratios 1/4, 3/8, and 5/16 at 2-, 3-, and 4-bit resolutions, respectively.

The proposed topology minimizes cascaded losses by maximizing the number of input voltage, V_{in}, and ground, 0, connections. Specifically, each 2:1 stage Ci has at least one input port connected to either the input voltage V_{in} or the converter ground 0, and thus each stage loads half of its output charge q_i on the input supply, V_{in}, or ground, 0, instead of loading such charge on a previous cascaded stage. For example, Figure 6.2 illustrates two different configurations that both realize an 11/16 ratio. In Figure 6.2a, the last stage $C4$ loads half of the output charge q_{out} on the second stage, $C2$, which in turn loads the first stage, $C0$, with $3q_{out}/8$. The third stage, $C3$, loads the first stage by an additional $q_{out}/4$, and thus the total charge delivered by the first stage is $5q_{out}/8$. In contrast, the RSC converter employs the configuration shown in Figure 6.2b, where the IN_{top} of $C4$ and the IN_{bottom} of $C3$ are directly connected to the converter input, V_{in}, and the ground, 0, respectively, and therefore, the loaded charge on $C2$ and $C1$ are both reduced by $q_{out}/2$. For an arbitrary recursion depth N, each stage is loaded with a charge q_i that is divided by a binary weight of the total output charge, q_{out}, such that $q_i = q_{out}/2^{N-i}$, where i is the stage order in the cascade.

It is known that the intrinsic loss mechanisms in an SC converter can be modeled by a finite output resistance, R_{out}, in either the slow-switching limit (SSL) or fast-switching limit (FSL): R_{SSL}, where the charge-sharing loss dominates, and R_{FSL}, where the switches' on-resistance dominates the losses [11,28]. In the SSL, the total energy loss through the converter can be found by adding the charge-sharing loss across each capacitor C_i, $(q_i/2)^2/C_i$, and by normalizing the charge-sharing power loss by the squared output current I_L^2, that is, $(q_{out}f_{sw})^2$, the equivalent output resistance R_{SSL} can be calculated as

$$R_{SSL} = \sum_{i=1}^{N} \left(\frac{1}{2^{N-i+1}} \right)^2 \frac{1}{f_{sw}C_i}, \tag{6.1}$$

where C_i is the total capacitance of the two flying capacitors per stage. The derived R_{SSL} is for a symmetric RSC, where each cell consists of two oppositely phased 2:1 SC, which eliminates any charge-balance DC capacitor between the cascaded stages. Similarly, at the FSL, the current through each switch becomes the delivered current by that stage, which is a binary weighted fraction of the load current, I_L (i.e., $I_L/2^{N-i}$). Thus, the equivalent output resistance R_{FSL}, for a 50% duty-cycle converter clock, is

$$R_{FSL} = \sum_{i=1}^{N} \sum_{j=1}^{4} \frac{1}{2} \left(\frac{1}{2^{N-i}} \right)^2 R_{i,j}, \tag{6.2}$$

where
  the summation over j accounts for the four switches per stage i
  each switch resistance $R_{i,j}$ results from two parallel switches in a symmetric RSC
  of eight switches

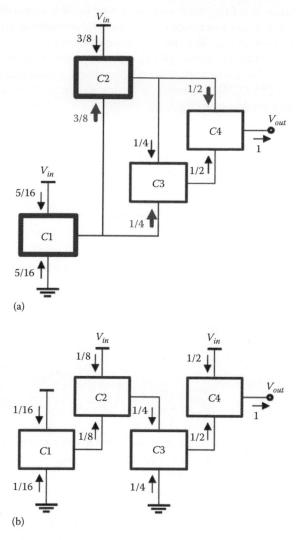

FIGURE 6.2   Charge flow through two inter-cell connections to realize the same ratio 11/16 (a) nonoptimal cascading (b) proposed RSC connection. Bold blocks are loaded with extra charge than the corresponding blocks in (b) with RSC connection. Bold arrows represent the extra loading charge.

The total equivalent output resistance R_{out} at a given switching frequency, f_{sw}, occurring between the two asymptotes can be approximated by the Euclidean norm of the two limits, R_{SSL} and R_{FSL} [28]. From Equations 6.1 and 6.2, the RSC equivalent output resistance R_{out} only depends on N and does not change across the resolution ratios.

Allocating a larger capacitance, C_i, for each stage results in a lower voltage swing, ΔV_i, and lower charge-sharing loss, as dictated by Equation 6.1. Given the limited available capacitance in a fully integrated SC converter, it is important to find the relative sizing of each stage capacitance C_i from the total available on-die capacitance C_{tot} to realize the minimal R_{SSL}. For fully integrated capacitors with single-voltage-rating and with no stacking of switches to block higher voltages, the optimal capacitance and conductance relative-sizing match the relative charge transferred through each capacitor or switch [25,27,28], and hence is binary weighted of the total available capacitance C_{tot} and conductance G_{tot}:

$$C_i = \left(\frac{2^{i-1}}{2^N - 1} \right) C_{tot}, \tag{6.3}$$

$$G_i = \frac{1}{4} \left(\frac{2^{i-1}}{2^N - 1} \right) G_{tot}. \tag{6.4}$$

With such optimal sizing, the equivalent output impedance at the two asymptotes can be found as

$$R_{SSL}^* = \frac{1}{f_{sw} C_{tot}} \left(1 - \frac{1}{2^N} \right)^2, \tag{6.5}$$

$$R_{FSL}^* = \frac{2}{G_{tot}} \left(1 - \frac{1}{2^N} \right)^2. \tag{6.6}$$

To realize the highest possible efficiency for a given silicon area, it is desired to select the SC topology that incurs the lowest charge-sharing loss, R_{SSL}, to deliver the same q_{out} and conversion ratio. The power available from an SC converter normalized by the power available from a 2:1 SC, using the same silicon area, can be used as a metric to compare various SC topologies in the SSL and FSL. After assigning the capacitors appropriate optimal relative sizing, the SSL normalized power available from a topology at a conversion ratio m/n becomes $M_{SSL} = (m/n)/\Sigma_i a_{c,i})^2$, where $m < n$ and $a_{c,i}$ is the fraction of the output charge q_{out} that flows through the capacitor C_i.

Figure 6.3 compares five conventional SC topologies [6,9,19], as well as a successive approximation (SAR) SC converter [4] and the proposed RSC converter, using the established SSL metric, M_{SSL}, where the capacitors of each topology are assigned the optimal relative sizing. The charge multiplier vectors of the various topologies can be found in [29] and through the analysis in [28]. The topologies are compared up to 5-bit binary conversion ratios. The SP topology M_{SSL} is also shown at the ratios 1/6, 1/5, 2/7, 1/3, 2/5, 3/7, while the *Fibonacci* topology M_{SSL} is shown for the *Fibonacci* series ratios 1/21, 1/13, ..., 1/2. All topologies, with the exception of the ladder topology, have the same M_{SSL} at the minimum and maximum conversion ratios within each resolution, for example, 1/2, 1/4, 1/8, 1/16, 1/32 and 3/4, 7/8, 15/16, 31/32, respectively. As shown in Figure 6.3b, due to the

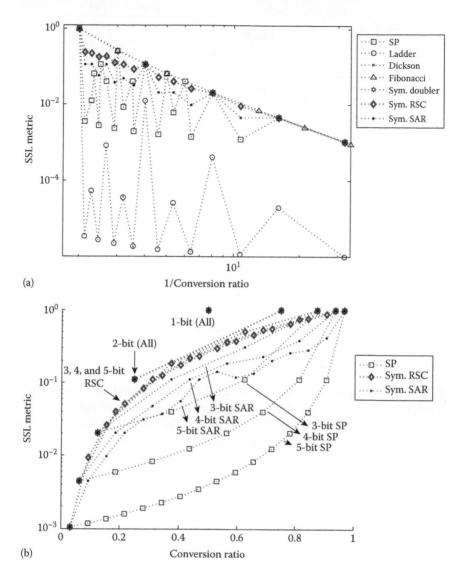

FIGURE 6.3 The SSL power-available metric, M_{SSL}, for the seven topologies at binary ratios up to 5-bit resolution. The topology of the highest power available at certain ratio incurs the lowest charge-sharing loss for a given silicon area. (a) Power-available metric for the seven topologies. (b) Detailed power-available metric for SP, symmetric RSC, and symmetric SAR topologies.

binary division of the output charge across the various stages, the RSC cascading loss converges to an upper limit, $1/(f_{sw}C_{tot})$, at large resolutions N, without further M_{SSL} degradation. The other topologies exhibit an M_{SSL} eye opening with higher resolutions N for ratios $m_{odd}/2^N$, where the SSL loss becomes the summation of a divergent series.

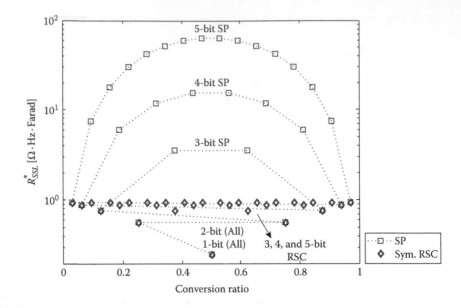

FIGURE 6.4 The R^*_{SSL} for the SP and symmetric RSC versus the binary ratios using a 1 F total capacitance and for an SC converter operated at 1 Hz.

Figure 6.4 shows the R_{SSL}, using a 1F total capacitance and at 1 Hz switching frequency for the SP and the RSC topologies, with capacitors of optimal relative sizing, across binary ratios up to 5-bit resolutions. The RSC normalized R^*_{SSL} saturates at an upper limit of $4 \times R_{SSL}$ of a 1/2 ratio. Figure 6.5 shows the FSL optimal voltage metric [28] for the seven topologies at the same binary ratios as previously discussed. In general, for fully integrated converters, capacitors consume most of the die area, and thus topologies that achieve the lowest SSL loss for a given silicon area (i.e., topologies with the highest M_{SSL}) are desired.

6.2.2 OPEN-LOOP POWER STAGE OPTIMIZATION

After defining the optimal relative sizing of individual RSC components, it is critical to select the total switch area A_{sw} and switching frequency f_{sw} that result in the maximum efficiency for a given load I_L and input voltage V_{in}. In a fully integrated SC, the charge-sharing SSL loss constitutes the major loss component. To decrease the SSL loss, either the available capacitance or the switching frequency, and hence switching parasitics, should be increased. In integrated converters, capacitance is not typically considered as a variable in the optimization process, and the maximum available capacitance for a given silicon area is implemented. The maximum efficiency over the design space (A_{sw}, f_{sw}) can be found by minimizing the total losses arising from the intrinsic SC R_{out}, and the switching losses that result from the power switches gate drive as well as the capacitor bottom-plate losses. The drain parasitics of the switches are treated as part of the capacitors bottom-plate parasitics.

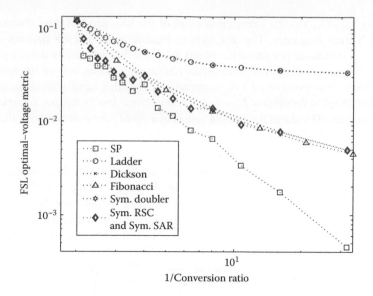

FIGURE 6.5 The FSL performance metric M_{FSL} of the seven topologies at binary conversion ratios up to 5-bit resolution.

Since an RSC consists of individual 2:1 SC cells that provide binary-weighted currents $I_i = I_L/2^{N-i}$, it can be shown that the optimal switching frequency f_{sw}^* and total conductance G_{tot}^* are given by*

$$f_{sw}^* = \frac{1}{4\sqrt[3]{2}} \sqrt[3]{\frac{G_{on}}{C_{gate}V_{gate}^2}\left(\frac{I_L}{C_{tot}}\right)^2} \cdot \sqrt[3]{\left(\frac{2^N-1}{2^{N-1}}\right)^2}, \tag{6.7}$$

$$\frac{G_{tot}^*}{C_{tot}} = 4\sqrt[3]{4} \sqrt[3]{\frac{G_{on}}{C_{gate}V_{gate}^2}\left(\frac{I_L}{C_{tot}}\right)^2} \cdot \sqrt[3]{\left(\frac{2^N-1}{2^{N-1}}\right)^2}, \tag{6.8}$$

where
  G_{on} and C_{gate} are the switch conductance density in S/m and the switch gate
    capacitance per unit width F/m, respectively
  V_{gate} is the gate drive voltage
  G_{tot}^*/C_{tot} is the optimal total conductance per unit capacitance

---

* A simple addition of the two loss limits, R_{SSL} and R_{FSL}, is used to express the intrinsic RSC loss that overestimates the total R_{out}. A negligible bottom plate parasitics are assumed, besides the equivalent load resistance R_L is assumed to be larger than R_{out}, to obtain simple intuitive expressions. The formula for a 2:1 SC optimal switch width and frequency in [16] are used in the derivation.

Essentially, G_{tot}^*/C_{tot} sets the intersection point of the SSL and FSL loss components, or the SC corner frequency. The first term in Equations 6.7 and 6.8 depends on the technology conductance per gate drive energy loss, and the load current density per unit capacitance. The second term depends on the resolution N, where at 1-bit resolution the optimal values correspond to a 2:1 SC converter. On the other hand, with larger number of cascaded stages N, the optimal f_{sw} and total conductance density reaches an upper limit of approximately 60% above the optimal values of a 2:1 SC converter utilizing the available C_{tot}. Essentially, the allocated capacitance of the last stage at large N becomes $C_{tot}/2$ while supplying I_L load current, and thus the optimal design point shifts by $\sqrt[3]{4}$. From Equation 6.8, the optimal total switch area does not change from one ratio to another within a given resolution N, simplifying the implementation of a reconfigurable SC. However, a small change in the optimal total conductance results when the bottom-plate parasitics are significant, and an average total switch width across the various ratios slightly affects the optimal efficiency. The optimal total loss per unit ampere becomes

$$\frac{P_{loss}^*}{I_L} = 3\sqrt[3]{2}\sqrt[3]{\frac{I_L/C_{tot}}{G_{on}/C_{gate}V_{gate}^2}} \cdot \sqrt[3]{\left(\frac{2^N-1}{2^{N-1}}\right)^4}. \tag{6.9}$$

The minimum loss at the optimal design point depends on the ratio of the current density I_L/C_{tot} to the switch conductance per gate loss, and the required resolution N. However, the efficiency $(1 + P_{Loss}/I_L V_{out})^{-1}$ depends on the desired ratio and increases with larger output voltages V_{out}. For arbitrarily large resolutions N, the loss per ampere in Equation 6.9 saturates at about 2.5× the loss of a 2:1 SC that utilizes the same available C_{tot}.

6.3 RECURSIVE RESOLUTION-RECONFIGURATION ARCHITECTURE

In order to achieve the highest possible efficiency for a given silicon area, the various ratios must be realized while ensuring 100% utilization of the available on-die capacitance. Additionally, the optimal relative sizing of the constituent capacitors and switches should be guaranteed. Unlike conventional topologies, the proposed RSC topology inherently enables recursive inter-cell connection and recursive binary slicing that can simultaneously achieve both conditions with low complexity.

6.3.1 RECURSIVE INTER-CELL CONNECTION

The proposed recursive inter-cell connection brings individual cells in parallel instead of disabling them when realizing lower-resolution ratios. Figure 6.6 summarizes the challenge of lowering the resolution in a 4-bit RSC. The converter consists of four 2:1 SC cells connected in succession $C1$, $C2$, $C3$, $C4$ to realize $m_{odd}/2^4$ ratios. As shown, the cells are allocated optimal binary sizing of the total available capacitance, C_{tot}, and conductance, G_{tot}. One method to realize a 1/2 ratio from the 4-bit RSC is to route the output from the first stage using an output selection multiplexer

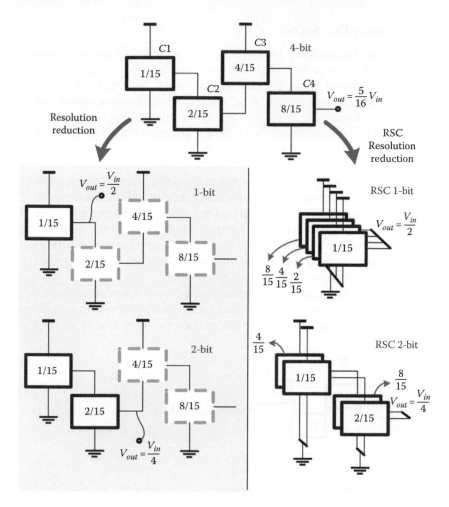

FIGURE 6.6 Resolution reduction from 4-bit to 1-bit and 2-bit using output selection multiplexer (left) and recursive inter-cell connection (right). The dashed cells are disabled when realizing lower resolutions.

and disabling all other stages. While this will produce the correct output voltage, such an approach wastes the available capacitance in the last three cells $C2$, $C3$, and $C4$, resulting in a 14/15 (93.33%) reduction in the available capacitance for charge transfer, thereby incurring a 15× penalty in R_{SSL}.

On the other hand, the *recursive* implementation connects the four 2:1 SC cells in parallel when a 1/2 ratio is desired, as shown in Figure 6.6, which results in 100% capacitance usage and the minimum possible ΔV for a given output charge and silicon area. Similarly, to lower the resolution from 4-bit to 2-bit, the cascade of the last two cells $C3$ and $C4$ is brought in parallel to the cascade of the first two cells $C1$ and $C2$, as shown in Figure 6.6, ensuring optimal relative sizing, that is, $\frac{1}{3}:\frac{2}{3}$, and 100% capacitance usage.

6.3.2 RECURSIVE CELL SLICING

Recursive slicing breaks down the largest cell in a cascade into binary weighted subcells to enable even-to-odd, and odd-to-odd, resolution reconfiguration, all the while satisfying optimal sizing. For example, instead of disabling the fourth cell $C4$ to realize a 3-bit resolution in a 4-bit SC converter, which wastes more than half of the total capacitance, one or more of the four available cells is sliced to realize six cells in total, and then the resulted cells are arranged in two parallel cascades of three cells each. In general terms, it can be shown that recursively slicing the last cell in the cascade CN into $(N-1)$ binary weighted cells results in the optimal solution. Such slicing achieves the optimal relative sizing when lowering the resolution, with a minimum number of sliced subcells and thus complexity. The resulted binary sliced subcells are connected in cascade, while operating in parallel with the cascade of the original $(N-1)$ stages. For example, in the 4-bit converter shown in Figure 6.7, the fourth cell $C4$ is sliced into three subcells of binary weights $(1/7, 2/7, 4/7)$, and arranged in parallel to the original cascade of the stages, $C1, C2, C3$ to achieve $m_{odd}/8$ ratios.

Similarly, when lowering the resolution further from three bits to two bits for $m_{odd}/4$ ratios, the last cells $C3$ and $C4_3$, which in parallel represent the last stage in the 3-bit cascade, are each binary sliced into two subcells, $(C3_1, C3_2)$, and

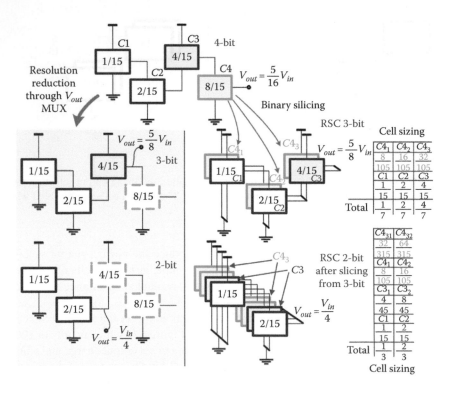

FIGURE 6.7 Resolution reduction from 4-bit to 3-bit and from 3-bit to 2-bit using output selection multiplexer (left) and recursive slicing with recursive inter-cell connection (right).

$(C4_{31}, C4_{32})$, respectively. Figure 6.7 shows the resulted eight cells sizing and connections of the topology implemented in this chapter. The relative sizing should be as close as possible to the illustrated weighting to achieve the peak performance; however, the optimal efficiency is not critically sensitive to mismatches between the various charge–transfer capacitors. It should be noted that four cells are only technically needed in order to realize all resolutions up to 4-bits; however, in order to guarantee 100% total capacitance utilization among all the possible resolutions while achieving optimal relative sizing, eight cells in total are instead employed.

6.3.3 Inter-Cell Reconfiguration Switches

This section discusses the implementation details to generate the desired ratios with a minimum set of programming switches, and hence minimum added parasitics. The required inter-cell reconfiguration switches can be divided into two main categories: switches to implement ratio programming within a specific recursion depth N and switches for resolution reconfiguration.

6.3.3.1 Ratio-Reconfiguration Switches

Figure 6.8 illustrates a simplified schematic of two 2:1 SC cells connected in parallel. By operating the four switches in each 2:1 cell from the nonoverlapped clock phases, Φ_1 and Φ_2, the 1/2 ratio is realized. In order to realize 1/4 and 3/4 conversion ratios in a 2-bit RSC, the two cells in Figure 6.8 are either connected in cascade or in stack through the added four reconfiguration switches r_1, r_2, r_3, and r_4. To realize a 1/4 conversion ratio, the second cell is connected between the output port MID_1 of the first cell and the converter ground, 0. This is accomplished through the three reconfiguration switches r_2, r_3, and r_4. The first cell output side (i.e., V_{out})

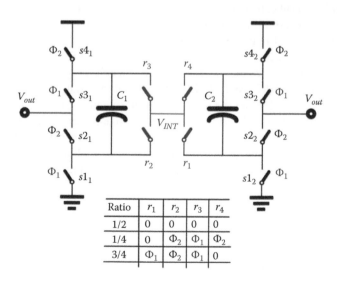

Ratio	r_1	r_2	r_3	r_4
1/2	0	0	0	0
1/4	0	Φ_2	Φ_1	Φ_2
3/4	Φ_1	Φ_2	Φ_1	0

FIGURE 6.8 Two 2:1 SC cells interconnection through ratio-reconfiguration switches. V_{INT} is the inter-cell intermediate node.

switches $s2_1$ and $s3_1$ are disabled and replaced by the reconfiguration switches r_2 and r_3, and hence r_2, r_3 are operated through Φ_2 and Φ_1, respectively. As a result, the first cell output charge is routed to the intermediate node V_{INT} between the two cells instead of the converter output V_{out}. To cascade both cells, the second cell input port IN_{top2} is reconfigured to the intermediate node V_{INT} between the two cells instead of the converter input voltage V_{in}. The switch $s4_2$ is disabled and the reconfiguration switch r_4 is operated in its place through the same clock phase Φ_2. Similarly, to realize the 3/4 conversion ratio, the first cell charge is routed to the intermediate node V_{INT} through the switches r_2 and r_3, and the reconfiguration switch r_1 is operated in place of $s1_2$. With such inter-cell connection, no extra series reconfiguration switches are required.

The proposed inter-cell reconfiguration switches are scalable. By replicating the same four connections between each pair of consecutive cells in an N-stage cascade, reconfiguration among the various ratios with a resolution of $m_{odd}/2^N$ can be realized. The conductance of the right half switches, r_1 and r_4, is double the conductance of the left half switches, r_2 and r_3, for optimal binary sizing.

6.3.3.2 Resolution-Reconfiguration Switches

Reconfiguration of the recursion depth (i.e., resolution) can be implemented through the same four ratio-reconfiguration switches; no additional programming switches are required. During resolution reconfiguration, the function of the reconfiguration switch pair r_2 and r_3 in Figure 6.8 is changed from routing the cell output charge to V_{INT}, to instead extracting charge from the intermediate node. Figure 6.9 illustrates the operation of the ratio-reconfiguration switches to reduce the resolution from 3-bit to 2-bit in an RSC. As shown in Figure 6.9a, the converter connects three 2:1 cells in cascade through the reconfiguration switch blocks $R_{1,2}$ and $R_{2,3}$. The 3-bit converter employs two subcells $C3_1$ and $C3_2$ to realize the third cell $C3$ in the cascade, for maximum resource utilization. The reconfiguration switch pairs $(r1_{3_1}, r1_{3_2})$ and $(r4_{3_1}, r4_{3_2})$ are operated in parallel, to connect the two subcells $C3_1$ and $C3_2$ as one cell in series or stack with the second cell $C2$. As shown in Figure 6.9b, to connect the sub cells $C3_1$ and $C3_2$ in cascade, the inter-cell switches $r1_{3_1}$ and $r4_{3_1}$ are operated in place of the switches $s2_{3_1}$ and $s3_{3_1}$ in order to route the output of cell $C3_1$ to the intermediate node V_{INT2}, while the reconfiguration switch $r1_{3_2}$ or $r4_{3_2}$ is operated in place of the switch $s1_{3_2}$ or $s4_{3_2}$, respectively, to realize 3/4 or 1/4 ratios. A similar procedure is followed for the reconfiguration block $R_{1,2}$ to connect the cells $C1$ and $C2$ in cascade. Finally, the second cell $C2$ output-side switches $s2_2$ and $s3_2$ are operated in place of the reconfiguration switches $r2_2$ and $r3_2$, and a 2-bit resolution is realized as shown in Figure 6.9b.

6.4 CIRCUIT IMPLEMENTATION

In order to validate the performance of the proposed RSC topology, a 4-bit RSC converter that realizes 15 ratios is implemented in 0.25 μm bulk CMOS process. Importantly, the RSC topology is inherently modular. Thus, the design of the converter requires custom implementation of only two SC building blocks.

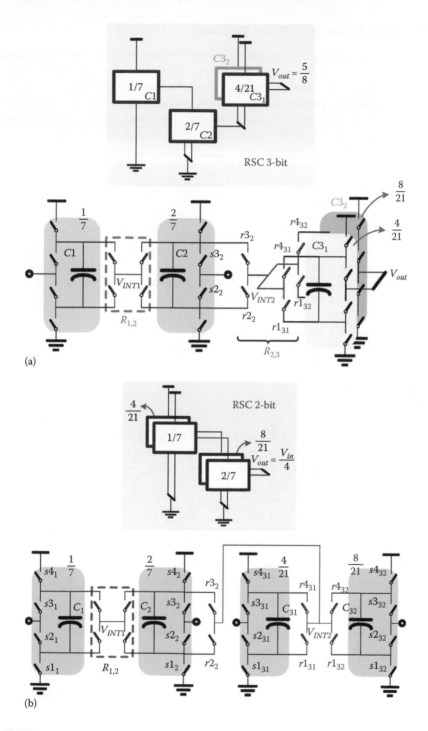

FIGURE 6.9 Realization of 2-bit resolution from 3-bit resolution RSC using the same ratio-reconfiguration switches. (a) 3 bit RSC. (b) 2 bit RSC.

6.4.1   4-Bit Power Stage Block Diagram

Figure 6.10 shows the recursive block diagram of the implemented 4-bit power stage, consisting of the two basic 2:1 building blocks: boundary and transfer cells. These two building blocks are connected together to implement four reconfigurable stages: $C1$, $C2$, $C3$, and $C4$. The capacitance and conductance of the last two stages, $C3$ and $C4$, are recursively binary-sliced to achieve 100% capacitance utilization and optimal relative sizing across the various ratios at any resolution. The fourth cell, $C4$, consists of three binary-sized subcells $C4_1$, $C4_2$, and $C4_3$, while the subcell $C4_3$ is further sliced into two subcells, $C4_{3_1}$ and $C4_{3_2}$. Similarly, the third cell $C3$ comprises two binary weighted subcells, $C3_1$ and $C3_2$. The eight total cells are interconnected at four intermediate nodes, V_{INT1}, V_{INT2}, V_{INT3_1}, and V_{INT3_2}, through four reconfiguration blocks, $R_{1,2}$, $R_{2,3}$, $R_{4_1,4_2}$, and $R_{4_2,4_3}$, along with a half reconfiguration block $R_{3,4}$.

As shown in Figure 6.10, two reconfiguration-switch blocks $R_{1,2}$, $R_{2,3}$ are employed between the three stages $C1$, $C2$, and $C3$ to realize recursive interconnection across the various resolutions until 3-bit operation. Similarly, another two reconfiguration-switch blocks, $R_{4_1,4_2}$, and $R_{4_2,4_3}$ are used to interconnect the subcells of the fourth stage, $C4$, for 3-bit resolution or lower. Instead of using the typical four-switch reconfiguration block, a two-switch reconfiguration block $R_{3,4}$ is used to cascade the third and fourth stages, $C3$ and $C4$. The two-switch reconfiguration block includes only the two switches that deliver charge to an intermediate node, and hence can be considered

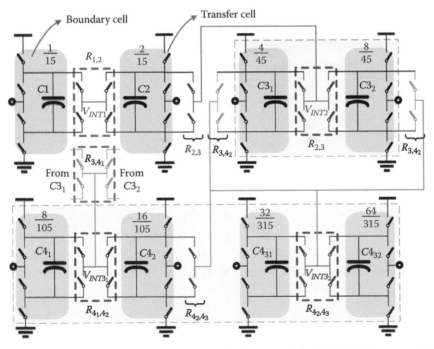

FIGURE 6.10   Recursive implementation block diagram of the 4-bit RSC converter. The implemented RSC comprises four stages of eight cells Ci and five reconfiguration switch blocks $R_{i,i+1}$.

as a half reconfiguration block. Since the nodes V_{INT3_1} and V_{INT3_2} should be separate when cascading the subcells $C4_1$, $C4_2$, and $C4_3$ to realize the 3-bit resolution, the reconfiguration block $R_{3,4}$ is further sliced into two sub-blocks, $R_{3,4_1}$ and $R_{3,4_2}$, to enable node isolation as illustrated in Figure 6.10. The two sub-blocks $R_{3,4_1}$, $R_{3,4_2}$ have relative conductance of, $\frac{1}{3} : \frac{2}{3}$, respectively, to match the relative sizing between the subcells $(C4_1, C4_2)$, and $(C4_3)$. Each switch in the implemented five reconfiguration blocks is assigned the optimal binary sizing of the total available conductance G_{tot}, which matches the relative charge that it routes.

6.4.2 Reconfiguration Costs

In the implemented 4-bit converter, boundary cells extract charge from the converter input voltage V_{in} (e.g., $C1$ and $C4_1$) or deliver charge to the converter output V_{out} (e.g., $C4_{31}$ and $C4_{32}$) across all the ratios. Therefore, these boundary cells only need an extra reconfiguration switch pair to deliver charge to a neighboring cell or shuttle the charge from a neighboring cell to the converter output V_{out}. On the other hand, transfer cells perform charge displacement from one stage to the next (e.g., $C2$, $C3_1$, $C3_2$, and $C4_2$). Thus, transfer cells employ four reconfiguration switches to extract the charge from one stage and deliver it to the next. Since all the switches are binary weighted to match the relative charge shuttled through a cell, the contribution of the extra four reconfiguration switches in a transfer cell to the flying capacitor bottom-plate parasitics matches the contribution of the original four switches of the 2:1 cell. In a boundary cell, such contribution is divided by two in relation to the original switches contribution.

In total, four cells contribute a normalize added-drain-parasitics of 1/2, while the remaining cells add 100%. The average normalized added-drain-parasitics from the used reconfiguration switches is less than unity or approximately 77.6% of the original switches drain parasitics. It should be noted that, in general, the drain parasitics constitute a small percentage of the gate capacitance.

6.4.3 Programmable-Port SC Boundary and Transfer Cells

In Figure 6.10, each 2:1 SC cell is represented with a single capacitor and four switches. However, in the actual implementation, each cell includes two capacitors and eight switches to implement two out-of-phase 2:1 cells. A port state can be defined for a cell (IN_{top}, IN_{bottom}, MID). A boundary cell operates in one of the four port-states: (V_{in}, 0, V_{out}), (V_{in}, 0, V_{INT}), (V_{INT}, 0, V_{out}), and (V_{in}, V_{INT}, V_{out}), where INT represents an inter-cell node. The first state is the typical case where the cell divides the converter input V_{in} by two. In the second state, the cell extracts charge from V_{in} to a neighboring cell. On the other hand, for a boundary cell to deliver charge to the output V_{out} from a neighbor, the cell input or ground ports are routed from the intermediate node, INT, instead of V_{in} or 0, which results in the last two states (V_{INT}, 0, V_{out}), and (V_{in}, V_{INT}, V_{out}).

Figure 6.11 illustrates the implemented standard boundary cell. Two 180° phase-shifted 2:1 SC cells are used to guarantee continuous input current through the cell input port, eliminating the need for a bypass capacitance. Since the intermediate node DC level is reconfigured at binary ratios of the input voltage, a transmission gate is used to implement the switches, with the exception of the V_{in}

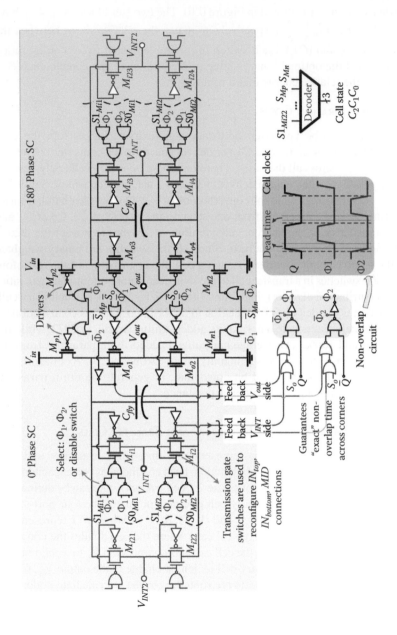

FIGURE 6.11   Boundary and transfer cells schematic.

and ground, 0, switches. The switches $M_{n1,2}$, $M_{o2,4}$, $M_{o1,3}$, and $M_{p1,2}$ are the original switches of the 2:1 SC converter that implement the typical port state (V_{in}, 0, V_{out}). A pair of reconfiguration switches can be operated as output-side switches or input-side switches by controlling their driving phases. For instance, by operating the switches M_{i1}, M_{i2}, in Figure 6.11, from the nonoverlapped clock phases Φ_1, Φ_2, respectively, the switches M_{i1}, M_{i2} act as output port switches. On the other hand, by driving M_{i1} from Φ_2, and disabling M_{i2} and M_{p1}, the switch M_{i1} is operated as an input-side switch and hence the cell input port becomes connected to V_{INT}. A similar explanation can be followed to connect the cell ground port IN_{bottom} to V_{INT} using M_{i2}. Figure 6.12 illustrates the four states of a boundary cell and the implemented cell decoder functional table.

The transfer cell is designed using the boundary cell as a starting point. At lower-resolution ratios, a transfer cell acts as a boundary cell and hence incorporates the same port states of the boundary cell. On the other hand, a transfer cell requires two additional states to shuttle charge from one stage to the next. In such cases, the transfer cell input or ground port is connected to the previous cell output port, which is connected to an intermediate node denoted as V_{INT} in Figure 6.11, while the transfer cell output port is connected to the next stage input/ground port intermediate node V_{INT2}. Thus, two additional port-states, (IN_{top}, IN_{bottom}, MID), are required for a transfer cell, (V_{INT}, 0, V_{INT2}) and (V_{in}, V_{INT}, V_{INT2}), respectively. Figure 6.12 illustrates the additional two states and the selection signals generated from the transfer cell decoder.

6.4.4 OUTPUT VOLTAGE REGULATION

Figure 6.13a shows the overall block diagram of the implemented 4-bit RSC converter chip. Two control loops are implemented in the proposed converter: an inner fine-grain loop and an outer coarse-grain loop. The inner loop, working within a single conversion ratio, should modulate either the switching frequency, f_{sw}, or the switched capacitance (i.e., digital capacitance modulation [22]) for fine-grain linear output voltage regulation and adaptation under load variations. Frequency modulation is chosen in this work to simplify the implementation complexity, as individual control of split subcells is not required in this case. The outer loop, implemented in an all-digital fashion, reconfigures the unloaded conversion ratio to minimize the range over which linear regulation is performed, thereby minimizing efficiency degradation.

6.4.4.1 Inner Fine-Grain Controller

The T flip-flop employed in Figure 6.13a guarantees a 50% duty-cycle input clock to the nonoverlap phase generator. A strong-arm comparator running at f_{comp} is used to provide the clock input to the T flip-flop, as shown in Figure 6.13a. The comparator sampling clock is produced by an on-chip current-starved oscillator that is set to twice the maximum switching frequency of the power stage; since the power stage switching frequency across all the 15 ratios does not exceed 8 MHz, the current starved oscillator is set to 16 MHz through an external bias, V_B.

	2:1 Cell state							M_{i1}		M_{i2}		M_{f21}		M_{f22}	
	IN_{top}	IN_{bottom}	MID	$C_2C_1C_0$	S_o	S_{Mn}	S_{Mp}	S0	S1	S0	S1	S0	S1	S0	S1
Boundary and transfer cell	V_{in}	0	V_{out}	011	1	1	1	0	0	0	0	0	0	0	0
	V_{in}	0	V_{INT}	000	0	1	1	1	0	1	0	0	0	0	0
	V_{INT}	0	V_{out}	010	1	1	0	0	1	0	0	0	0	0	0
	V_{in}	V_{INT}	V_{out}	001	1	0	1	0	0	0	1	0	0	0	0
Transfer cell	V_{INT}	0	V_{INT2}	110	0	1	0	0	1	0	0	1	1	1	0
	V_{in}	V_{INT}	V_{INT2}	101	0	0	1	0	0	0	1	1	0	1	0

FIGURE 6.12   Boundary and transfer cells decoder truth table.

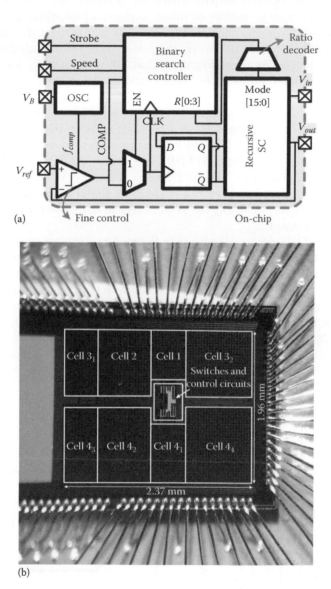

(a)

(b)

FIGURE 6.13 *Recursive* switched-capacitor voltage regulator implementation, comprising eight cells of binary weights and two control loops. (a) Block diagram of the overall RSC chip with binary search controller. (b) 4 bit Recursive SC test chip photo.

6.4.4.2 Outer Coarse-Grain Controller

Coarse-grain control in reconfigurable SC converters typically switch between discrete ratios by using a resistor string to generate ratio threshold levels [4,10]. However, a large number of ratios requires a prohibitively large resistor string, which takes into account R_{out} variation across the different ratios in order to avoid deadlock. In this work, the power stage itself is used to produce the threshold levels. By operating the SC at the maximum f_{sw} and scanning through the available ratios using binary search, the optimal ratio (i.e., the ratio that provides the required output level V_{ref} with minimum resistive voltage drop) can be located. The block diagram of the implemented binary search controller is shown in Figure 6.14. A 4-bit shift register, which is supplied to the ratio-decoder, is used to hold the current ratio state of the SC power stage as shown in Figure 6.13a. Once *STROBE* is asserted, *RST* is triggered and the power stage is reconfigured into the 1/2 ratio. Then, *EN* is asserted, initiating the binary search procedure. As a result, the *CLK* signal is routed directly from the on-chip oscillator, switching the power stage at 8 MHz to provide the minimal output resistance, R_{out}.

The proposed ratio-state code, shown in Figure 6.14, registers consecutive comparison decisions and enables a recursive implementation of the binary controller. Once the counter overflows (*OVR* is asserted), the 4-bit shift-register stores the present fine-grain controller comparison decision with V_{ref}. If the comparator output, *COMP*, is zero, the present power stage output is lower than the desired level, V_{ref}, and the SC is reconfigured into a larger binary ratio at the next resolution configuration, $(1 + R_{i-1})/2$, once the comparison decision 0 is registered at the *OVR* edge.

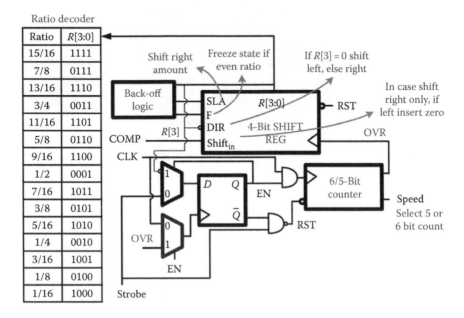

FIGURE 6.14 Recursive binary search controller block diagram.

On the other hand, when *COMP* is 1, the 4-bit register shifts in 1 and the power stage is reconfigured to the lower next-resolution binary ratio $(R_{i-1})/2$, where $i-1$ is the previous search iteration.

6.5 EXPERIMENTAL VERIFICATION

The proposed 4-bit *recursive* SC converter was fabricated in a 0.25 μm bulk CMOS process using 0.9 fF/μm² MIM capacitors and thin-oxide 2.5 V MOS transistors; a die photo is shown in Figure 6.13b. The RSC occupies 4.645 mm² for a total capacitance of 3 nF. A three-ratio (1/3, 1/2, 2/3) series-parallel (SP) SC converter was fabricated using the same technology to enable normalized performance comparison with the prototyped RSC. The implemented three-ratio SP is optimized for the same current density 0.5 mA/mm² as the prototyped 4-bit RSC.

Figure 6.15 shows the measured efficiency of the developed RSC and three-ratio SP converters, along with the results of a numerical model developed for the RSC, three-ratio SP, and 7-bit SAR topologies, with models based on the work in [29]. In addition, an ideal LDO is included for comparison. All converters are shown for a 2.5 V input voltage V_{in} and a 2 mA constant load current, except for the SP, which has a 1.86 mA load current to ensure equal current density. The efficiency of the RSC is measured for the following 14 ratios (1/8, 3/16, 1/4, 5/16, 3/8, 7/16, 1/2, 9/16, 5/8, 11/16, 3/4, 13/16, 7/8, 15/16) over an output voltage ranging from 0.1 to 2.2 V. Interestingly, the efficiency of the RSC at the 9/16 ratio falls below the RSC 1/2 ratio efficiency since the 9/16 ratio R_{SSL} is 3.5× larger than the 1/2 ratio. The RSC and SP SC converters both achieve a peak efficiency of 85%, and the numerical models are each within 1% of measurement results across the output voltage range. The large number of ratios afforded by the RSC topology enables a 38% expanded output voltage range (0.1–2.2 V in contrast to 0.2–1.6 V for the SP), while achieving 6.4% and 3.5% higher efficiency at 0.79 and

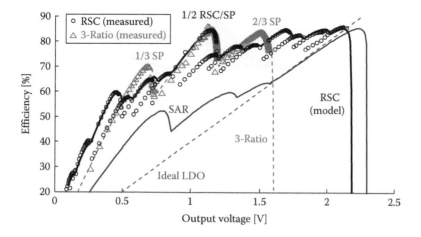

FIGURE 6.15 Measured and model-predicted efficiency, at 2 mA fixed load current, of the fabricated 4-bit RSC versus the output voltage at an input voltage of 2.5 V. The measured three-ratio efficiency is at 1.86 mA current and the same input voltage.

1.2 V output voltages, respectively, compared to the SP converter. The measured RSC also achieves 17.7% higher efficiency than an ideal LDO at 1.6 V. On the other hand, the SP peak efficiencies at the 1/3 and 2/3 ratios (at 0.68 and 1.5 V output voltages) exceed the RSC by 5.6% and 5.3%, respectively. The implemented RSC essentially takes the average of the three-ratio efficiency over the 0.52–1.6 V output range, filling the gaps between the three ratios (1/3, 1/2, 2/3) and maintaining a flatter efficiency profile. The 4-bit RSC achieves greater than 70% efficiency over the 0.9–2.2 V output range with an efficiency improvement of 28% over the 7-bit SAR.

Figure 6.16 shows the measured and numerically modeled efficiency given a 940 Ω resistive load for the RSC and 1 KΩ load for the SP in order to mimic the operation of a CMOS digital load under DVS conditions. At 0.8 and 1.2 V output voltages, the three-ratio SC achieves 59% and 68.7% efficiencies while the 15-ratio RSC achieves 8% and 7.6% higher efficiencies at the same voltages, respectively. The RSC delivers a dynamic voltage operating range from 0.04 to 2.16 V, which is 40.4% larger than the three-ratio SC output range from 0.09 to 1.6 V, thereby enabling wider-range DVS operation. The measured operating frequency of the RSC and SP with the external resistive load is shown in Figure 6.17. The RSC is switched over a 45× dynamic range, from 200 kHz to 9 MHz, to realize the 0.04–2.16 V output voltage range. In contrast, the SP requires a 100× frequency dynamic range, from 100 kHz to 10 MHz, to produce V_{out} from 0.09 to 1.6 V.

Figure 6.18 shows the measured efficiency of the 1/2 RSC conversion ratio versus the load current at an output voltage of 1.15 V. In this case, greater than 80% efficiency is achieved for load currents ranging from 30 µA to 1 mA. These results illustrate the primary advantage of a frequency modulation control, where the switching frequency, as well as switching parasitics loss, scales with the load current.

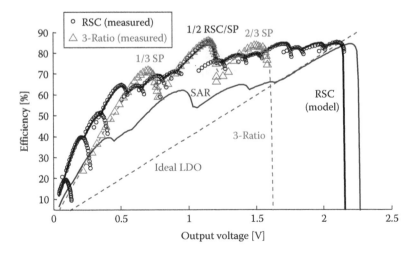

FIGURE 6.16 Measured and model-predicted efficiency with external resistive load, modeling a digital load under DVS operation, of the three-ratio SP and the 4-bit RSC across the output voltage, at an input voltage of 2.5 V.

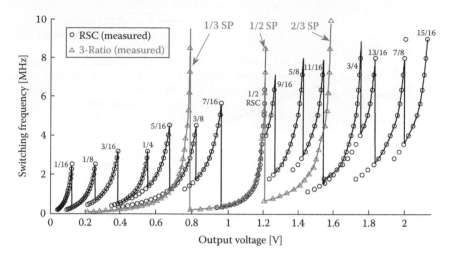

FIGURE 6.17 Measured RSC and three-ratio SC switching frequency f_{sw} across V_{out} using the same external resistive load in Figure 6.16.

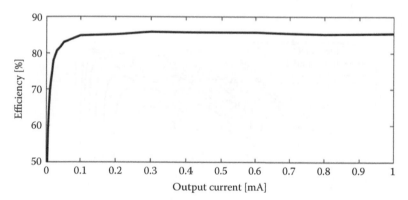

FIGURE 6.18 Measured RSC efficiency versus the load current at 1/2 ratio, while supplying 1.15 V output voltage V_{out}.

The peak efficiency of the RSC and the three-ratio SP for various power/current densities are essentially identical since both deliver the same 1/2 ratio. In DVS applications, system battery life is a key parameter, and for a digital load of uniform probability power states, the system energy efficiency is essentially the weighted-average efficiency of the converter over the output voltage range. The weighted-average efficiency is given by $\int P(V_{out}) \cdot V_{out}\eta(V_{out})dV_{out}$, where $P(V_{out})$ is the probability of a given power state and the integration is over the achievable converter range. Figure 6.19 shows the measured and numerically modeled weighted-average efficiencies across the output voltage range, plotted versus current density. As shown in Figure 6.19a, the measured weighted-average efficiency of the RSC exceeds the SP weighted-average by 6.9% at the same current density of 0.23 mA/mm². The modeled efficiency of the RSC maintains higher

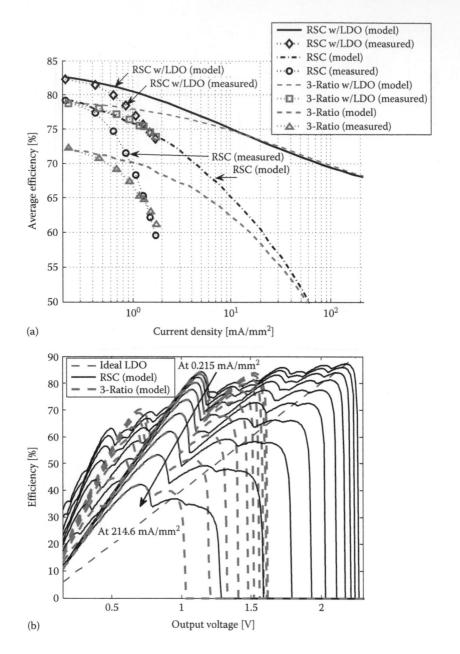

FIGURE 6.19 Measured and predicted weighted-average efficiency versus the load current density, from 0.215 to 215 mA/mm², for the fabricated RSC and SP in 0.25 μm bulk CMOS. A load of equal probability power state is assumed. When indicated, an ideal LDO is assumed to fill the efficiency gaps over the 0–2.5 V output range. (a) Average efficiency of RSC and SP versus current density. (b) Model-predicted RSC and SP efficiency across V_{out}, versus different current densities.

weighted-average efficiency across different current densities and approaches a 2.5% higher average than the SP at 16 mA/mm². Note that the modeled and measured results diverge after the nominal current density of 0.5 mA/mm² as the model assumes optimal total switch width given the increased current density, while the fabricated chips have fixed total conductance.

Since the SP converter can only deliver voltages up to 1.6 V, another weighted-average efficiency metric is calculated assuming that an ideal LDO is used to fill any efficiency gap. With an LDO, the RSC still exceeds the SP measured weighted-average by 3.3% at 0.23 mA/mm². At 16 mA/mm² and above, the LDO performance dominates the RSC and the SP efficiency and both converge to the same value. As shown in Figure 6.19b, the RSC maintains superior performance than the SP converter at higher power densities until the LDO performance dominates.

All presented numerically modeled results employ MIM capacitors with a 1.4% bottom-plate parasitic capacitance ratio. If MOS capacitors were employed in place of MIM capacitors, the 10% bottom-plate parasitics in this technology would degrade the efficiency by 12.5% across the output voltage range for a 3 nF of total flying capacitance. On the other hand, if a higher density MIM capacitance were available, for example, with a MIM density of 4 fF/μm² and bottom-plate ratio of 4× lower, the efficiency of both the RSC and SP converters would increase at each discrete ratio. However, due to severe linear regulation away from the nominal three ratios in the SP topology, the efficiency between these ratios only marginally improves. On the other hand, the RSC converter has explicit ratios between these gaps, and thus the efficiency of the RSC topology at these voltages is increased. For example, with 4 fF/μm² MIM capacitors, the weighted-average efficiency of the RSC exceeds the three-ratio SP by 9% at 0.23 mA/mm² or by 6.8% when including an ideal LDO. In this example, the RSC and SP weighted-averages converge at 60 mA/mm², which is 3.8× larger than the 0.9 fF/μm² MIM capacitor case. Migrating to a more modern technology node with higher-density MIM [14,32], MOS [15,20,22], ferroelectric [10], or deep-trench capacitors [1,2,8] and lower parasitic switches will thus enable improved performance of the RSC over the SP topology at larger current densities.

Figure 6.20a shows the control response to a variable staircase voltage reference, V_{ref}. The control voltage V_{ref} is changed every 500 μs with variable step sizes of 650 mV maximum value. Figure 6.20b details the transient coarse controller response when the *STROBE* signal is activated while the SC is initially producing a 2 V output voltage. Here, the SC power stage phase clock, *CLK*, is switched at the maximum frequency while the coarse controller cycles through the various binary ratios until the output reaches the desired level after 8 μs. In the third cycle of this example, the coarse controller reaches the 13/16 ratio, which cannot produce the desired level V_{ref} = 2 V, given the converter R_{out}. Thus, a fourth correction cycle automatically results and the *Back–Off* logic returns the power stage to the correct 7/8 ratio. Finally, the coarse controller hands off the regulation operation to the fine-level frequency controller where *clk* goes back to a normal frequency. Table I summarizes the results of the presented design, and compares to prior work (Table 6.1).

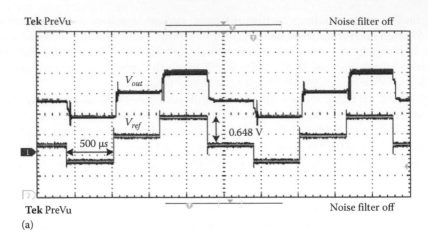

(a)

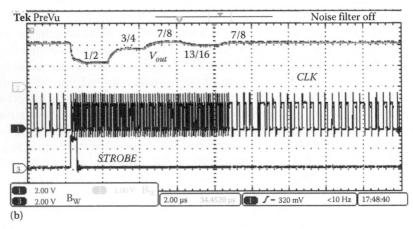

(b)

FIGURE 6.20 (a) Coarse-controller measured transient response to a stair control voltage V_{ref}. (b) Controller transient response when *STROBE* is activated while $V_{ref} = 2$ V, showing the detailed ratio binary search operation.

TABLE 6.1
Comparison with Previously Published Fully Integrated SC Converters

Work	[15]	[4]	3-Ratio SP	4-Bit RSC
Technology	65 nm	180 nm	0.25 μm	0.25 μm
Capacitor	Bulk PMOS	On-chip	MIM	MIM
Area [mm²]	0.64	1.69	4.33	4.645
Capacitance [nF]	3.88	2.24	2.8	3
Topology	1/3, 2/5 SP	7-bit SAR	2/3, 1/2, 1/3 SP	4-bit RSC
V_{in} [V]	3–4	3.4–4.3	2.5	2.5
V_{out} [V]	1	0.9–1.5	0.2–1.6	0.1–2.2
Efficiency (η) [%]	74	72	85	85
I_L at (η) mA	32 mA	0.01 mA	1.86 mA	2 mA

6.6 CONCLUSION

A *recursive* SC converter topology is presented that achieves a flattened efficiency profile over a wide voltage range by employing $2^N - 1$ ratios in an intelligent and modular manner. Compared to a co-fabricated three-ratio SP converter, the proposed 4-bit RSC achieves a wider operating range and achieves a higher weighted average efficiency. To achieve high efficiency with a large number of ratios, the RSC topology maximizes the number of connections to the converter input supply and ground in order to minimize both the charge shuttled through the converter flying capacitors and the cascaded losses. Unlike conventional SC topologies, the RSC SSL loss converges to an upper limit $1/(f_{sw}C_{tot})$ and becomes fixed for arbitrarily high-resolution N. The RSC loss for large resolutions N thus saturates at approximately 2.5× the loss of a 2:1 SC that utilizes the same available C_{tot}. By employing both recursive inter-cell connection and recursive slicing, all possible resolutions, N, and hence their ratios, can be realized without disconnecting a single capacitor and while satisfying optimal relative sizing of the constituent capacitors and switches, thereby ensuring high efficiency even at larger values of N. The inherent regularity and modularity of the RSC topology simplifies the implementation of arbitrarily large resolutions with 2^{N-1} possible ratios, resulting in opportunities to achieve greater than 15 ratios in future work.

REFERENCES

1. T.M. Andersen, F. Krismer, J.W. Kolar, T. Toifl, C. Menolfi, L. Kull, T. Morf et al. A 4.6 W/mm² power density 86% efficiency on-chip switched capacitor DC-DC converter in 32 nm SOI CMOS. In *2013 28th Annual IEEE Applied Power Electronics Conference and Exposition (APEC)*, Long Beach, CA, March 17–21, 2013, pp. 692–699.
2. T.M. Andersen, F. Krismer, J.W. Kolar, T. Toifl, C. Menolfi, L. Kull, T. Morf et al. A sub-ns response on-chip switched-capacitor DC-DC voltage regulator delivering 3.7 W/mm² at 90% efficiency using deep-trench capacitors in 32 nm SOI CMOS. In *2014 IEEE International Solid-State Circuits Conference Digest of Technical Papers (ISSCC)*, San Francisco, CA, February 9–13, 2014, pp. 90–91.
3. S. Bandyopadhyay, Y.K. Ramadass, and A.P. Chandrakasan. 20 µA to 100 mA DCDC converter with 2.8–4.2 V battery supply for portable applications in 45 nm CMOS. *IEEE Journal of Solid-State Circuits*, 46(12):2807–2820, December 2011.
4. S. Bang, A. Wang, B. Giridhar, D. Blaauw, and D. Sylvester. A fully integrated successive-approximation switched-capacitor DC-DC converter with 31 mV output voltage resolution. In *2013 IEEE International Solid-State Circuits Conference Digest of Technical Papers*, San Francisco, CA, February 2013, pp. 370–371.
5. T. Van Breussegem and M. Steyaert. A 82% efficiency 0.5% ripple 16-phase fully integrated capacitive voltage doubler. In *2009 Symposium on VLSI Circuits*, Kyoto, Japan, June 16–18, 2009, pp. 198–199.
6. J.S. Brugler. Theoretical performance of voltage multiplier circuits. *IEEE Journal of Solid-State Circuits*, 6(3):132–135, June 1971.
7. B.H. Calhoun and A.P. Chandrakasan. Ultra-dynamic voltage scaling (UDVS) using sub-threshold operation and local voltage dithering. *IEEE Journal of Solid-State Circuits*, 41(1):238–245, January 2006.
8. L. Chang, R.K. Montoye, B.L. Ji, A.J. Weger, K.G. Stawiasz, and R.H. Dennard. A fully-integrated switched-capacitor 21 voltage converter with regulation capability and 90% efficiency at 2.3 A/mm². In *2010 Symposium on VLSI Circuits*, Honolulu, HI, June 2010, pp. 55–56.

9. J.F. Dickson. On-chip high-voltage generation in MNOS integrated circuits using an improved voltage multiplier technique. *IEEE Journal of Solid-State Circuits*, 11(3):374–378, June 1976.

10. D. El-Damak, S. Bandyopadhyay, and A.P. Chandrakasan. A 93% efficiency reconfigurable switched-capacitor DC-DC converter using on-chip ferroelectric capacitors. In *2013 IEEE International Solid-State Circuits Conference Digest of Technical Papers*, San Francisco, CA, February 17–21, 2013, pp. 374–375.

11. M. Evzelman and S. Ben-Yaakov. Average-current-based conduction losses model of switched capacitor converters. *IEEE Transactions on Power Electronics*, 28(7):3341–3352, July 2013.

12. P. Hazucha, G. Schrom, B.A. Bloechel, P. Hack, G.E. Dermer, S. Narendra, D. Gardner, T. Karnik, V. De, and S. Borkar. A 233-MHz 80%–87% efficient four-phase DC-DC converter utilizing air-core inductors on package. *IEEE Journal of Solid-State Circuits*, 40(4):838–845, April 2005.

13. C. Huang and P.K.T. Mok. A 100 MHz 82.4% efficiency package-bondwire based four-phase fully-integrated buck converter with flying capacitor for area reduction. *IEEE Journal of Solid-State Circuits*, 48(12):2977–2988, December 2013.

14. R. Jain, B.M. Geuskens, S.T. Kim, M.M. Khellah, J. Kulkarni, J.W. Tschanz, and V. De. A 0.451 V fully-integrated distributed switched capacitor DC-DC converter with high density MIM capacitor in 22 nm tri-gate CMOS. *IEEE Journal of Solid-State Circuits*, 49(4):917–927, April 2014.

15. H.-P. Le, J. Crossley, S.R. Sanders, and E. Alon. A sub-ns response fully integrated battery-connected switched-capacitor voltage regulator delivering 0.19 W/mm² at 73% efficiency. In *2013 IEEE International Solid-State Circuits Conference Digest of Technical Papers*, San Francisco, CA, February 2013, pp. 372–373.

16. H.-P. Le, S.R. Sanders, and E. Alon. Design techniques for fully integrated switched-capacitor DC-DC converters. *IEEE Journal of Solid-State Circuits*, 46(9):2120–2131, September 2011.

17. K.N. Leung and P.K.T. Mok. A capacitor-free CMOS low-dropout regulator with damping-factor-control frequency compensation. *IEEE Journal of Solid-State Circuits*, 38(10):1691–1702, October 2003.

18. P. Li, L. Xue, P. Hazucha, T. Karnik, and R. Bashirullah. A delay-locked loop synchronization scheme for high-frequency multiphase hysteretic DC-DC converters. *IEEE Journal of Solid-State Circuits*, 44(11):3131–3145, November 2009.

19. M.S. Makowski and D. Maksimovic. Performance limits of switched-capacitor DC-DC converters. In *Proceedings of PESC'95—Power Electronics Specialist Conference*, Atlanta, GA, Vol. 2, 1995, pp. 1215–1221.

20. Y.K. Ramadass and A.P. Chandrakasan. Voltage scalable switched capacitor DC-DC converter for ultra-low-power on-chip applications. In *2007 IEEE Power Electronics Specialists Conference*, Orlando, FL, June 17–21, 2007, pp. 2353–2359.

21. Y.K. Ramadass and A.P. Chandrakasan. Minimum energy tracking loop with embedded DCDC converter enabling ultra-low-voltage operation down to 250 mV in 65 nm CMOS. *IEEE Journal of Solid-State Circuits*, 43(1):256–265, January 2008.

22. Y.K. Ramadass, A.A. Fayed, and A.P. Chandrakasan. A fully-integrated switched-capacitor step-down DC-DC converter with digital capacitance modulation in 45 nm CMOS. *IEEE Journal of Solid-State Circuits*, 45(12):2557–2565, December 2010.

23. G.A. Rincon-Mora and P.E. Allen. A low-voltage, low quiescent current, low drop-out regulator. *IEEE Journal of Solid-State Circuits*, 33(1):36–44, 1998.

24. L.G. Salem and P.P. Mercier. An 85%-efficiency fully integrated 15-ratio recursive switched-capacitor DC-DC converter with 0.1-to-2.2 V output voltage range. In *2014 IEEE International Solid-State Circuits Conference Digest of Technical Papers (ISSCC)*, San Francisco, CA, February 9–13, 2014, pp. 88–89.

25. S.R. Sanders, E. Alon, H.-P. Le, M.D. Seeman, M. John, and V.W. Ng. The road to fully integrated DCDC conversion via the switched-capacitor approach. *IEEE Transactions on Power Electronics*, 28(9):4146–4155, September 2013.

26. G. Schrom, P. Hazucha, F. Paillet, D.J. Rennie, S.T. Moon, D.S. Gardner, T. Kamik et al. A 100 MHz eight-phase buck converter delivering 12 A in 25 mm² using air-core inductors. In *APEC 07—Twenty-Second Annual IEEE Applied Power Electronics Conference and Exposition*, Anaheim, CA, February 25–March 1, 2007, pp. 727–730.

27. M.D. Seeman, V.W. Ng, H.-P. Le, M. John, E. Alon, and S.R. Sanders. A comparative analysis of switched-capacitor and inductor-based DC-DC conversion technologies. In *2010 IEEE 12th Workshop on Control and Modeling for Power Electronics (COMPEL)*, Boulder, CO, June 28–30, 2010, pp. 1–7.

28. M.D. Seeman and S.R. Sanders. Analysis and optimization of switched-capacitor DC-DC converters. *IEEE Transactions on Power Electronics*, 23(2):841–851, March 2008.

29. M.D. Seeman. A design methodology for switched-capacitor DC-DC converters. PhD thesis, EECS Department, University of California, Berkeley, CA, May 2009.

30. D. Somasekhar, B. Srinivasan, G. Pandya, F. Hamzaoglu, M. Khellah, T. Karnik, and K. Zhang. Multi-phase 1 GHz voltage doubler charge pump in 32 nm logic process. *IEEE Journal of Solid-State Circuits*, 45(4):751–758, April 2010.

31. N. Sturcken, M. Petracca, S. Warren, P. Mantovani, L.P. Carloni, A.V. Peterchev, and K.L. Shepard. A switched-inductor integrated voltage regulator with nonlinear feedback and network-on-chip load in 45 nm SOI. *IEEE Journal of Solid-State Circuits*, 47(8):1935–1945, August 2012.

32. T.M. Van Breussegem and M.S.J. Steyaert. Monolithic capacitive DC-DC converter with single boundary multiphase control and voltage domain stacking in 90 nm CMOS. *IEEE Journal of Solid-State Circuits*, 46(7):1715–1727, July 2011.

7 GaAs Power Devices and Modules

Vipindas Pala and T. Paul Chow

CONTENTS

7.1   GaAs TECHNOLOGY FOR INTEGRATED HIGH-FREQUENCY POWER CONVERTERS

Innovations in low-voltage (<50 V) power electronic systems are driven by the requirements of high-energy efficiency, bandwidth, size, and cost, mainly in the portable electronics domain. In a typical handheld system, like a smartphone, there are different domains with varying power requirements like audio, displays, lighting, communication circuits, and processors. As the output voltages of the integrated circuit reduce, the power conversion circuit needs to have tighter specifications for the voltage set point, lower voltage ripple, faster transient response, and reduced noise. Routing the power to subsystem from a central power conversion unit results in higher line voltage variations and increased noise. Increasingly, the power conversion circuits are placed in close proximity to the load to cater to the load requirements (point of load [POL]). It is advantageous to integrate the POL converter along with the load as a system-on-chip or a system-in-package (SiP) implementation to reduce the form factor. The POL approach, therefore, requires the size of the power conversion circuit to be kept to a minimum for the sake of integration. The key factor to integrating a switch mode voltage regulator on chip or on package is the size of the energy storage components. In a typical switch mode DC–DC converter, the bulkiest component is typically the inductor. The key to reducing the inductor size is increasing the switching frequency of the DC–DC converter.

The second advantage of increasing the switching frequency is that of improved bandwidth. As the battery life becomes more critical, complex power management strategies are employed in portable systems to extend the time between charging. Most of these strategies require the power management circuit to respond quickly to changing load conditions. From a DC–DC converter point of view, a high regulated bandwidth is needed to meet these. The regulated bandwidth of the converter is limited ultimately by the switching frequency of the converter. In a pulse width modulated (PWM) DC–DC converter, the bandwidth is limited by the sampling rate. This is equal to half of the switching frequency [1] according to the Nyquist criterion. In the real world, ripple requirements at the output limits the bandwidth to anywhere between one-fifth and one-tenth of the switching frequency.

For DC–DC converters powering microprocessors in laptops, the future requirements for slew rate specifications may exceed 100 A/μs [2]. The regulated bandwidth in today's state-of-the-art DC–DC converters is not high enough to meet the slew rate and prevent large voltage swings from appearing at the load without the use of bulky capacitors with ratings of up to 1 mF [3]. These capacitors have to be placed in close proximity to the processor die, increasing the overall form factor. A high-frequency DC–DC converter operating at a regulated bandwidth of 1 MHz or greater can cater to the fast load transients of the processor and integrated in close proximity to the die without using such large capacitors at the output.

A high bandwidth DC–DC converter can also be used to improve the efficiency of radio-frequency (RF) power amplifiers (PAs) using envelope tracking shown in Figure 7.1. PAs are among the largest consumers of power in a

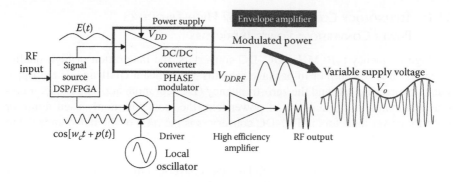

FIGURE 7.1   Envelope tracking DC–DC converter powering a radio-frequency amplifier.

typical handheld wireless device [4]. In today's mobile phones and other wireless devices, saturated GSM PA modules require up to 2 A of input current when driven from an input voltage VCC of 3.3 V, while for linear WCMDA PAs the peak current requirement is close to 0.7 A [5,6]. This requires downconversion of the typical Li–ion battery voltages (>4.2 V) using a linear or a switched mode regulator. For saturated PAs for GSM standards, V_{CC} is adjusted to control the output power, and high efficiencies are maintained at backed off power by reducing V_{CC}. In GSM time-division multiplexing, the power supply also needs to be able to slew V_{CC} from zero to full power in 10–20 μs at the beginning of each time step [7], which cannot be done by low-frequency DC–DC converters alone due to their slower response. For this purpose, a low drop-out (LDO) linear regulator is typically used in parallel with a switching converter to provide a fast transient response while retaining high steady-state efficiency. When using nonlinear PAs for standards like EDGE, WCDMA, or LTE that require linear amplification, ideally the power supply needs to be able to track the envelope of the RF signal, which imposes a much larger requirement on the power supply bandwidth [7,8]. One proposed solution to this problem is to use a low-MHz DC–DC converter that tracks the average of the envelope signal in conjunction with a linear regulator or a class AB amplifier that provides the bandwidth. The hybrid supply modulator solution can meet the bandwidth requirement, but has a low efficiency especially when the peak-to-average ratio of the envelope signal is low. To meet the bandwidth requirements, a PWM switch mode converter alone needs to be operated at least 5.1 times the bandwidth of the envelope signal [9]. Today's wireless standards like LTE have an envelope bandwidth of up to 20 MHz, which means that the peak switching frequency of the converter has to be higher than 100 MHz to meet the requirements of the modulator. This is more than an order of magnitude higher than that of state-of-the-art power conversion circuits in portable systems. The wide adoption of switching modulators for envelope tracking PAs is limited by the poor efficiency of switch mode power converters when operating above tens of MHz. This is mainly due to two reasons—lack of fast switches with low switching losses when operating at tens of MHz and lack of low loss power inductors at these frequencies.

7.1.1 Technology Considerations for High-Frequency Power Conversion: Power Inductors

For high-frequency (>10 MHz) DC–DC converters, the integration of high-quality inductors either on chip or in package remains a significant challenge. If the ripple inductor current is low and the zero-frequency (DC) component of the inductor current contributes to the majority of power loss, the most commonly used figure of merit for power inductors in DC–DC converters is the ratio of the inductance (L) to DC resistance (DCR). Inductors that have a high permeability magnetic core have extremely good figure of merits at kilohertz frequencies due to their low L/DCR ratio. However, they are not suited to DC–DC converters that operate beyond a few MHz due to hysteresis and core losses that arise due to harmonics. In monolithic or SiP implementations, air-core spiral inductors are the most common solution. These inductors are available in compact form factors up to the hundreds of nano-Henry range. L/DCR ratio of these components is much lower than ferrites. Below 10 MHz or so, the inductor losses are high and limit their adoption in DC–DC converters. However, as the switching frequencies increase beyond 50 MHz or so, the size of the inductor required to minimize current harmonics decreases to the point where the series resistance losses in the inductor are not prohibitive.

At these frequencies, low-height, planar inductors also become attractive. If integrated on the PCB laminate, planar inductors, fabricated on a single layer of metal, can now be placed in the substrate below the semiconductor components. In high-frequency converter implementations that use planar inductors, the total output inductance is constrained by the area consumed by the spiral, and therefore, it is desirable to use a small inductance at the cost of a high-current ripple. As the harmonics of the inductor current increase, the loss due to harmonics also starts to come into play, which means that the quality factor of the inductor starts to become important. Up to few hundreds of MHz, the quality factor of typical chip inductors increases with frequency and is mainly determined by the series resistance of the inductor. *Off*-chip inductor solutions include inductors on high-resistivity substrates that are stacked on top of the active die [10], SMD components soldered on laminate [11–13], flip-chip bonding of the active die on top of the passive silicon die [14], and inductors using bond wires [15].

Ultimately, for switching frequencies around 100 MHz, it is possible to design the inductors on the metal layers on the semiconductor chip itself. On-chip spiral inductors are common in nm-CMOS-based DC–DC converters that operate at hundreds of MHz [16–19]. Planar inductors on CMOS processes typically use multiple thick metal spirals stacked on top of each other to increase the total cross-sectional area and reduce the DCR. However, the thickness of interconnect metal layers is limited, and therefore, cross-sectional area of CMOS cannot be achieved without increasing chip area. Based on published data, DC–DC converters with on-chip inductors implemented on silicon CMOS show a significantly lower performance compared to off-chip solutions. This is mainly because silicon CMOS substrates have higher conductivity and energy is lost due to circulating substrate currents induced by the inductor, which is in close proximity to the chip. Wide bandgap semiconductors like GaAs and gallium nitride (GaN), which typically employ lower conductivity substrates, also have the benefit of higher-quality *on*-chip inductors.

In one example [20], spiral inductors are fabricated on a 65 µm thick Cu metal layer electroplated on top of a field metal on semi-insulating GaAs substrate. The inductor spirals are then etched on copper. Even with a thick copper layer, feature exclusions of up to 30 µm are obtained using a directional vertical etch. The planar spirals have a large cross-sectional area due to the large vertical sidewalls reducing the DCR and maximizing the quality factor. Contacts to the ends of the spiral inductor are made using solder bumps with a thickness of close to 25 µm, and the die is then flip-chipped to a laminate. The 25 µm separation between the copper metal and traces on the laminate ensures that not too much energy is coupled into the copper traces below, and thus reduces eddy current losses.

Figure 7.2 shows an example of inductor built-in thick copper and its measurement results. The winding width and spacing are 80 µm. The inductor size is 1.5×1.6 mm^2. The four spare bumps at the corner of the die are for mechanical stability considerations. Two-port S-parameter measurement using a network analyzer is used to characterize the thick copper inductors once it is mounted on the laminate with signal and ground access by sweeping the frequency from 40 MHz to 2.5 GHz. The measured DCR is 86.97 mΩ. At 100 MHz, the inductance is 11.94 nH and the peak inductance of 22 nH appears at 2 GHz as shown in Figure 7.3. A quality factor of 22.1 at 100 MHz and almost 80 at 2 GHz is measured, which is much higher than similar inductor values built in 10 µm thick copper on silicon [16] with $Q > 20$ at 2 GHz.

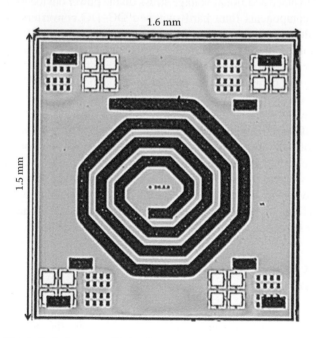

FIGURE 7.2 Thick Cu inductor on GaAs die.

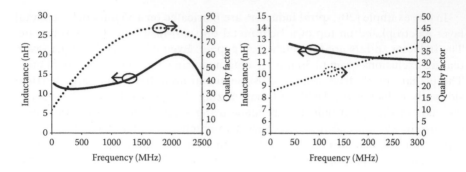

FIGURE 7.3   Measured inductance and quality factor of the GaAs thick Cu *on*-chip spiral inductor on two different scales.

7.1.2 Technology Considerations for High-Frequency Power Conversion: Power Switches

At present, silicon vertical or lateral NMOSFETs are the technology of choice for state-of-the-art switch mode converters. Efficiency degradation due to frequency-dependent power losses in the switching transistor imposes a fundamental switching frequency limit for semiconductor power switch. Some of these limits can be overcome by circuit topologies that employ soft switching. Techniques such as charge recycling [21], class E type, or resonant switching [22,23] have been adopted for converters operating at extremely high switching frequencies. Resonant topologies, however, tend to place additional voltage stress on the power device itself and require more passive components than hard-switched DC–DC converters. These factors make it difficult to implement in highly integrated solutions.

When monolithic implementations of power conversion circuits become preferable, lateral devices become more attractive as they can integrate multiple devices on the same die. Power conversion circuits built using nanometer CMOS processes have been demonstrated to switch at extremely high frequencies. In published reports of partially or fully integrated high-frequency DC–DC converters operating at frequencies of above 10 MHz [21], efficiencies higher than 85% have been obtained using CMOS nodes of 0.18 μm or below. In these implementations, multiple transistors are cascaded to meet the ruggedness requirements in the portable environment where typical battery voltages of up to 4.2 V need to be supported using low-voltage transistors. The disadvantage of this approach is that the overall resistance of the cascaded power device is now higher due to extrinsic components like the ohmic contact resistance, and access resistance of each transistor connected in series.

7.1.3 Device Technology Candidates for Integrated Low-Voltage High-Frequency Switching

A variety of considerations determine the optimal choice of a device technology for power switching at high frequencies. For minimizing the power losses in the switch, it is desirable to have a low *on* resistance, which determines conduction losses, as well as a low gate charge, which results in low transition times as well as low losses in the driver circuit. A commonly used figure of merit to evaluate high-frequency switching transistors is the product of the *on* resistance and gate charge [24], which is independent of the gate periphery and die area to the first order:

$$FOM = R_{ON}Q_G. \tag{7.1}$$

Generally, bipolar devices are not preferred for high-frequency applications because the minority charges in the base of the transistor have to be removed at each switching transition, which reduces the maximum switching speed. In addition, bipolar devices are current driven, and power consumption in the driver circuits considerably reduces the system efficiency in low-voltage applications.

7.1.3.1   Silicon FETs

Power FETs, most commonly silicon MOSFETs, have been conventionally used for high-frequency devices since they are unipolar and require less driving power since they are driven by a gate voltage as opposed to a base current. They are also less susceptible to a second breakdown as compared to bipolar transistors. In traditional low-voltage switching applications, silicon power MOSFETs are the technology of choice. Power MOSFETs can be implemented in both vertical and lateral structures. Vertical MOSFETs like the DMOS [25] or the UMOS [25] have a higher current rating and a lower *on* resistance as compared to lateral MOS devices. However, because the high-voltage contact is on the back of the wafer, the vertical MOSFETs are not suited to monolithic power IC implementation when the power switch is integrated with other low-voltage analog and digital circuits on the same chip.

Lateral power MOSFETs are more suited to monolithic integration with analog and digital components as all the terminals of the device are on the top side of the wafer. They generally have lower FOM as compared to the vertical MOSFETs [26] and are optimized according to the reduced surface field (RESURF) [27] principle to achieve the optimal breakdown voltage-*on* resistance trade-off. The lateral DMOS (LDMOS) shown in Figure 7.4 is traditionally the most important lateral power MOSFET and has been prevalent in power ICs since the 1980s [28]. LDMOS transistors have been integrated along with bipolar and CMOS transistors for analog and logic functions in the BCD technologies [28]. On bulk substrates, this is done using junction isolation (JI). BCD technologies are also available on silicon-on-insulator (SOI) wafers, which utilize dielectric isolation. Driven by lateral scaling and improvements in the LDMOS figure of merit, various generations of BCD generations have dominated in the power IC industry. A survey of latest BCD technology generations is shown in Table 7.1.

As the VLSI technology has evolved, various lateral MOS structures have been introduced to be compatible with advanced CMOS technologies to integrate power MOSFETs. All of these structures have a lightly doped drain (LDD) region to support the breakdown voltage, while the channel length has been scaled according to VLSI scaling rules to achieve the maximum figure of merit. For low voltages (<30 V), lateral MOSFETs based on a horizontal gate structure [31] have shown superior

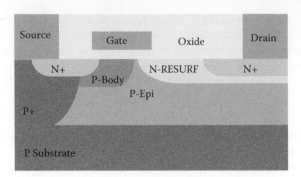

FIGURE 7.4　30 V silicon reduced surface field N-LDMOS structure. (From Yasuhara, N. et al., *Low gate charge 30V N-Channel LDMOS for DC-DC converters*, in *International Symposium on Power Semiconductor Devices and ICs*, Cambridge, U.K., April 14–17, 2003, pp. 47–51.)

TABLE 7.1

Evolution of BCD Generations

	SOIBCD	BCD4	BCD5	BCD6	BCD8
CMOS L_G (μm)	1.0	0.8	0.6	0.35	0.18
DMOS (lateral/vertical)	L	L and V	L	L	L
DMOS (rated voltage, V)	30–200	30–90	18–80	5–70	5–45
Metal levels	1–2	2–3	2–3	3–5	4–6

Source:　Adapted from Conteiro, C. et al., Roadmap differentiation and emerging trends in BCD technology, in *European Solid-State Research Conference*, Florence, Italy, September 24–26, 2002, pp. 275–282.

performance, whereas for rated breakdown voltages of 80 V or above, lateral trench MOSFETs [32] have been shown to have the best FOM among integrable structures. A few examples of device structures that have been used for low-voltage (<30 V) integrable lateral MOSFETs are shown in Figure 7.5. Compatible with CMOS IC fabrication, these power devices are usually separated from low-power analog and digital circuits using shallow trench isolation.

Finally, as gate lengths in the advanced CMOS nodes have shrunk to under 100 nm for logic applications, low-voltage silicon MOSFETs optimized for digital applications have been used for power switching by stacking them in series in a cascode configuration to support the breakdown voltage [34]. This design is suboptimal due to the additional channel resistance of transistors stacked in series, and the feasibility of such an approach has been demonstrated only for voltages under 5 V.

7.1.3.2　Gallium Arsenide–Based FETs

Although GaAs semiconductor technology has been around since the 1970s, their commercial use has been mostly limited to RF, optoelectronics, and solid-state lighting. The advantages of GaAs over silicon are its high electron mobility, high critical

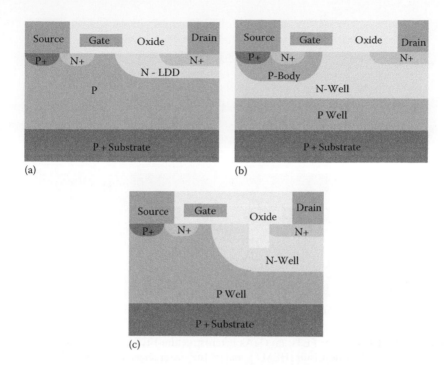

FIGURE 7.5 Various low-voltage silicon lateral power MOSFET structures compatible incorporating a lightly doped N-drain region. (From Ng, W.T. and Yoo, A., Advanced lateral power MOSFETs for power integrated circuits, in *IEEE International Conference on Solid-State and Integrated Circuit Technology*, Shanghai, China, November 1–4, 2010, pp. 859–862.)

field strength, larger bandgap, higher operating temperature, and the availability of a semi-insulating material to provide isolation and minimize parasitics [35,36]. Because GaAs does not possess a native oxide with a good interface, fabricating high-quality MOSFETs is difficult. The field effect transistors that show good performance in the GaAs system are depletion mode devices like junction field effect transistor (JFETs), metal semiconductor field effect transistor (MESFETs), and high-electron-mobility transistor (HEMT). The first power GaAs FETs were vertical JFETs, fabricated in the 1980s [37,38]. In the 1990s, lateral high-voltage MESFETs with breakdown voltages up to 130 V were demonstrated [39,40]. In these devices, the breakdown voltage is supported using field plating or by depleting the drift region using unpassivated surface charges.

The mobility of the GaAs FETs was improved using a heterostructure in the 35 V heterostructure FET (HFET) device proposed first in [41]. In these devices, a high-mobility 2D electron gas (2DEG) layer was induced on the AlGaAs/GaAs interface. A breakthrough in HEMT technology occurred when it was demonstrated that GaAs could be replaced by high-quality pseudomorphically grown InGaAs for the high-mobility channel [42,43]. In pHEMTs, since then, high-voltage pHEMT devices with breakdown voltages of up to 40 V have been extensively used for power RF applications [44,45].

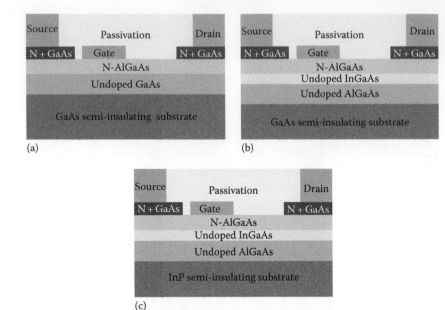

FIGURE 7.6 III–V power FETs: (a) GaAs heterostructure FET, (b) GaAs pseudomorphic high-electron-mobility transistor (HEMT), and (c) InP metamorphic HEMT.

High-power metamorphic HEMTs have also been developed using the AlInAs/InGaAs material system fabricated on InP substrates [46,47]. This system has higher band offsets compared to pHEMTs and also higher 2DEG density (Figure 7.6).

7.1.3.3 GaN FETs

The GaN material system has emerged in recent years as one of the most promising candidates for power switching applications. GaN MESFETs came first, in the early 1990s [48], followed by AlGaN/GaN HFETs [49]. GaN heterojunction FETs achieve very high carrier densities without the use of dopants due to their inherent polarizing fields [50]. GaN also has a large critical electric field and a large bandgap, which makes it suitable for high-temperature and high-voltage applications. Research in GaN system for power electronics has sense taken *off*. Low-voltage discrete GaN FETs have been commercialized [51,52]. Most recently, integrated circuits based on enhancement/depletion mode GaN FETs have also been reported [53] (Figure 7.7).

7.1.4 Physics-Based Comparison between Device Technologies

A comparison of three different material systems—silicon, GaAs, and GaN—is presented in this section to assess their performance in high-speed power switching. As candidates for integrated power devices in the 5–100 V range, three representative

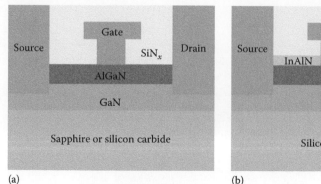

FIGURE 7.7 (a) AlGaN/GaN high-electron-mobility transistor (HEMT) and (b) enhancement mode GaN HEMT. (From Tang, Y. et al., High-performance monolithically integrated E/D mode InAlN/AlN/GaN HEMTs for mixed signal applications, in *IEEE International Electron Devices Meeting*, San Francisco, CA, December 6–8, 2010, pp. 30.4.1–30.4.4.)

device structures are also chosen: silicon LDD NMOS as shown in Figure 7.11a, AlGaAs/InGaAs/AlGaAs pHEMT as shown in Figure 7.12b, and AlGaN/GaN HEMT as shown in Figure 7.13a.

7.1.4.1 Power FET *on* Resistance

For a low-voltage lateral FET, the various components of the *on* resistance are shown in Figure 7.8. Apart from the drift resistance (R_D), the channel resistance (R_{CH}) becomes significant as the gate length L_{CH} becomes comparable to the drift length L_D. Also significant in non-self-aligned FETs (like HEMTs) is the lateral resistance between the source and the channel (R_S). The source and drain access resistances (R_{AS} and R_{AD}) represent the region of transition between vertical and lateral flow of carriers. For HEMT-type device structures, where the charge carriers have to flow across a wide bandgap donor layer to reach the channel layer (Figure 7.8), this component of the resistance is higher than that of MOS-type structures. The contact resistances between the metallization and the heavily doped layers (R_{CS} and R_{CD}) also become crucial for low-voltage transistors. Another significant factor is the resistance in the metal interconnects (R_{METAL}) between semiconductor and package, which become prominent for devices with a large gate periphery and current rating. So we have the specific *on* resistance written as

$$R_{ON,SP} = R_D + R_{CH} + R_{EXT},$$

$$R_{EXT} = R_S + R_{AS} + R_{AD} + R_{CS} + R_{CD} + R_{METAL}.$$

(7.2)

where each resistance term is normalized by multiplying with the gate periphery width. For lateral flow of current in the semiconductor, in the drift region or

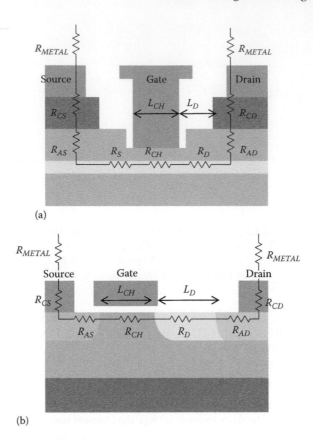

FIGURE 7.8 *On* resistance components of a power FET: (a) high-electron-mobility transistor–type structure; (b) MOS-type structure.

the channel, the specific *on* resistance is determined by the mobility of carriers in the material. The specific *on* resistance can then be expressed in terms of mobilities and sheet carrier concentrations in each region:

$$R_{ON,SP} = \frac{L_{CH}}{\mu_{CH}Q_{CH}} + \frac{L_D}{\mu_D Q_D} + R_{EXT}, \tag{7.3}$$

where
  μ is the carrier mobility
  Q is the sheet charge concentration per unit area
  the subscripts represent the channel or drift region

In the MOS structure, the peak carrier mobility in the channel region is limited by scattering in the Si/SiO$_2$ interface, where the physical mechanism that reduces the mobility is columbic scattering due to ionized dopants in the channel. However, in an

HEMT-type structure, where a 2D electron gas is responsible for conduction in both the channel and drift region, the degradation of mobility due to the columbic scattering is very low even at room temperature because the channel region is usually not doped. Hence, the achievable electron mobility is much higher than the MOS structure and very close to the phonon scattering limit.

It is observed that the GaAs pHEMT system has the highest carrier mobility of over 6000 cm²/V/s that can be obtained at room temperature for a sheet carrier concentration of 3×10^{12} cm^{-2}. By comparison, in AlGaN/GaN, mobilities are of the order of 1500 cm²/V/s with sheet carrier concentrations exceeding 10^{13} cm^{-2}. In silicon inversion layers, the mobility is dependent on the effective vertical electric field, which depends on applied gate voltage, the oxide thickness, and the background dopant concentration. For a device with an oxide thickness of 5 nm, a threshold voltage of 1 V, the inversion mobility at a carrier concentration of 4×10^{12} cm^{-2} is 486 cm²/V/s. In the drift region, assuming that the doping is 10^{16} cm^{-3} a peak mobility of 1200 cm²/V/s can be obtained.

In addition to the drift and channel resistances, the contact and access resistances become crucial in estimating the total *on* resistance of low-voltage lateral devices.

The specific contact resistivity of a metal–semiconductor contact is expressed in units of Ω cm², indicating that the contact resistance scales down in proportion to the area of the contact (Figure 7.9). This is indeed the case when the direction of current in the device is perpendicular to the contact. However, for lateral devices, the effect of increasing the contact length beyond a certain length (called the transfer length L_T) has little effect on the contact resistance.

The contact resistance to a planar semiconductor layer can be expressed as

$$R_C = \frac{R_{SK} L_T}{W} \coth\left(\frac{d}{L_T}\right), \qquad (7.4)$$

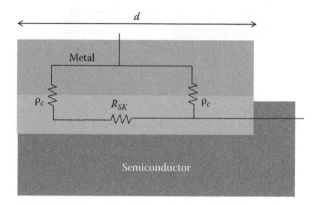

FIGURE 7.9   Resistance of a metal–semiconductor contact.

where L_T is the transfer length

$$L_T = \sqrt{\frac{\rho_c}{R_{SK}}}$$ (7.5)

R_{SK} is the sheet resistivity of the region under the contact
ρ_c is the specific contact resistivity
W is the width of the contact area
d is the contact length as shown in Figure 7.10

It is common, therefore, to normalize the contact resistance of lateral devices by device width rather than device area.

In silicon NMOS devices, the best contacts made to the N+ source or drain region use a metal silicide like $TiSi_2$ [54]. A specific contact resistivity of less than 10^{-7} Ω cm^2 can be obtained using silicided contacts. Also, if self-aligned processes

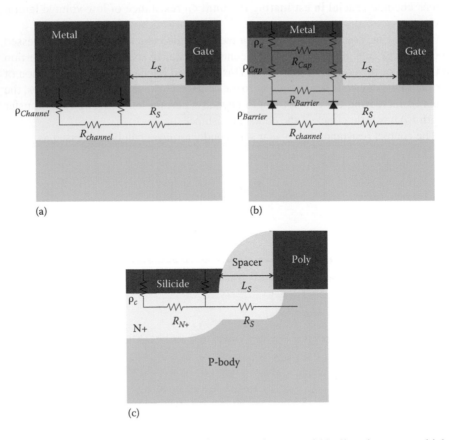

(a)   (b)

(c)

FIGURE 7.10 Illustration of contact and access resistances of (a) alloyed contact to high-electron-mobility transistor (HEMT) structure, (b) nonalloyed contact to HEMT structure using a cap layer, and (c) silicided contact to MOS structure.

are used in defining the source regions, the distance between the source and gate (L_S) can be reduced to 10% of the minimum feature size used in the NMOS process. In submicron CMOS nodes, a contact resistance less than 0.06 Ω mm can be obtained for NMOS devices [55].

For buried channel devices like HEMTs, two types of contacts are possible. In alloyed contacts, the metal is allowed to sink all the way down to the channel layer using solid-phase sintering or diffusion. Using alloyed AuGe/Ni/Au contacts, a contact resistance of up to 0.06 Ω mm has been reported on AlGaAs/GaAs HEMTs [56]. It has also been shown that a low-contact resistance can also be achieved by using a nonalloyed contact formed on a heavily doped cap layer [57]. The actual contact resistance obtained for nonalloyed contacts for multilayer devices is mainly dependent on the contact resistance between the high bandgap barrier and channel layer [58]. In GaN-based HEMTs, a low-contact resistance is harder to obtain due to the wider bandgap. Using Ti/Al/Pt/Au or Ti/Al/Ni/Au contacts, a resistance lower than 0.6 Ω mm was reported in [59]. More recently, a contact resistance of 0.23 Ω mm was obtained [60] using selective regrowth of GaN on AlGaN, and 0.023 Ω mm was obtained [61] on a GaN MIS-HEMT using selective regrowth of InGaN on GaN.

Additionally, the GaAs pHEMTs and GaN HEMTs use metal gates and their fabrication is not conducive to self-aligned contacts to source and drain when using CMOS-type process steps. Hence, the gate-source distance L_S is close to the minimum feature size used in the process and this increases the access resistance in these structures. More recently, InGaAs and GaN MIS HEMTs [58,61] have been demonstrated using self-aligned gates by using refractory metal gates.

7.1.4.2 Gate Charge

The gate charge of an FET is mainly determined by the channel charge Q_{CH} and drift charge Q_D. In a lateral power FET, the area of the channel that needs to be depleted in the blocking state is illustrated in Figure 7.11. During switching, the charge needed to deplete this region is supplied through Q_{GS}, Q_{GD}, and Q_{DS}, so that

$$Q_{CH} + Q_D = Q_{GS} + Q_{GD} + Q_{DS}. \tag{7.6}$$

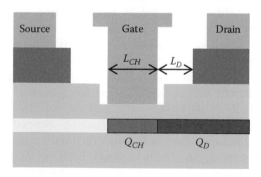

FIGURE 7.11 Illustration of the gate charge components of a power high-electron-mobility transistor.

Partitioning these charges and identifying the individual components is a complex problem. However, for devices with no substrate contact, we can make the assumption that all the charges in the drift region are imaged at the gate, so that Q_{DS} is small. Therefore, we can represent the gate charge normalized by width as

$$Q_{CH} + Q_D = Q_{GS} + Q_{GD} = Q_G, \tag{7.7}$$

$$Q_G = L_{CH} Q_{CH} + L_D Q_D. \tag{7.8}$$

Q_{CH} scales down as the channel length is reduced. As the voltage rating of the device decreases, the contribution of the channel charge to the gate charge becomes significant. The FOM from Equation 7.1 can then be written as the product of the specific *on* resistance and gate charge from Equation 7.8:

$$FOM = \left(\frac{L_{CH}}{\mu_{CH} Q_{CH}} + \frac{L_D}{\mu_D Q_D} + R_{EXT} \right) \times \left(L_{CH} Q_{CH} + L_D Q_D \right). \tag{7.9}$$

Equation 7.9 illustrates the important device parameters of a lateral power device, which determine the optimum technology choice.

7.1.4.3   Breakdown Voltage

For a lateral power device, the breakdown voltage is primarily dependent on the drift length as long as the drift region is engineered to maintain a low peak electric field. Since the breakdown voltage is extremely sensitive to the peak electric field, it is desirable that a uniform electric field is maintained in the drift region, so that the blocking voltage achievable for a given drift length is maximized. If the drift region is fully depleted before the blocking voltage is reached, then it behaves like an insulator and a uniform electric field can be maintained in this way.

In lateral devices, this is achieved by compensating for the drift charges using charges of the opposite type. This technique is known as reduced surface field [27]. The positively charged N-type drift region can be compensated by using P-type dopants in the substrate (JI RESURF) [27] by imaging the depleted dopants on a field plate [62] on a backgate when separated by an insulator (SOI RESURF) [63] or using alternating pillars of opposing dopants (superjunction or 3D RESURF) [64].

An ideal limit for the 2D RESURF structure can be obtained by considering the inclined field plate structure shown in Figure 7.12. First suggested in [65], in this structure, when the drift region is fully depleted, both the lateral and the vertical electric field in the semiconductor are uniform. Breakdown occurs when peak field in the entire drift length reaches the critical electric and the drift length required to support a certain blocking voltage is minimized.

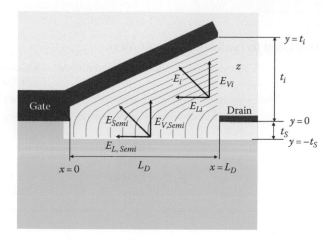

FIGURE 7.12 Illustration of the ideal electric field distribution in the drift region of a power high-electron-mobility transistor in the blocking state.

In the ideal case, to minimize the *on* resistance for a particular breakdown voltage, it can be shown that the vertical and horizontal electric fields in the semiconductor are equal, so that

$$E_{V,Semi} = E_{L,Semi} = \frac{E_{Crit}}{\sqrt{2}}. \tag{7.10}$$

where E_{Crit} is the breakdown electric field of the semiconductor. This yields an expression for the minimum length of the drift region:

$$L_D = \frac{\sqrt{2}BV}{E_{Crit}}. \tag{7.11}$$

The minimum specific drift resistance is given by

$$R_D = \frac{2BV}{\epsilon_S \epsilon_0 \mu_D E_{Crit}^2}. \tag{7.12}$$

Therefore, in the ideal case, the critical electric field has a big impact on the drift resistance of lateral power devices. Wide bandgap devices, therefore, have a lower drift resistance for the same breakdown voltage owing to their higher critical electric field.

7.1.4.4 Figure-of-Merit Comparison

The figure of merit ($R_{ON} \times Q_G$) for the three material systems discussed in the previous sections can be computed based on the equations described in the previous sections. Table 7.2 summarizes the parameters used for the calculations. In all cases,

TABLE 7.2

Parameters Used to Evaluate Figure of Merit

	Silicon LDD MOS	AlGaAs/InGaAs/AlGaAs pHEMT	AlGaN/GaN HEMT
E_{Crit} (V/cm)	3×10^5	4×10^5	2.4×10^6
μ_D (cm²/V/s)	1200	6000	1800
ε_s	11.7	12.9	8.9
RESURF dose (cm⁻²)	1.37×10^{12}	2.01×10^{12}	8.3×10^{12}
μ_{CH} (cm²/V/s)	488	6000	1800
Q_{CH} (cm²/V/s)	4×10^{12}	3×10^{12}	1×10^{13}
R_C (Ω cm)	0.034	0.05	0.05

a self-aligned technology is assumed so that the minimum possible spacing between gate to source or gate to drain contact is 10% of the minimum feature size in the process.

The comparison of the *on* resistance, gate charge, and the switching figure of merit for silicon LDD NMOS, AlGaAs/InGaAs/GaAs pHEMT, and AlGaN/GaN HEMT devices as a function of breakdown voltages is shown in Figure 7.13. It can be seen that the GaN-based system has the lowest *on* resistance among the three. This is primarily due to the high carrier concentration that is achievable in the GaN channel and drift region without compromising for the breakdown voltage. As the voltage rating of the device is reduced, however, this advantage diminishes. This is because the contact and access resistances start to dominate the total *on* resistance. GaAs system has intermediate *on* resistance, but because the GaAs system has the highest electron mobility, the low *on* resistance is achieved using a much smaller charge in the drift region. Therefore, the GaAs system has the lowest gate charge.

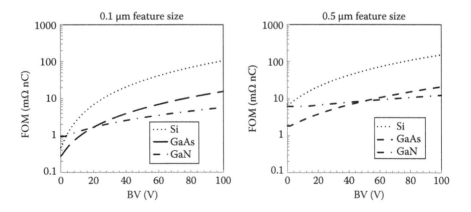

FIGURE 7.13 Comparison of the power switching figure of merit (in mΩ nC) between silicon lightly doped drain NMOS, GaAs pHEMT, and GaN HEMT for 0.1 and 0.5 μm minimum feature size processes.

If we compare the product of the *on* resistance and the gate charge, the GaAs-based system is seen to have the lowest FOM for $BV < 50$ V for 0.5 μm feature size processes, and <20 V for 0.1 μm feature size processes. This is again due to the superior carrier mobility in the GaAs system that enables a reduction in the *on* resistance without increasing the channel or drift charge. From the calculations, it follows that unless sub-100 nm feature sizes are used, GaAs is the most optimal material system for high-frequency switching power conversion from an efficiency point of view.

7.1.5 EVALUATION OF GaAs pHEMT FOR POWER SWITCHING

To evaluate the feasibility of pHEMTs for power switching applications, a 0.5 μm commercially available pHEMT process (TriQuint TQPED) was chosen [66]. The process integrates depletion mode (D-pHEMT) and enhancement mode (E-pHEMT) devices on the same wafer. The availability of low-standby current enhancement mode devices makes the pHEMT process suited to fabrication of integrated power management circuits.

The higher-level cross section and the device structure of the pHEMT in the TQPED process are shown in Figures 7.14 and 7.15. The process utilizes a semi-insulating substrate as the starting material. A superlattice buffer layer consisting of alternating layers of AlGaAs and GaAs are deposited on the substrate for isolation. Unintentionally doped spacer layers of AlGaAs are placed on either sides of an InGaAs channel with silicon delta–doped layers placed on both sides of the channel-spacer stack. A Schottky AlGaAs layer is placed on top of the spacer on which the gate is deposited. The process utilizes a thin InGaP etch stop layer on top of the Schottky layer for definition of the gate recess. A heavily doped cap layer of GaAs is deposited on top to reduce the ohmic resistance to the contacts. The device fabrication steps are found in [69].

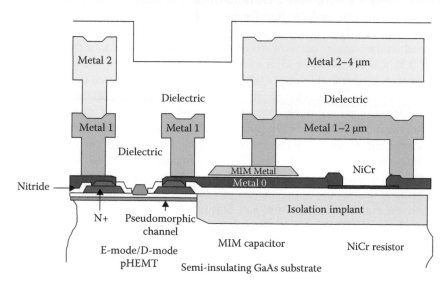

FIGURE 7.14 0.5 μm pHEMT cross section. (From TriQuint Semiconductor Inc., TriQuint Semiconductor website, http://bit.ly/Qbx8l9, March 2010.)

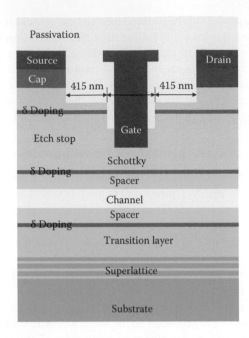

FIGURE 7.15 0.5 μm enhancement mode GaAs pHEMT structure and epi stack.

7.1.5.1 Enhancement Mode pHEMT Characteristics

An enhancement mode pHEMT device with a gate width of 100 μm was characterized to estimate the basic device characteristics. The output curve of the device is shown in Figure 7.16, and the transfer curve of the device in the linear regime is shown in Figure 7.17.

Based on the transfer curve of the device in the linear regime, a threshold voltage of 0.36 V is observed, which is the gate voltage needed for an output current of 1 mA/mm. A subthreshold slope of 64 mV/decade is observed when the gate voltage is below this value. The device clearly has a Schottky-type gate as the gate current starts to increase exponentially when the gate voltage is greater than 0. This can be observed in Figure 7.18. When operated as a switch, beyond a certain value of gate voltage, the driver circuits have to supply a high steady-state current and this is undesirable in terms of keeping the driver power consumption low. We can choose a maximum permissible gate bias depending on this power dissipation limit. If the ratio between the output current and the gate current has to be limited to less than a hundred, the maximum permissible gate bias is 0.88 V.

7.1.5.2 Specific *on* Resistance

The measured specific *on* resistance of the 100 μm enhancement mode pHEMT as a function of the gate voltage is shown in Figure 7.19. For this device, the specific *on* resistance at a gate voltage of 0.88 V is 1.46 Ω mm when drain–source voltage is 0.1 V.

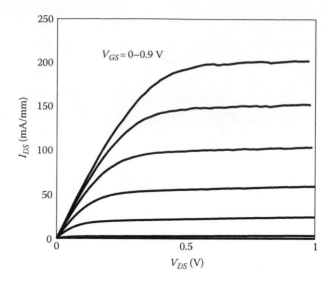

FIGURE 7.16   Output characteristics of the enhancement mode pHEMT device.

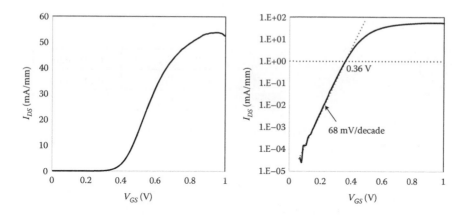

FIGURE 7.17   Transfer curve of the enhancement mode pHEMT device in linear and log scale. The drain voltage is 0.1 V.

7.1.5.2.1   TLM Measurements

To identify the various components of *on* resistance, transmission line measurements (TLMs) were performed on each layer in the HEMT structure. These measurements, therefore, allow us to identify the components of the *on* resistance as shown in Figure 7.19. It is seen that the vertical components of the resistance, including contact and access resistances, account for more than 64% of the total *on* resistance. The drift resistance L_D accounts for 8% of the total resistance and the channel resistance accounts for 24%.

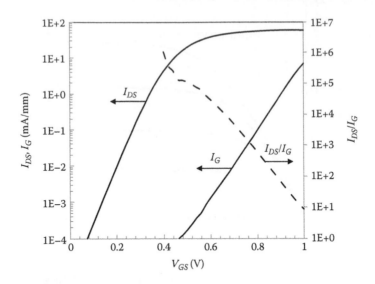

FIGURE 7.18 Gate current, drain current, and their ratio as a function of gate voltage.

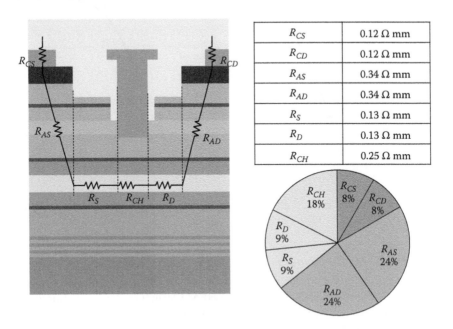

R_{CS}	0.12 Ω mm
R_{CD}	0.12 Ω mm
R_{AS}	0.34 Ω mm
R_{AD}	0.34 Ω mm
R_S	0.13 Ω mm
R_D	0.13 Ω mm
R_{CH}	0.25 Ω mm

FIGURE 7.19 Components of *on* resistance of the enhancement mode pHEMT extracted from transmission line measurements.

7.1.5.3 Interconnect Metallization

To design large area devices, a multilayer metallization scheme needs to be used. An illustration of the interconnect scheme is shown in Figure 7.20. In this scheme, three metal layers run perpendicular to each other conducting current through to copper bumps that are placed in a grid on the large area device. Placing of Cu bumps directly on top of the device reduces the resistance considerably up to the level of the circuit board. Since the board metal layers are much thicker than the ones on the chip, further resistance increase is minimal. The minimum distance between the copper bumps is limited by the minimum feature size in the PCB. In state-of-the-art PCBs, this limits the bump to bump spacing to 100 μm or above. Using this approach, for any device with dimensions larger than 200 μm × 200 μm, there is very little increase in the total *on* resistance. The top metal lines run diagonally to the copper bumps, so that there are no vias required on the PCB pads. Measured *on* resistance of large area devices is shown in Table 7.3.

7.1.5.4 Gate Charge

The gate charge measurement was done using a switching circuit as shown in Figure 7.21. The E-pHEMT S1 used has a total gate width of 5 mm.

The extracted gate charge waveform, plotted as a function of gate voltage, is shown in Figure 7.22.

When switching to a gate voltage of 0.63 V, the gate charge for the 5 mm device is 22.5 pC. The miller charge is extracted to be 10.4 pC.

7.1.5.5 Breakdown Voltage

In the TQPED enhancement mode pHEMT, when the gate voltage is set to zero, the maximum operating voltage is determined by the leakage current rather than avalanche breakdown. For a current ratio of 10^4 between the *off* state and the *on* state, the drain voltage cannot exceed 11 V, and the leakage current is limited to 10 μA/mm. The leakage characteristic of the device is shown in Figure 7.23. The leakage current has two components, the first of which is the drain-source current, which is primarily determined by drain-induced barrier lowering (DIBL). The second component is the reverse gate current through the Schottky barrier between the gate and drain. It is seen that only a small fraction of the drain current flows through the gate, which implies that the leakage is dominated by the drain–source current.

Upon applying a negative gate bias, the potential barrier between the gate and source can be increased, which leads to the increase of the breakdown voltage as shown in Figure 7.24. However, a negative bias also applies a larger stress on the gate-drain Schottky junction, and therefore, the leakage current through the gate increases. When a voltage of −0.6 V is applied, half of the drain current flows through the gate and the rest flows through the source. When a lower potential bias is applied, we can say that the breakdown is no longer limited by DIBL but by the current through the gate Schottky junction. The reverse current through the gate is primarily determined by thermionic and field emission [67]. A peak BV_{DS} of up to 19.4 V can be obtained when a gate bias of −0.9 V is applied, as shown in Figure 7.24.

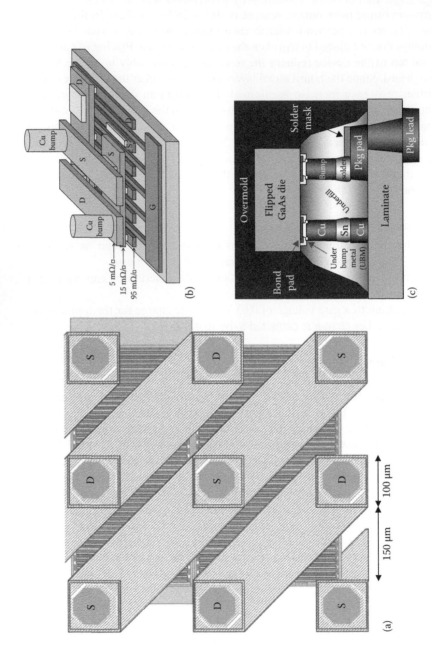

FIGURE 7.20 Illustration of layout of copper bumps for flip-chip packaging. (a) Die layout for a 20 mm device, (b) 3D schematic of copper bumps, and (c) flipped die on laminate showing Cu bumps soldered to PCB pad.

TABLE 7.3

Layout Details and Measured *On* Resistance of the Large Devices

Gate width (mm)	5	10	20
Current rating (A)	0.5	1	2
Area (µm)	240 × 500	480 × 500	600 × 800
R_{ON} (mΩ)	318.9	167.5	91.7
$R_{ON} \times W$ (Ω mm)	1.59	1.68	1.83
$R_{ON} \times$ Area (mΩ mm²)	38.26	40.21	44.02

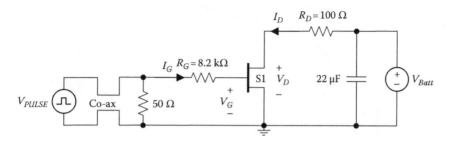

FIGURE 7.21 Circuit used to measure gate charge.

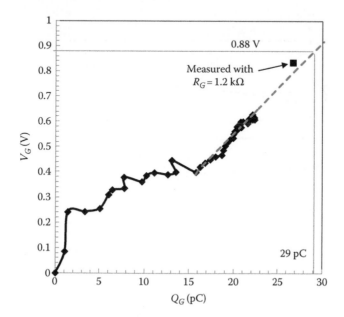

FIGURE 7.22 Gate voltage versus gate charge for switching with $R_G = 8.2$ kΩ, $V_D = 5$ V, and $R_D = 100$ Ω.

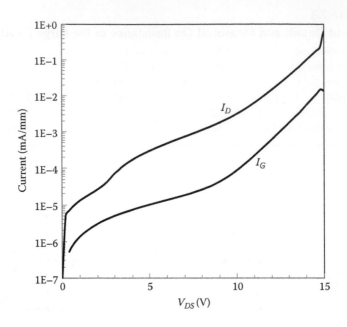

FIGURE 7.23 Leakage current of the TQPED enhancement mode pHEMT when V_{GS} is 0 V.

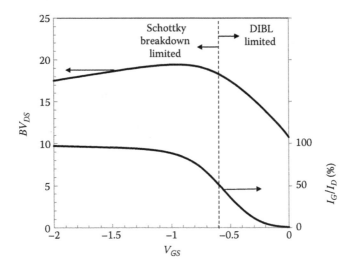

FIGURE 7.24 Breakdown voltage as a function of gate voltage and the fraction of drain current flowing through the gate at breakdown.

7.1.5.6 Figure-of-Merit Comparison

The 0.5 μm enhancement mode TQPED GaAs pHEMT device is nominally rated at 11 V at room temperature. The *on* resistance and the gate charge of the enhancement mode pHEMT device are shown in Figure 7.25. The details of best-in-class commercially available discrete silicon NMOS devices in the 8–12 V breakdown voltage range are also shown in the same figure.

Since in the GaAs pHEMT device the measurements are made on wafer level, the effect of packaging is not included in the overall FOM measurement. In Figure 7.25, not discounting packaging effects, an order of magnitude improvement can be seen in GaAs over a silicon commercial device with the same *on* resistance. When comparing devices of different current ratings, the effect of packaging cannot be neglected. Especially in cases where the total *on* resistance of the device is low, as in [68], interconnect and package resistances and extrinsic capacitances can cause a significant deterioration of the figure of merit. From [68], we see that for a 1.25 mΩ silicon device with a chip scale flip-chip packaging, about 54% of the total resistance comes from the interconnect, bump, and PCB. Assuming that these components remain in the same proportion, and that the intrinsic resistance and gate charge scale with device width, we can compare the device from [68] with the GaAs pHEMT device as shown in Table 7.4. With these effects, the GaAs pHEMT improvement in FOM is 30% over silicon.

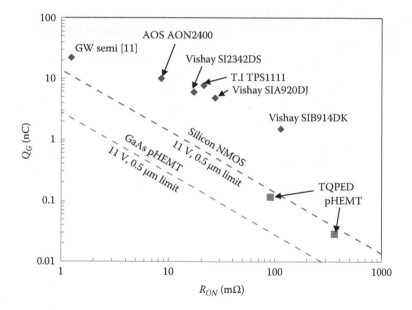

FIGURE 7.25 The comparison of *on* resistance and gate charge for TQPED enhancement mode pHEMT and commercially available silicon devices in the 8–12 V range. Also shown is the limit for 11 V lateral devices in the two technologies with a minimum feature size of 0.5 μm.

TABLE 7.4

Comparison of GaAs pHEMT with a High Current Silicon
MOSFET [68] Including Packaging Effects

Total Resistance	GW Si MOSFET [17]	GaAs pHEMT
Total R_{ON} (mΩ)	1.25	1.25
Intrinsic R_{ON} (mΩ)	0.56	0.56
Interconnect (mΩ)	0.27	0.27
Packaging (mΩ)	0.4	0.4
Q_G (nC)	22	16
FOM (mΩ nC)	27.5	19.7

7.1.6 EXTENDED DRAIN DEVICES

Power switches in the 20–50 V range are used today in power supplies of servers, notebooks, telecom, brushless DC motors, etc. Silicon power MOSFETs are the most commonly used switches in this voltage range today [62]. More recently, 40 V enhancement mode GaN pHEMTs have also emerged as candidates to replace silicon in this range [49]. We have shown that GaAs pHEMTs can be highly efficient switches in this range for these applications due to their superior figure of merit. In addition, GaAs pHEMTs have the advantage of the semi-insulating GaAs substrate, which provides natural electrical isolation and minimizes parasitics for power ICs. Extending the voltage range of commercially available pHEMT devices to 20–50 V range can, therefore, potentially replace silicon in these applications.

The first high-voltage GaAs transistors that were proposed were vertical MESFETs and JFETs [35,36]. However, the design of lateral GaAs power devices is more complex. This is because lateral GaAs FETs on semi-insulating substrate are not suited to JI [25] and SOI [57] RESURF techniques unlike silicon. GaAs on semi-insulating substrate is similar to silicon on DI technology realized with poly-silicon handle wafer on other similar substrates [63]. In well-designed RESURF devices, a linear relationship is obtained between the drift length and the breakdown voltage. In low-voltage GaAs devices, the drift region charge is depleted in the blocking state by charges on the interface between the semiconductor and the passivating layer [64]. However, these charges are not controlled easily and their magnitudes vary widely depending on the exact process conditions. If the surface charge does not adequately compensate for the drift charge, a high electric field can result at the gate end of the drift region and the breakdown voltage is low. In this case, no improvement in the breakdown voltage is obtained even if the drift length is increased. If there is no adequate charge compensation, to limit the high electric field at the gate during reverse bias, other means of compensating the drift charge are required. Some workers have reported breakdown enhancement in GaAs lateral devices using field plates [65,69].

Using an optimized version of the pHEMT process, we can increase the breakdown voltage of the devices. The optimized process uses a double recessed structure,

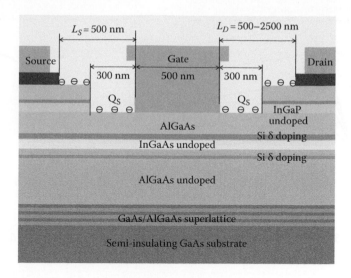

FIGURE 7.26 Structure of the double-recessed pHEMT.

shown in Figure 7.26, to reduce the field crowding near the gate. Upon optimizing the double recess to improve the field profile, the breakdown voltage increases in proportion to the separation between the gate and drain (L_D), as shown in Figure 7.27. For L_D of 2.5 μm, which was the largest drift length in the experiment, a breakdown voltage of 47 V was obtained. The output and transfer characteristics of the pHEMT with $L_D = 500$ nm are shown in Figure 7.28.

Using TLM measurements, the intrinsic *on* resistance of the device was characterized. As shown in Figure 7.29, the major components of the *on* resistance are the contact and access resistances. Therefore, increasing the drift length L_D does not contribute to a significant increase in the *on* resistance of the pHEMT. The *on* resistance of the optimized structure as a function of drift length is shown in Figure 7.30.

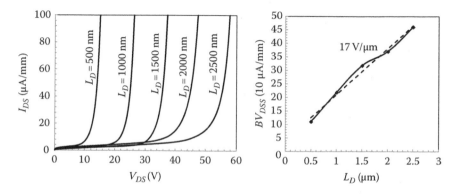

FIGURE 7.27 Breakdown characteristics of the optimized devices with varying L_D at $V_{GS} = 0$ V.

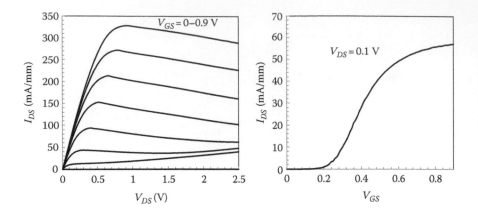

FIGURE 7.28 The output and transfer characteristics of the optimized device with $L_D = 500$ nm.

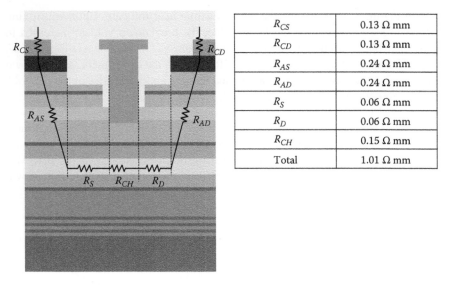

R_{CS}	0.13 Ω mm
R_{CD}	0.13 Ω mm
R_{AS}	0.24 Ω mm
R_{AD}	0.24 Ω mm
R_S	0.06 Ω mm
R_D	0.06 Ω mm
R_{CH}	0.15 Ω mm
Total	1.01 Ω mm

FIGURE 7.29 Components of the *on* resistance of the intrinsic optimized device.

The gate charge was measured for 22 and 32 V devices (with drift lengths equal to 1000 and 1500 nm, respectively), each with a total gate width of 5 mm. The 22 V device as a gate charge of 44 pC and the 32 V device has a gate charge of 58 pC. The switching characteristics and figure of merit of the 22 and 32 V devices are summarized in Table 7.5.

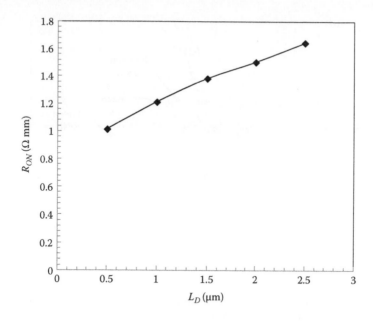

FIGURE 7.30 The *on* resistance of the optimized device as a function of L_D.

TABLE 7.5
Device Characteristics of 22 and 32 V pHEMT Devices

BV (V)	Gate Width (mm)	Current (A)	R_{ON} (mΩ)	Q_G (pC)	FOM (mΩ nC)
22	5	0.5	268	44	12
32	5	0.5	303	58	18

7.1.6.1 Comparison to Commercial Devices

Both the 22 and 32 V pHEMTs show an order of magnitude improvement over commercially available 20 V silicon NMOSFETs as shown in Figures 7.31 and 7.32. The switching figure of merit ($R_{ON} \times Q_G$) of GaAs pHEMTs have been compared to silicon NMOSFETs and GaN HEMTs at various voltage ratings in Figure 7.33. It is seen that the GaAs pHEMTs compare extremely favorably with commercially available silicon parts. Also shown in the figure is the 40 V GaN HEMT released by the EPC [49]. Although the measured FOM data for the 47 V GaAs pHEMT were not available, it is seen that the GaN fits in line with the FOM trend of the GaAs pHEMT.

7.1.7 GaAs pHEMT Integrated Power Converters

Single-chip DC–DC converters, where the output stage, control elements, and filters are integrated on chip, have relatively lower efficiency due to the lossy passives. The GaAs-based high-frequency DC–DC converter is conceived as a small

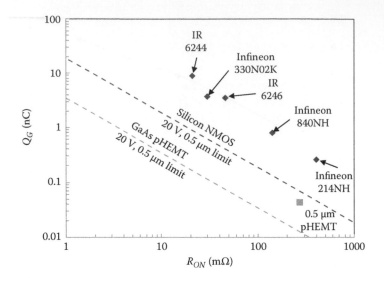

FIGURE 7.31 Performance of the 22 V pHEMT compared to silicon NMOSFETs. Also shown are the intrinsic limits for a 0.5 μm feature size.

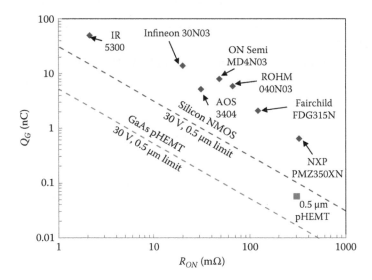

FIGURE 7.32 Performance of the 32 V pHEMT compared to silicon NMOSFETs. Also shown are the intrinsic limits for a 0.5 μm feature size.

form factor flip-chip multichip module, as shown in Figure 7.34. Multichip modules are advantageous because they enable the use of the best technology for each component of the circuit at the cost of larger interconnect parasitics compared to single-chip solutions. In buck-type high-frequency DC–DC converters, the current through the high-side switch (the input current) can vary by as much as

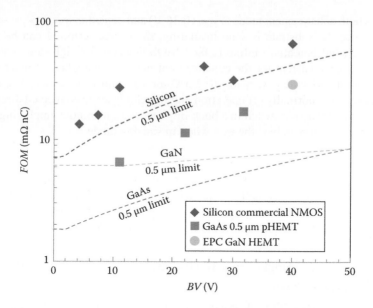

FIGURE 7.33 Comparison of switching FOM between silicon, GaN, and GaAs pHEMT devices.

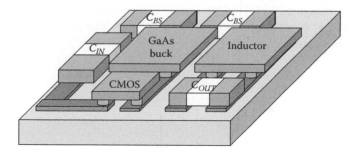

FIGURE 7.34 Multichip module for a pHEMT-based buck converter.

10 A/ns during the switching transition. Any wire resistance between the decoupling capacitors and the FET will, therefore, cause very large voltage spikes, which can force the switch to function outside its safe operating area and also reduce the power efficiency. To implement a high-frequency DC–DC converter using wire-bond packaging, it might be therefore necessary to place additional decoupling capacitors *on* chip, consuming valuable real estate on the die. With flip-chip inter-connects, the series inductance from chip to decoupling capacitors can be made much smaller, eliminating the need for additional decoupling capacitors on chip. In addition, flip chip has a smaller package size, higher reliability and throughput, and lower cost of assembly.

The device technology chosen for the converter prototypes is a commercially available 0.5 μm pHEMT process on semi-insulating GaAs substrate. The process integrates both enhancement mode ($V_T = +0.36$ V) and depletion mode ($V_T = -0.5$ V) devices. Since the substrate is semi-insulating, the source terminal can be raised above the ground potential (unlike LDD NMOS devices) for high-side switching. The drain leakage current for the enhancement mode device when connected to a 4.5 V rail in the off state is very low (0.2 µA) for a 10 mm wide device. The device thus achieves truly normally off operation and therefore can be connected directly to the battery as a high-side switch in a buck implementation without employing negative voltage generators to bias the gate when in standby mode.

The fabricated pHEMT output stage integrates the switches and gate drivers and is flip-chip bonded on the laminate. The capacitors for the output filter, input decoupling, and charge pumps are SMD type that are soldered on the laminate. For the closed-loop control circuits, pHEMT technology is not ideal for the required logic and analog circuits due to its high-power consumption, and a CMOS chip that can be flip-chip mounted on the package would be the optimal solution.

7.1.7.1   Prototype 1: Single-Phase Buck Converter with Integrated Drivers

7.1.7.1.1   Power Switches

The prototype pHEMT power IC is a buck converter that consists of an enhancement mode high-side switch P1, a diode-connected pHEMT P2 and a high gate driver circuit as shown in Figure 7.35. The gate width of P1 is optimized using SPICE simulations such that the sum of frequency-dependent losses and conduction losses in the switch is minimized at 100 MHz. The chip is designed for a rated output current of 500 mA, and it is determined using SPICE simulations that the optimal gate width for this load is 10 mm.

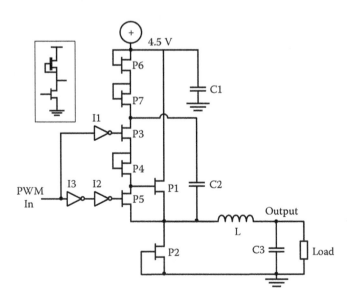

FIGURE 7.35   Schematic of the fabricated buck power IC along with external passives. The E/D inverting buffer is shown in the inset.

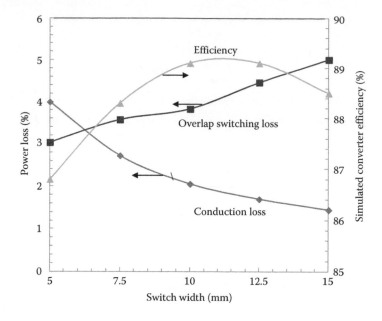

FIGURE 7.36 Simulated converter efficiency, conduction, and overlap switching loss for the high-side switch P1 as a function of device width.

A 10 mm diode-connected pHEMT P2 (gate shorted to drain) is used as the rectifier for the buck circuit. The diode-connected pHEMT is used instead of a Schottky diode because of the small threshold voltage of the enhancement mode pHEMT (0.36 V), which gives a forward voltage drop of 0.5 V when the current is 0.5 A. In comparison, the forward drop of the Schottky diode in the GaAs pHEMT process is 0.65 V (Figure 7.36).

7.1.7.1.2 Gate Driver

The gate driver circuit takes an externally generated PWM signal at the input and drives the high-side switch P1. Enhancement-/depletion-type inverters are used as buffer stages in the driver. Since there is no PMOSFET-like pull-up device in GaAs pHEMT technology, these buffer stages consume more power than static CMOS inverters. In the final driver stage, only enhancement mode devices are used so as to minimize power consumption. Only one of P3 and P5 are *on* at the same time, depending on which the gate of P1 is pulled high or low. To turn P1 fully *on*, the gate voltage of P1 needs to be raised above the supply voltage by 0.85 V. This is accomplished using a bootstrap capacitor C2, which is a 200 pF MIM structure integrated on chip. When P1 is *off*, the source of P1 is pulled below ground by the inductor current flowing through P2. At this point, C2 is charged through the level-shifting transistors P6 and P7. When P1 starts turning *on* through P6, P7, P3, and P4, the source voltage of P1 is raised above ground. The charge stored in C2 then starts elevating the drain voltage of P3 above the supply voltage. Since the gate Schottky diode of P1 starts conducting when VGS exceeds 0.89 V, tight control on the gate

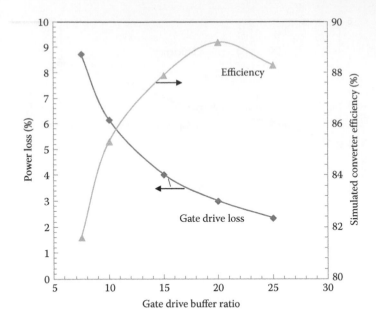

FIGURE 7.37 The gate drive power loss and simulated converter efficiency as a function of E/D inverter buffer sizing ratio.

voltage is necessary to minimize the losses in the driver. Transistors P6, P7, and P4 enable adequate voltage level shifting such that the gate overvoltage does not occur once the converter operation reaches steady state. Optimization of the size of the driver transistors is performed using SPICE simulations to minimize the power loss in the gate driver while achieving minimum possible overlap time for the switching of the power transistor P1. The optimization of the E/D inverter sizing ratio is shown in Figure 7.37.

7.1.7.1.3 Passives and Board-Level Design

When switching at 100 MHz, the transition times for switching in P1 are of the order of 250 ps. During such sharp transitions in current, parasitic inductances in PCB traces and even wire-bonds can cause a large voltage spike in the drain of P1, thus causing unsafe operating conditions for the transistor. To mitigate this problem, a 25 nF chip capacitor C1 is soldered on the PCB within a distance of 200 μm from the chip. The die is connected to the PCB using flip-chip bonding to avoid the effect of bondwire inductances. The inductor used was a 0603 15 nH air core surface mount part [70]. The specifications of the inductor are shown in Figure 7.38. An SMT output capacitor C3 (25 nF) is also soldered on the PCB in close proximity of the die to complete the converter circuit. The die photo of the GaAs pHEMT die is shown in Figure 7.39. The printed circuit board consisting of the converter module is shown in Figure 7.40.

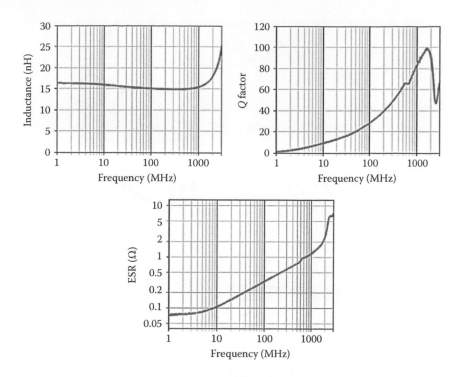

FIGURE 7.38   Inductance, quality factor, and ESR of the 15 nH air-core inductor.

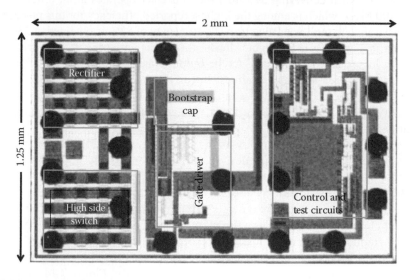

FIGURE 7.39   Die photo of the pHEMT power IC showing circuit elements and Cu bumps for flip-chip bonding.

FIGURE 7.40   Photograph of the evaluation board showing the converter module.

7.1.7.1.4   Power Converter Measurements

The fabricated buck circuit is operated in open-loop conditions by using a low-rise time variable duty pulse source to control the output voltage. For an output voltage of 3.3 V, when delivering 500 mA, the converter operates at an efficiency of 87% at 100 MHz, which is among the highest reported efficiency for this power and frequency range to our knowledge. A peak efficiency of 88% is obtained at 330 mA as shown in Figure 7.41. For lighter loads, the efficiency drops as gate drive and switching losses start to dominate. As the output voltage is lowered, the efficiency of the circuit is seen to decrease almost linearly as shown in Figure 7.42. This is due to the increased conduction loss in the rectifying diode connected transistor at lower duty cycles. The efficiency at lower voltages can be improved by adopting a synchronous buck design.

The gate drive and switching losses in the circuit increase linearly as the frequency is increased. The efficiency of the circuit drops by about 3% as the frequency is increased to 200 MHz as shown in Figure 7.43. The gate driver and the power transistor are connected to the supply through separate wires. Hence, the power consumed by the gate driver can be independently monitored. By extrapolation of power loss as a function of frequency, the frequency-independent and frequency-dependent losses are isolated. Apart from gate driver losses, other frequency-dependent losses include output switching losses in the power transistor and frequency-dependent losses in the inductor. Figure 7.44 shows the distribution of various power loss components in the circuit, with the simulated values shown for comparison.

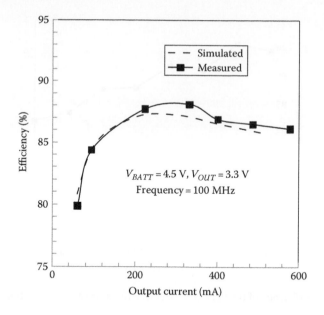

FIGURE 7.41 Efficiency of the buck converter as a function of load current at 100 MHz.

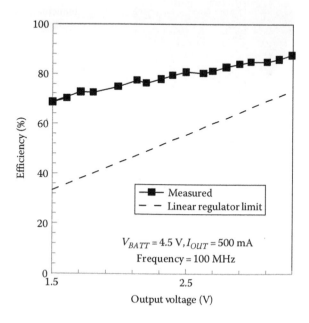

FIGURE 7.42 Efficiency of the buck converter as a function of output voltage at 100 MHz.

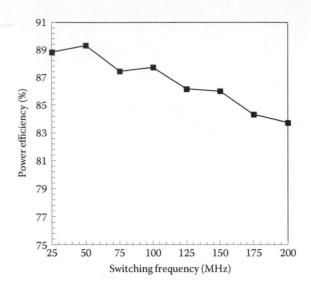

FIGURE 7.43   Efficiency of the buck converter as a function of switching frequency.

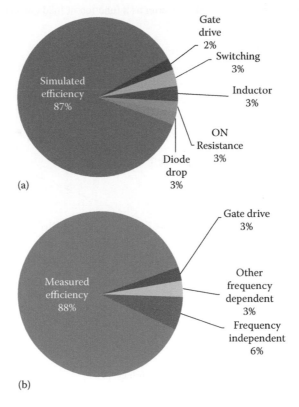

FIGURE 7.44   Comparison between the (a) simulated and (b) measured power loss components in the converter at 100 MHz, 330 mA, 3.3 V output.

7.1.7.2 Prototype II: High-Bandwidth pHEMT DC–DC Converter for Envelope Tracking

In the second example, a single-phase GaAs buck DC–DC converter designed to drive PA loads is implemented with SMD inductors on laminate. A closed-loop hysteric control network is implemented externally using a GaAs IC to estimate the loop bandwidth and transient response of the power supply. The converter is designed to provide an output power of up to 37 dBm from a 4.2 to 4.5 V supply. Both linear and saturated PA loads are approximated by a fixed resistance across a wide range of output power. For typical PA loads, therefore, the output power drops in a quadratic fashion as V_{CC} is reduced. The DC–DC converter is optimized at peak load conditions, which is at the maximum output voltage.

7.1.7.2.1 Output Stage

The output stage of the buck converter, consisting of switches and drivers integrated on the GaAs pHEMT die, is shown in Figure 7.45. The high-side switch is an enhancement mode pHEMT (Q1) that is connected directly to the battery voltage (V_{BATT}). The low side consists of a synchronous rectifier Q2 in parallel with a diode connected pHEMT Q3. After Q1 is turned *off*, Q3 supplies the bulk of the inductor current for a short dead time till when Q2 is turned *on*. For a voltage conversion ratio from 4.5 to 3.3 V, and an output current of 1 A, the optimal gate periphery is 20 mm. In the low-side synchronous rectifier, the overlap switching losses are negligible because it undergoes zero volt switching transitions. In the optimization of Q2, the trade-off is only between the *on* resistance and gate driver losses. For maximizing the efficiency at the peak conversion ratio of 4.5:3.3, the optimal size of the synchronous rectifier is also close to 20 mm.

7.1.7.2.2 Gate Driver Design

The gate driver circuits consist of multiple E/D inverting buffer stages whose function is to provide the gate charge required for Q1 and Q2 during the switching transition. E/D inverters are inherently lossier than CMOS inverters due to lack of true pull-up (PMOS like) transistor. In the high-side driver shown in Figure 7.46, the first buffer stage is made up of E-pHEMTs Q4 and Q5, driven by complementary gate signals to minimize power consumption. The driver sizing ratio is optimized such that the switching time of Q1 is minimized while the static conduction losses in the drivers is kept low. A sizing ratio between 15:1 and 20:1 between Q1 and the final buffer stage gives the maximum efficiency for this configuration.

The high-side transistor Q1 needs a floating driver that can raise the gate potential above the V_S node. The driver circuit, therefore, needs a charge pump that can boost the gate voltage above the battery voltage when Q1 is conducting. In the circuit shown in Figure 7.46, this is accomplished through a bootstrap capacitor (C_{BS}). When Q1 is *off*, and the voltage at V_S is low, C_{BS} gets charged to three diode voltages and one V_{SAT} below V_{BATT} through the pHEMT Q13 and the string of diodes D1, D2, and D3. After the high-side pulse is turned *on*, the gate of Q1 is pulled up by the EFET Q4, which elevates the gate to one threshold voltage below V_{BATT}. The current is switched to Q1 and thus correspondingly raises V_S. The stored charge in C_{BS} now discharges through D4 and Q4, eventually raising the potential at the gate of Q1 above V_{BATT}. From this point *on*, the remainder of the gate charge required to fully turn *on* Q1 is supplied by

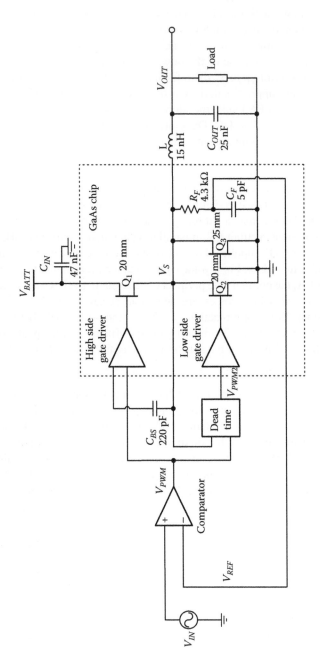

FIGURE 7.45 Schematic of the single-phase converter with hysteric control.

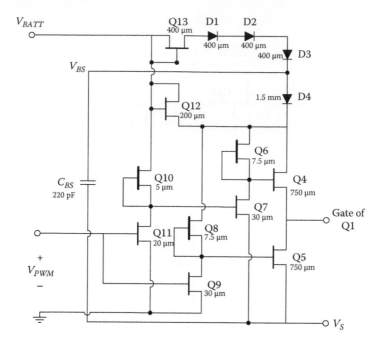

FIGURE 7.46 High-side gate driver circuit.

the capacitor C_{BS}. The number of level shifting stages is chosen such that the gate voltage of Q1 is raised to only 0.85 V above V_{BATT}. This is necessary to eliminate forward conduction in the gate Schottky diode of pHEMT Q1, which would increase the drive power consumption and the size of C_{BS}. Using this configuration, the optimal value of C_{BS} for a 20 mm E-pHEMT at 100 MHz is 220 pF. The bootstrap capacitors are implemented off-chip using SMD components that are soldered on the module.

The driver for the synchronous rectifier has a similar topology to that of the high-side driver, except the supply voltage has to be level shifted to a smaller value instead, so that the forward conduction state of the Schottky gate of Q2 is avoided (Figure 7.47). To down shift the supply voltage, two D-mode pHEMTs Q22 and Q23 are used. The gate of Q22 is grounded, so that the source potential is close to the pinch-off voltage of the D-pHEMT. The gate of Q23 is now connected to a voltage equal to the pinch-off voltage of the D pHEMT, and therefore, the final inverting buffer stage sees a virtual rail voltage of twice the pinch-off voltage of the D-pHEMT. When Q2 is being charged, the E-pHEMT Q14 will shift the potential down by another V_{SAT}, thus bringing the bias voltage of Q2 such that the Schottky diode forward conduction does not drastically increase the gate driver power consumption.

7.1.7.2.3 Controller Design

While the optimal technology for implementing the control and logic circuits is CMOS, to estimate the loop bandwidth and transient performance of the DC–DC output stage, a hysteric control loop was designed as shown in Figure 7.45. This control strategy is selected to ensure fast response and a large bandwidth.

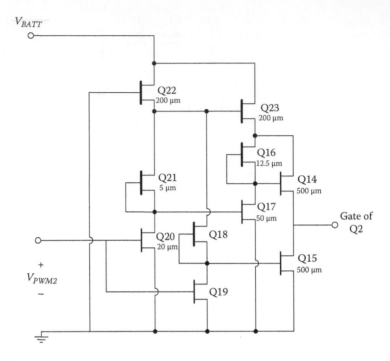

FIGURE 7.47 Synchronous rectifier gate drive circuit.

A separate CMOS emulator pHEMT die was used, which incorporated a hysteric comparator and a dead-time signal generator. The schematic of the comparator is shown in Figure 7.48. The comparator compares a reference voltage with a ramp signal generated by the R_F and C_F at the switching node V_S. The dead-time circuit is a NOR operation between V_{PWM} and a delayed version of V_{PWM} from the switching node V_S so that Q1 and Q2 are not turned *on* at the same time (Figure 7.49).

No additional phase compensation is included in the design. The self-starting converter utilizes the voltage generated at the switching node, which is fed back into a voltage comparator that generates the control pulse for the gate driver by comparing the fed back voltage with a reference voltage. The switching frequency of the converter is dependent upon the time constant of the feedback network, which is designed so that the converter switches at 100 MHz at full power. The control pulse is then fed to a dead-time circuit that generates nonoverlapping signals for the high- and low-side switches. The power losses in this chip are much higher than what would be for a CMOS implementation of control and logic circuits and therefore are not included in power efficiency measurements presented.

7.1.7.2.4 Measurements

The design prototype was flip-chip assembled on a 5 mm × 5 mm laminate on which the GaAs die and passives were integrated. A chip micrograph of the output stage die, with integrated drivers, is shown in Figure 7.50. A 0402 size 15 nH wire-wound surface mount inductor from Coilcraft [70] was used as the output

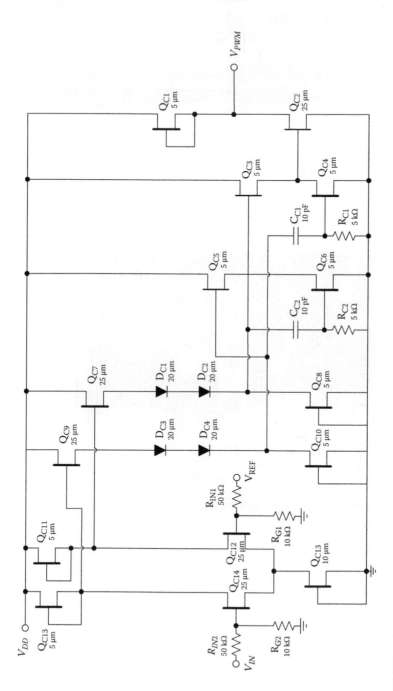

FIGURE 7.48 GaAs pHEMT comparator.

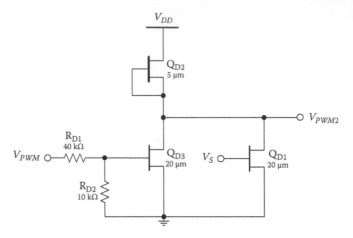

FIGURE 7.49 Dead-time generator block.

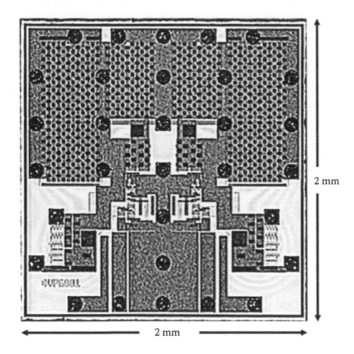

FIGURE 7.50 Die photo of the GaAs pHEMT DC–DC converter chip.

filter at the filtering capacitor. The low-voltage efficiency is improved by synchronous rectification as shown in Figure 7.51. The open-loop power efficiency of the circuit for various voltage conversion ratios for an input voltage of 4.5 V is shown in Figure 7.52. When switched at 100 MHz, the peak measured efficiency of the converter at a steady-state output voltage of 3.375 V is 88% for 34 dBm power output. The converter can operate up to 37 dBm output power at the peak voltage.

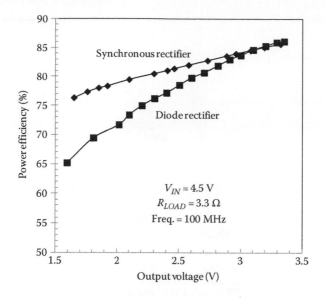

FIGURE 7.51 Effect of synchronous rectifier on power conversion efficiency.

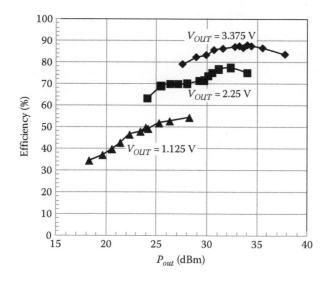

FIGURE 7.52 Measured power efficiency of the converter as a function of output power at 75%, 50%, and 25% voltage conversion ratios.

At lower output voltages, the efficiency degrades. The major reason for this degradation at low voltages in our design is the static gate driver losses in the GaAs driver, which do not scale as a function of output power. Simulated power loss components in the circuit for three different voltage conversion ratios are shown in Figure 7.53.

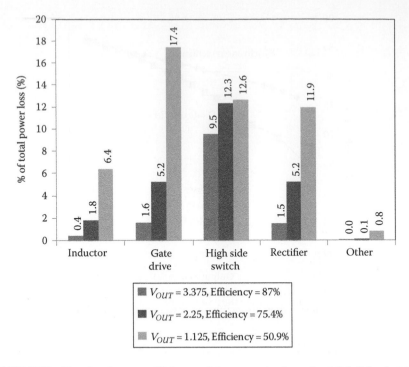

FIGURE 7.53 Simulated power efficiency of the converter for a fixed 3.3 Ω load at three different output voltages.

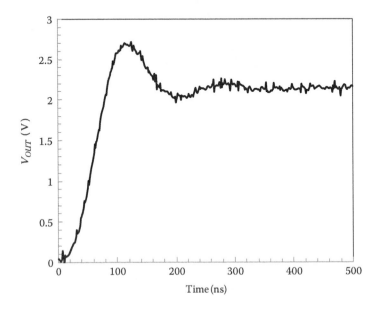

FIGURE 7.54 Measured step response of the converter for a 100 kHz input pulse.

The dynamic performance of the converter for a reference voltage step is shown in Figure 7.54. The voltage output can rise to steady state in 160 ns for the full power case. Figure 7.55 shows the amplitude response of the converter when driven by a single tone signal. The amplitude response remains flat till about 3 MHz and then peaks due to the underdamped double pole of the output LC network. The 3 dB bandwidth of the converter is 14.5 MHz. The tracking response of the converter to a 1 MHz sinusoid is shown in Figure 7.56.

The power efficiency of the converter is compared to other high-frequency converters in published literature in Figure 7.57. The efficiency of the pHEMT converter is among the highest reported for the frequency and power range.

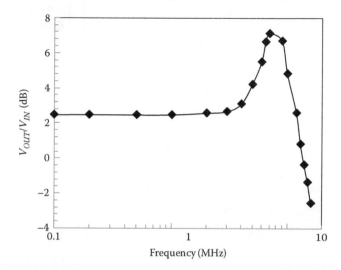

FIGURE 7.55   Closed-loop amplitude response of the converter.

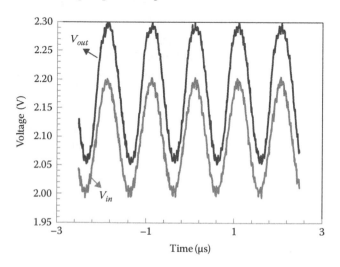

FIGURE 7.56   Tracking response of the converter to a 1 MHz sinusoid.

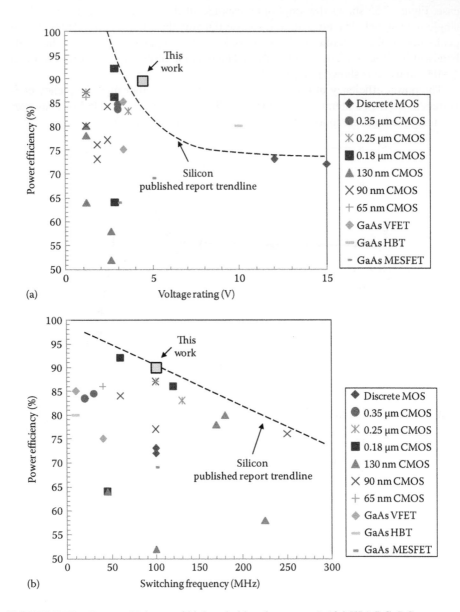

FIGURE 7.57 Power efficiency of high switching frequency (≥10 MHz) DC–DC converters from published sources plotted against (a) input voltage rating and (b) against switching frequency. Device technology is shown in legend.

7.2 SUMMARY

The use of GaAs pHEMTs as a platform for efficient high-frequency switching regulators is demonstrated. It is shown that the E-Mode pHEMT devices can demonstrate superior switching performance due to its inherent material properties. For portable applications, DC–DC converters consisting of GaAs switching devices, along with passives and control elements that are assembled as a flip-chip module can enable form factor reduction and improvement in battery life. Design approaches are presented and a suitable gate drive topology is demonstrated for single-chip integration of the output stage of a GaAs-based converter. The prototype hysteric-controlled single-phase synchronous buck regulator achieves a peak efficiency of 88% at 100 MHz switching frequency and can provide a 37 dBm peak power output. The efficiency of the presented converter is among the highest report for switching converters in this power and frequency range. The converter has a 3 dB loop bandwidth of 14.5 MHz and is suited to envelope-tracking applications.

GaAs pHEMTs are shown to be a promising alternative to silicon NMOS for power switching applications. As the trends of increasing switching frequencies and miniaturization continues, the GaAs alternative looks promising for power IC applications that require high efficiency while maintaining a small form factor, especially in portable electronics. The advantage comes from the inherent material properties of GaAs pHEMTs like the superior electron mobility and bandgap, which translates to a low *on* resistance and gate charge for switches. Due to the superior figure of merit for the GaAs pHEMTs, they have much smaller power dissipation even in hard switching at high frequencies compared to silicon. Furthermore, in GaAs the improvement in switching performance can be obtained without resorting to sub-100 nm feature size scaling, which makes it a low-cost alternative to GaN, and competitive to nanometer CMOS-based power switches. This allows the use of simple system architectures and low passive counts to meet efficiency goals at switching frequencies beyond 100 MHz. This is a 10-fold improvement in frequency and bandwidth when compared to the state of the art. This will also reduce the inductor size required in power electronic circuits to the nano-Henry range, which opens up possibilities of integrating the whole power supply on a single chip. A large bandwidth allows for special techniques to be employed to reduce load power consumption. For example, the efficiency of a linear RF PA in a 3G or 4G wireless transceiver can be improved by using a technique called envelope modulation, which requires that the power supply tracks an RF envelope with a bandwidth of 10–20 MHz using a DC–DC converter that switches above 100 MHz. GaAs pHEMTs could potentially be utilized to fabricate power supplies on a chip scale package or as a system on-chip implementation, and potentially enable a form factor reduction of up to 10–15 times, while improving the bandwidth and maintaining a high-energy efficiency. The improvement in bandwidth could potentially enable up to 2× improvement in battery life for wireless transceivers when compared to using a fixed or a linear regulated power supply.

A normally *off* GaAs pHEMT device fabricated using a commercially available process (TriQuint TQPED) shows excellent characteristics for power switching, although the device itself was optimized for RF applications. The baseline device has a rating of 11 V.

Using a three metal layer layout and flip chip packaging, the power device properties were seen to scale fairly well with the device area. A switching figure of merit of 8.5 mΩ nC was measured for a 2 A device in the baseline process. The figure of merit represents an order of magnitude improvement over commercially available silicon NMOS power devices and a twofold improvement over submicron CMOS-based NMOS devices in the same voltage range. GaAs pHEMTs also show excellent ruggedness properties since the voltage limit of operation in power switching is limited by the *off* state leakage current rather than avalanche initiated second breakdown.

Using the baseline process, a platform for high-frequency compact power supplies for portable electronics was developed and shows excellent performance. The platform integrates an output stage power IC consisting of pHEMT switches and gate drives, along with passives and control circuits in a multichip module. Prototype power converters show acceptable performance at specifications for typical battery-powered applications, at switching frequencies of 100 MHz and above. A prototype converter achieved 88% conversion efficiency at 100 MHz, which is among the highest report for switching converters in the power and frequency range. Moreover, the high bandwidth achieved makes such a design attractive to envelope tracking applications, leading to a marked improvement in system efficiency in handset applications.

On a broader scale, GaAs pHEMTs could also be utilized to fabricate DC–DC converters for microprocessors, base band amplifiers, and in other applications, some not conceived where wide bandwidth power supplies could offer huge gains. In a scenario where innovations in silicon-based low-voltage transistors have saturated, this approach gives a new way of breaking the paradigm and making large leaps in performance. An optimized process can achieve breakdown voltages of up to 47 V in pHEMTs without making major changes in the epi structure and process flow. The fabricated figure of merit of the devices shows an order of magnitude improvement over comparable silicon NMOSFETs and is of the same range as commercially available GaN FETs. This thesis also shows that for under 50 V, due to extrinsic resistances, GaN might be an inferior option to GaAs for power devices.

REFERENCES

1. RW Ericksson and D Maksimovic, *Fundamentals of Power Electronics*, 2nd ed. Norwell, MA: Kluwer Academic Publishers, 2004.
2. A Lidow and G Sheridan, Defining the future for microprocessor power delivery, in *Applied Power Electronics Conference and Exposition*, February 9–13, 2003, pp. 3–9.
3. Freescale Semiconductor. (September 2004). Freescale Corporation website [Online]. http://www.freescale.com/files/32bit/doc/app_note/AN2747.pdf. Accessed on August 1, 2012.
4. A Carroll and G Heiser, An analysis of power consumption in a smartphone, in *USENIX Annual Technical Conference*, Boston, MA, June 23–25, 2010, pp. 21–21.
5. Skyworks Inc. (2008, January). Skyworks website [Online]. http://www.skyworksinc.com/uploads/documents/200643a.pdf. Accessed on August 1, 2012.
6. TriQuint Semiconductor Inc. (March 2010). TriQuint Semiconductor website [Online]. http://www.triquint.com/products/p/TQM7M5012H. Accessed on August 1, 2012.
7. Y Li and D Maksimovic, High efficiency wide bandwidth power supplies for GSM and EDGE RF power amplifiers, in *IEEE International Symposium on Circuits Systems*, Kyushu, Japan, May 23–26, 2005, pp. 1314–1317.

8. V Pinon, F Hasbani, A Giry, D Pache, and C Garnier, A single-chip WCDMA envelope reconstruction LDMOS PA with 130 MHz switched-mode power supply, in *Proceedings of the International Solid State Circuits Conference*, 2008, pp. 564–566.

9. L Marco, A Poveda, E Alarcon, and D Maksimovic, Bandwidth limits in PWM switching amplifiers, in *IEEE International Symposium on Circuits Systems*, Island of Kos, Greece, May 21–24, 2006, pp. 5323–5326.

10. K Onizuka, K Inagaki, H Kawaguchi, M Takamiya, and T Sakurai, Stacked chip implementation of on-chip buck converter for distributed power supply system in SiPs, *IEEE J. Solid State Circuits*, 42(11), 2404–2410, 2007.

11. G Schrom et al., A 100 MHz eight-phase buck converter delivering 12 A in 25 mm² using air-core inductors, in *Proceedings of the Applied Power Electronics Conference*, 2007, pp. 727–730.

12. P Hazucha et al., A 233 MHz 80%–87% efficient four-phase DC-DC converter utilizing air-core inductors on package, *IEEE J. Solid State Circuits*, 40(5), 838–845, 2005.

13. G Schrom et al., A 480 MHz multi-phase interleaved buck DC-DC converter with hysteretic control, in *Proceedings of the Power Electronics Specialists Conference*, 2004, pp. 4702–4707.

14. HJ Bergveld et al., A 65 nm CMOS 100 MHz 87% efficient DC-DC down converter based on dual-die system-in-package integration, in *Proceedings of RFIC*, 2008, pp. 3698–3705.

15. M Wens and M Steyaert, A fully-integrated 0.18 um CMOS DC-DC step-down converter using a bondwire spiral inductor, in *IEEE Custom Integrated Circuits Conference*, San Jose, CA, September 21–24, 2008, pp. 17–20.

16. S Abedinpour, B Bakkaloglu, and S Kiaei, A multi-stage interleaved synchronous buck converter with integrated output filter in 0.18 um SiGe process, *IEEE Trans. Power Electron.*, 22(6), 2164–2175, 2007.

17. J Wibben and R Harjani, A high efficiency DC-DC converter using 2 nH integrated inductors, *IEEE J. Solid State Circuits*, 43(4), 844–854, 2008.

18. M Alimadadi et al., A 660 MHz ZVS DC-DC converter using gate-driver charge-recycling in 0.18 μm CMOS with an integrated output filter, in *Power Electronics Specialists Conference*, 2008, pp. 140–146.

19. M Wens and M Steyaert, A fully-integrated 130 nm CMOS DC-DC step-down converter regulated by a constant ON/OFF time control system, in *Proceedings of the European Solid-State Circuits Conference*, 2008, pp. 62–65.

20. V Pala, H Peng, P Wright, MM Hella, and TP Chow, Integrated high-frequency power converters based on GaAs pHEMT: Technology characterization and design examples, *IEEE Trans. Power Electron.*, 27(5), 2644–2656, 2012.

21. M Alimadadi, S Sheikhaei, G Lemieux, S Mirabbasi, and P Palmer, A 3 GHz switching DC-DC converter using clock-tree charge recycling in 90 nm CMOS with integrated output filter, in *Proceedings of the International Solid State Circuits Conference*, 2007, pp. 532–540.

22. J Hu et al., High frequency resonant SEPIC converter with wide input and output voltage ranges, in *Proceedings of the Power Electronics Specialists Conference*, 2008, pp. 1397–1406.

23. JM Rivas, RS Wahby, JS Shafran, and DJ Perreault, New architectures for radio-frequency DC-DC power conversion, *IEEE Trans. Power Electron.*, 21(2), 380–393, 2006.

24. J Rabaey, AP Chandrakasan, and B Nikolic, *Digital Integrated Circuits*, 2nd ed. Ann Arbor, MI: Pearson Education, 2003.

25. M Shur, *GaAs Devices and Circuits*. New York: Springer, 1987.

26. ZJ Shen, DN Okada, F Lin, S Anderson, and X Cheng, Lateral power MOSFET for megahertz-frequency, high-density DC/DC converters, *IEEE Trans. Power Electron.*, 21(1), 11–17, 2006.

27. J Baliga, *Power Semiconductor Devices*. Boston, MA: PWS Publishing Company, 1996.
28. T Efland, C Tsai, and S Pendharkar, Lateral thinking about power devices, in *International Electron Devices Meeting*, San Francisco, CA, December 6–9, 1998, pp. 679–682.
29. JA Appels and HMJ Vaes, High voltage thin layer devices (RESURF devices), in *International Electron Devices Meeting*, Washington, DC, December 3–5, 1979, pp. 238–241.
30. B Murari, F Berlott, and GA Vignola, *Smart Power ICs: Technologies and Applications*. New York: Springer, 2002.
31. N Yasuhara et al., Low gate charge 30 V N-Channel LDMOS for DC-DC converters, in *International Symposium on Power Semiconductor Devices and ICs*, Cambridge, U.K., April 14–17, 2003, pp. 47–51.
32. C Conteiro, A Andreini, and P Galbiati, Roadmap differentiation and emerging trends in BCD technology, in *European Solid-State Device Research Conference*, Florence, Italy, September 24–26, 2002, pp. 275–282.
33. ZJ Shen, D Okada, and F Lin, Lateral power MOSFET for megahertz-frequency high-density DC/DC converters, *IEEE Trans. Power Electron.*, 21(1), 11–17, 2006.
34. KR Varadarajan, A Sinkar, and TP Chow, Novel integrable 80 V silicon lateral trench power MOSFETs for high frequency DC-DC converters, in *IEEE Power Electronics Specialists Conference*, June 17–21, 2007, pp. 1013–1017.
35. WT Ng and A Yoo, Advanced lateral power MOSFETs for power integrated circuits, in *IEEE International Conference on Solid-State and Integrated Circuit Technology*, Shanghai, China, November 1–4, 2010, pp. 859–862.
36. G Schrom et al., Feasibility of monolithic and 3D-stacked DC-DC converters for microprocessors in 90 nm technology generation, in *International Symposium on Low Power Electronics and Design*, 2004, pp. 245–248.
37. BJ Baliga, MS Adler, and DW Oliver, Optimum semiconductors for power field effect transistors, *IEEE Electron Dev. Lett.*, 2, 162–164, 1982.
38. AJ Atkinson, Power devices in gallium arsenide, *IEE Proc. Int. Solid-State Electron. Dev.*, 132, 264–271, 1985.
39. PM Campbell, RS Ehle, PV Gray, and BJ Baliga, 150 V vertical channel GaAs FET, in *International Electron Devices Meeting*, San Francisco, CA, December 13–15, 1982, pp. 258–260.
40. PM Campbell, W Garwacki, AP Sears, P Menditto, and BJ Baliga, Trapezoidal-groove Schottky-gate vertical-channel GaAs FET (GaAs static induction transistor), *IEEE Electron Dev. Lett.*, 6, 304–306, 1985.
41. LW Yin et al., Improved breakdown voltage in GaAs MESFETs by utilizing surface layers of GaAs grown at a low temperature by MBE, *IEEE Electron Dev. Lett.*, 11, 561–563, 1990.
42. CL Chen et al., High breakdown voltage MESFET with a low-temperature grown GaAs passivation layer and overlapping gate structure, *IEEE Electron Dev. Lett.*, 13, 335–337, 1992.
43. K Asano et al., Novel high power AlGaAs/GaAs HFET with a field-modulating plate operated at 35 V drain voltage, in *International Electron Devices Meeting*, December 6–9, 1998, pp. 59–62.
44. A Ketterson et al., Characterization of InGaAs/AlGaAs pseudomorphic modulation-doped field effect transistors, *IEEE Trans. Electron Dev.*, 33, 564–571, 1986.
45. T Henderson et al., Microwave performance of a quarter-micrometer gate low noise pseudomorphic InGaAs/AlGaAs modulation-doped field effect transistor, *IEEE Electron Dev. Lett.*, 7, 641–649, 1986.

46. M Miller, Design, performance and application of high voltage GaAs FETs, in *Compound Semiconductor Integrated Circuits Symposium*, Palm Springs, CA, October 30–November 2, 2005, pp. 236–239.

47. K Ishikura et al., A 28 V over 300 W GaAs heterojunction FET with dual field-modulating plates for W-CDMA base station, in *IEEE International Microwave Symposium*, Long Beach, CA, June 12–17, 2005, pp. 823–826.

48. CK Peng, MI Aksun, AA Ketterson, H Morkoc, and KR Gleason, Microwave performance of InAlAs/GaAs/InP MODFETs, *IEEE Electron Dev. Lett.*, 8, 24–26, 1987.

49. AS Brown, UK Mishra, JA Henige, and M Delaney, The impact of epitaxial layer design and quality on GaInAs/AlInAs high-electron mobility transistor performance, *J. Vacuum Sci. Technol.*, B6(2), 678–681.

50. MA Khan, JN Kuznia, AR Bhattarai, and DT Olson, Metal semiconductor field effect transistor based on single crystal GaN, *Appl. Phys. Lett.*, 62(15), 1786–1787, 1993.

51. MA Khan et al., Microwave performance of a 0.25 µm gate AlGaN/GaN heterostructure field effect transistor, *Appl. Phys. Lett.*, 65(9), 2164–2175, 1994.

52. H Morkoc, R Cingolani, and B Gil, Polarization effects in nitride semiconductor device structures and performance of modulation doped field effect transistors, *Solid State Electron.*, 43(10), 1909–1927, 1999.

53. EPC Corporation. (June 2012). EPC website [Online]. http://bit.ly/NTUaqn. Accessed on August 1, 2012.

54. International Rectifier. (February 2012). IR website [Online]. http://bit.ly/MKjY9b. Accessed on August 1, 2012.

55. Y Tang et al., High-performance monolithically integrated E/D mode InAlN/AlN/GaN HEMTs for mixed signal applications, in *IEEE International Electron Devices Meeting*, San Francisco, CA, December 6–8, 2010, pp. 30.4.1–30.4.4.

56. TriQuint Semiconductor Inc. (March 2010). TriQuint Semiconductor website [Online]. http://bit.ly/Qbx8l9. Accessed on August 1, 2012.

57. B Davari, Shallow junctions, silicide requirements and process technologies for sub 0.5 µm CMOS, *Microelectron. Eng.*, 19, 649–656, 1992.

58. SD Kim, CM Park, and JCS Woo, Advanced model and analysis of series resistance for CMOS scaling into nanometer regime—Part II: Quantitative analysis, *IEEE J. Electron Dev.*, 49(3), 467–472, 2002.

59. A Ketterson et al., Characterization of extremely low contact resistances on modulation doped FETs, *IEEE Trans. Electron Dev.*, 32(11), 2257–2261, 1985.

60. M Feuer, Two-layer model for source resistance in selectively doped heterojunction transistors, *IEEE Trans. Electron Dev.*, 32(1), 7–11, 1985.

61. N Waldron, DH Kim, and JA del Alamo, A self-aligned InGaAs HEMT architecture for logic applications, *IEEE Trans. Electron Dev.*, 57(1), 297–300, 2010.

62. SJ Cai et al., High performance AlGaN/GaN HEMT with improved ohmic contacts, *Electron. Lett.*, 34(24), 2354–2356, 1998.

63. S Heikman, S Keller, SP DenBaars, and UK Mishra, Mass transport regrowth of GaN for ohmic contacts to AlGaN/GaN, *Appl. Phys. Lett.*, 78(19), 2876–2878, 2001.

64. Nidhi et al., N-polar GaN-based highly scaled self-aligned MIS-HEMTs with state-of-the-art ft.LG product of 16.8 GHz-µm for mixed signal applications, in *CS MANTECH Conference*, Portland, OR, 2010, pp. 341–344.

65. N Fujishima et al., A 700 V lateral power MOSFET with narrow gap double metal field plates realizing low on-resistance and long term stability of performance, in *Proceedings of the International Symposium on Power Semiconductor Devices and ICs*, Osaka, Japan, 2001, p. 399. Accessed on August 1, 2012.

66. S Merchant et al., Realization of high breakdown voltage (>700 V) in thin SOI devices, in *Proceedings of the International Symposium on Power Semiconductor Devices and Circuits*, 1991, pp. 31–35.

67. F Udrea, A Popescu, and WL Milne, The 3D-RESURF double-gate MOSFET: A revolutionary power device concept, *Electron Lett.*, 34(8), 808, 1998.

68. W Saito, I Omura, T Ogura, and H Ohashi, Theoretical limit estimation of lateral wide band-gap semiconductor power-switching device, *Solid State Electron.*, 48, 1555–1562, 2004.

69. WA Wohlmuth et al., A 0.5 μm InGaP etch stop power pHEMT process utilizing multi-level high density interconnects, in *GaAs Mantech*, 2004, p. 7.1.

70. Infineon Inc. (June 2012) Power management selection guide [Online]. http://bit.ly/Q7CM96.

8 Silicon and III–V Technologies for High Switching Speed Monolithic DC–DC Power Converter ICs

Han Peng, Zemin Liu, and Mona Mostafa Hella

CONTENTS

8.1 INTRODUCTION

The recent advances in the development of multifunction system-on-chip (SOC) solutions for portable devices have increased the need for improved battery capacity and lifetime while maintaining small footprint and size. A typical mobile terminal includes data converters, memory chips, RF front-end circuits for the transceiver, frequency synthesizers, digital baseband cores, and a microcontroller. Each circuit block is optimized for maximum performance at a different supply voltage while the typical lithium ion battery voltage in a cell phone is 4.2–4.5 V and the ICs' internal supply voltage can vary from 3.6 to 0.8 V. A DC–DC converter is required to interface between the battery and the different functional blocks on chip. The employed interconnect technology for the different blocks can be anything from bond wires to flip-chip, future solutions of 3D integration, and SOC realization. Thus, to maximize the performance of the supply regulator, it has to be placed close to the functional blocks to limit the effect of interconnect parasitics, particularly when providing high current and voltage levels. Ideally, the supply modulator would be integrated in the same technology as the different building blocks. However, this poses another challenge for passives integration, regarding their size on-chip and the encountered losses. As shown in Figure 8.1, the trend for monolithic DC–DC converters is moving towards higher-frequency ranges (>10 MHz) with power limits of 5 W for modern analog and digital functional blocks.

The basic types of power supply modulators, shown in Figure 8.2, can be categorized as (1) low-dropout regulators, (2) switched-capacitor converters, (3) switch-mode DC–DC converters, and (4) hybrid-mode supply modulators (combined linear and switch-mode). Low-dropout regulators (LDOs) are primarily used for their ease of integration, wide control loop bandwidth, and low-area overhead. The pass elements in LDOs (Figure 8.2a) are usually common collector or common drain stages. The dropout voltage is regulated by adjusting the load current and the on-resistance of the pass devices. The efficiency of low-dropout regulators is limited by the ratio of the input to the output signal. As the voltage conversion ratio decreases, the efficiency of the LDO is drastically reduced. Another drawback of LDOs is the limitation on the output voltage levels, which can only be lower than the input.

Switched-capacitor DC–DC converters (Figure 8.2b) are easily integrated and can provide reasonable efficiency for large voltage differences but are typically limited to lower current loads. The peak efficiency of a capacitive converter is limited by its output resistance, given by $1/2\pi fC_L$, where f is the switching frequency and C_L is the load capacitance [2]. The efficiency can ideally reach 100% as C_L goes to zero, which indicates zero load current. Switched-capacitor DC–DC converters are a good

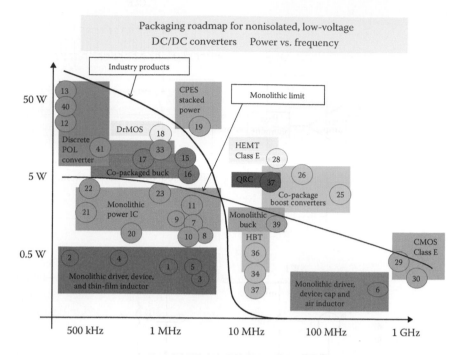

FIGURE 8.1 Roadmap of nonisolated low-voltage DC–DC converters: output power versus switching frequency. (From van Wyk, D., Power electronics integration technology (PEIT) thrust overview, CPES invited presentation, 2008.)

candidate for high-frequency, low-current voltage regulators. However, in order to realize different levels of output voltage, the control circuitry of the converter can be complicated, which can lead to increased power consumption.

Switch-mode DC–DC converters (Figure 8.2c) can be step down, step up, or step up and down depending on the different configurations for switches and inductors [2]. The efficiency of inductive load DC–DC converter mainly depends on the parasitic components of the inductors and the switch losses. Switch-mode DC–DC converters typically have a much higher efficiency over the output power range compared to LDOs and switch-capacitor DC–DC converters [3–5]. The main drawback of switch-mode converters is the limitation on the closed-loop bandwidth by either the switching frequency, closed-loop compensator, or the output filter depending on the type of the employed control technique. Hybrid power supply modulators as shown in Figure 8.2d are composed of switch-mode DC–DC converters and linear regulators. They have been proposed to meet the stringent requirements of wide-bandwidth modulation in high-performance polar loop transmitters [6,7]. In such systems, the high-frequency component in the baseband envelope signal will be fed into linear low-dropout circuit, and the low-frequency component is fed to a switch inductor DC–DC converter to improve the efficiency. In previously reported architectures, the switch-mode DC–DC converter is usually operating in the MHz range and thus does not suffer from very-high-frequency switching losses.

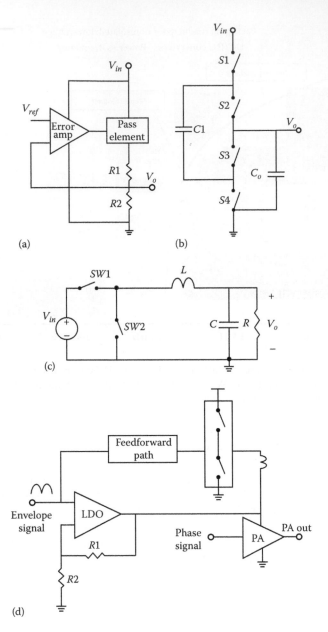

FIGURE 8.2 Power supply modules: (a) low-dropout circuit, (b) switched-capacitor converter, (c) switched-mode buck DC–DC converter, and (d) hybrid mode supply regulator.

While hybrid supply modulators have shown promising results compared to LDOs, their efficiency is still below that of the switching-mode converters.

In this chapter, switch-mode DC–DC converters are considered for high-efficiency, high-switching frequency supply modulators over a wide range of output voltages. Different implementation processes, passive device integrations, output inductor networks, and power converter topologies are investigated. The chapter is arranged as follows: Section 8.2 discusses the design of high-frequency interleaved DC–DC converter in GaAs/AlGaAs technology, including the core topology analysis, circuit implementation, measurement results, and power loss analysis. Section 8.3 discusses different light-load efficiency improvement techniques and focuses on resonant gate driver topology with a new output inductor network for phase shedding/segmentation applications.

8.1.1 HIGH SWITCHING SPEED SUPPLY MODULATORS

Traditional switch-mode DC–DC converters usually require inductors up to 100 µH and capacitors in the hundreds of micro-Farads [8]. With the continuous trend in integrating various components on the same die or the same package, increasing the switching frequency to the hundreds of MHz range would reduce the inductors' and capacitors' sizes to the nano Henry (nH) and pico-Farad (pF) ranges, where they can be implemented on-chip. Such integration level simplifies packaging and reduces the power loss from the interconnection of different dies and external components. In addition, the quality factor Q of on-chip inductors increases with frequency up to few GHz. Furthermore, in certain control techniques as voltage mode control and average current control technique, the closed-loop bandwidth is related to the switching frequency. To form a stable feedback loop with an adequate compensator network, the 3 dB bandwidth is typically selected as 1/10th to 1/15th of the switching frequency. Hence, the higher the switching frequency, the larger the control loop bandwidth. On the other hand, there are several frequency-dependent loss components in DC–DC converters. The switching loss (P_{SW}) and gate driver loss (P_{gate}) of the switching transistor will increase linearly with the frequency as given in Equations 8.1 and 8.2:

$$P_{SW} = \frac{1}{2} V_{DD} I_{OUT} (t_{ON} + t_{OFF}) f_{SW} \qquad (8.1)$$

where
t_{ON} and t_{OFF} are the turn-on and turn-off periods of the power devices
f_{SW} is the switching frequency
V_{DD} and I_{OUT} are the supply voltage and output current

$$P_{gate} = \left(C_{gd} + C_{gs} \right) V_{dd}^2 f_{SW} \qquad (8.2)$$

Here, C_{gd} and C_{gs} are the gate-drain and gate-source capacitances of the switching transistor.

As can be seen from the aforementioned discussion, increasing the switching frequency reduces the inductive losses, while increasing the transistor switching losses and the gate driver losses. Thus, it is clear that there is an optimum switching frequency at which the total loss is at a minimum. In addition, to achieve high DC–DC converter efficiency, optimization of the power stages, passive networks, and gate driver stages have to be done simultaneously.

8.1.2 Implementation Technologies for Integrated Supply Modulators

While improving the converter efficiency is an important design target for supply modulators, there are also other design criteria such as the accuracy of the output voltage, the speed of the transient response, the closed-loop control bandwidth, line and load regulations, power density, and system area/size. The transient response of supply regulators determines the speed by which the output can respond to the input or reference changes. The larger the closed-loop bandwidth, the less time it takes for the system to stabilize. The speed factor is extremely important for applications such as wireless systems. The accuracy of output, line, and load regulations is related to the noise rejection from the substrate, supply voltage changes, and load variation and are critical for applications that require a fixed output voltage such as microprocessors. The performance for supply modulators is generally dependent on the employed modulator architecture, as discussed in Section 8.1, and the implementation technology for both the active and passive devices involved.

8.1.2.1 Active Devices

The loss in the active devices is one of the main factors affecting the DC–DC converter efficiency. Conduction loss, switching loss, and gate driver loss are the three main loss mechanisms in the power transistors. A low ON resistance R_{ON} is necessary to minimize conduction losses. Reducing the gate charge Q_G is required to minimize power losses in the gate driver circuits, as well as to reduce the switching time, thereby reducing ($V_{DS} \times I_{DS}$) loss during turn-on and turn-off transitions. The gate capacitors C_{gd} and C_{gs} determine the required gate charge to turn on the transistor according to Equation 8.3. In addition, the power transistor has to have a high breakdown voltage since the switching noise caused by the parasitic drain inductance will generate voltage spikes higher than the supply voltage at high frequencies:

$$t_{ON} = t_{OFF} = \frac{\int C_{gd}dV_{gd} + \int C_{gs}dV_{gs}}{I_G} = \frac{Q_G}{I_G} \tag{8.3}$$

where
  V_{gd} and V_{gs} are the gate-drain and gate-source voltages
  I_G is the gate current

While various technologies have been used for DC–DC converters, this work is concerned with converters operating in the 5–15 V applications and in achieving high levels of integration for the various functional blocks. Two technology platforms will be discussed: GaAs pHEMT and Si-CMOS. The first is an example of III–V technologies that dominate the wireless transmitter and power amplifier markets. The second dominates the information technology and portable electronics market. While CMOS technology is the most popular technology for analog, digital, and low-complexity communication systems, GaAs pHEMTs dominate the landscape for radio frequency transmitters and high-performance transceivers.

TABLE 8.1

Device Comparison

Type	Silicon NMOS	GaAs pHEMT
Channel electron mobility	300–500 cm²/V·s	4000–6000 cm²/V·s
Drift region mobility	700 cm²/V·s	3500 cm²/V·s
$E_{C,Lat} = BV/L_D$	15–20 V/μm	25–30 V/μm

To compare GaAs technology with mainstream CMOS technology, channel electron mobility, drift region mobility, and lateral critical electric field $E_{C,Lat}$ are provided in Table 8.1. P-HEMT has a notably higher channel electron mobility, thus providing higher-frequency operation. It has a lower gate capacitor due to the Schottky gate structure and, therefore, a lower $R_{ON} \times Q_G$ compared to silicon NMOS in the same voltage range, which makes it more suitable for high-frequency power switching applications. GaAs also has higher-energy bandgap, so it allows higher breakdown voltages and higher-temperature operation. Furthermore, GaAs has a semi-isolated substrate that translates to better quality factor passive components.

The structure and epilayers of the enhancement mode pseudomorphic HEMT device fabricated on GaAs substrate is described in detail in Chapter 7. As an example of the implementation of power devices in CMOS technology, a 0.18 μm CMOS technology with 5 V high-voltage devices is employed. The threshold voltage for the 5 V NMOS transistor is 0.725 V and −0.894 V for PMOS transistors. The breakdown voltage of N+/Hi P-well junction and P+/Hi N-well junction is 6 V. There are six metallization layers with distribution metal layer for micro surface mount bumps in this process. The major advantage for CMOS converters is the design flexibility. With a PMOS transistor as high-side power switch in DC–DC converter, no boot-strapped gate driver is required, which significantly reduces the design complexity by removing level shift circuits and bootstrapped capacitor. With the help of complementary devices, the gate driver buffer stage can be more efficient with less static current and leakage current compared to GaAs pHEMT inverter with active resistor load. CMOS technology is also a good candidate for control circuit design, so that the whole converter system can be integrated on the same die to further reduce die area and improve power density. Moreover, CMOS devices always have parasitic body diodes that can protect the power switches during dead time. On the other hand, because CMOS devices have more parasitic capacitors, this tends to have higher switching loss. In the remaining sections, we will provide examples of both GaAs and CMOS technologies.

8.1.2.2 Passive Device Technology

For high-frequency (>10 MHz) DC–DC converters, the integration of high-quality inductors either on-chip or in-package continues to be a challenge. Micro-Henry inductors that have a high permeability magnetic core and extremely high-quality factors at kilohertz frequencies are not suited for multi-megahertz DC–DC converters due to hysteresis and core losses. In monolithic or system in package implementations, air core spiral inductors and recently bond-wire inductors [9–11] have been used. The inductor power loss in a DC–DC converter can be given by

$$P_{IND} = I_{LDC}^2 R_{LDC} + \frac{1}{12}\Delta I_{p-p}^2 R_{LAC} \tag{8.4}$$

where

I_{LDC} is the average current

ΔI_{p-p} is the peak-to-peak ripple

R_{LDC} and R_{LAC} are the resistances at DC and at the fundamental harmonic of current

If the ripple inductor current is low, the zero frequency (DC) component of the inductor current contributes to the majority of power loss. The most commonly used figure of merit for power inductors in DC–DC converters is the ratio of the inductance to DC resistance. However, in high-frequency converter implementations that employ planar inductors, the total output inductance is constrained by the area consumed by the spiral, and therefore, it is desirable to use a small inductance at the cost of a high current ripple. As the harmonics of the inductor current increase, the AC losses also start to come into play, which means that the quality factor of the inductor starts to become important.

To achieve higher load current, reduce inductor size and current ripple, multiphase interleaved DC–DC converters can be adopted. The normalized inductor ripple reduction is plotted in Figure 8.3. Complete ripple cancellation takes place at duty cycle equal to 0.5 for two-phase, duty cycle of 0.33 and 0.66 for three-phase, and duty cycle equal to 0.25, 0.5, and 0.75 for four-phase operation. To improve the ripple cancellation points over different duty cycles, more phases are required, which greatly reduces the inductor power loss and improves the efficiency at the expense of complicated phase generation circuitry and expanded passive network.

On-chip spiral inductors are common in nanometer-scale CMOS-based DC–DC converters that operate at hundreds of MHz [2,13–19]. Up to few hundreds of MHz, the quality factor of typical on-chip inductors increases with frequency and is mainly determined by the series resistance of the inductor. For switching frequencies around ~100 MHz, planar inductors on CMOS processes typically use multiple thick metal spirals stacked on top of each other to increase the total cross-sectional area and reduce the DC resistance as in [2,15]. However, in CMOS processes, the thickness of interconnect metal layers is still limited (e.g., 10 µm thick copper layer), and therefore, thick cross-sectional area is difficult to achieve. In general, based on published data for DC–DC converters with on-chip inductors implemented on silicon substrate, it shows a significantly lower performance compared to off-chip solutions. Hence, in order to have inductors in the nH range with low dc resistance, high-quality factor, and small area for supply modulators at hundreds of MHz, multiple thick metal layers are preferred.

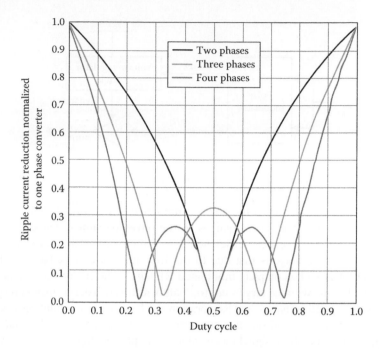

FIGURE 8.3   Normalized inductor ripple current for multiphase operation. (From Tuite, D., Reconciling power-factor correction standards leads to solutions. [Online], available: http://electronicdesign.com/article/engineering-essentials/reconciling power factor correction standards leads to solutions/5.aspx, accessed on September 9, 2008.)

8.2   TWO-PHASE INTERLEAVED DC–DC CONVERTER WITH NEGATIVE COUPLED INDUCTORS IN GaAs TECHNOLOGY

To illustrate the use of DC–DC converters in various applications, this section presents the design of a 150 MHz two-phase interleaved DC–DC converter with negative coupled inductors in GaAs p-HEMT 0.5 μm technology for power amplifiers (PA) in polar modulation systems. In wireless communication systems, the efficiency of the radio frequency power amplifiers (RF PAs) dominates the power consumption of the radio transceiver. In addition, to improve the spectral efficiency of communication standards, nonconstant envelope modulations with high peak to average signal variations are being adopted. This translates to strict requirements on the power amplifier linearity to avoid signal distortion. Such highly linear amplifiers are inefficient, particularly at "backed-off" power conditions, where the power amplifier typically operates for most of its operating time. There are basically three techniques to improve the PA efficiency based on the use of supply modulators [20]: power tracking (slow tracking), envelope tracking, and Envelope Elimination and Restoration (EER) or polar modulators.

In power tracking (slow tracking) system, the supply modulator will track the peak power in the baseband signal and supply the PA with a voltage slightly higher than the largest peak of the envelope. The input to the PA is the fully modulated

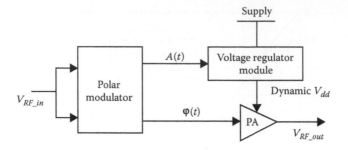

FIGURE 8.4 Block diagram of a polar modulation system.

RF signal, thus the linearity of the power amplifier is still required. The envelope-tracking technique restores any changes in the envelope signal. Thus, the whole system can recover the reduced efficiency due to modulation and power back off. However, the disadvantage is that due to the limitation in control bandwidth, the input of the PA is still a complete modulated RF signal and linear PAs are needed. The third technique with the largest control bandwidth and fastest transient response is called Envelope Elimination and Restoration (EER) or polar modulation as shown in Figure 8.4. The baseband signal is converted to an envelope signal $A(t)$ and phase signal $\varphi(t)$. Since the phase signal has a constant amplitude, high-efficiency switching power amplifier classes such as class E or class F can be utilized. The envelope-varying information is restored by modulating the supply voltage of the PA through the output of an envelope modulator. The whole system has highest efficiency, widest control bandwidth, and fastest transient response. Hence, polar modulation is an effective technique to alleviate the linearity-efficiency trade-off [21–23].

The selection of switching frequency for supply modulators for PA applications needs to consider the communication modulation schemes. For a given communication standard and its associated frequency band and channel separation, the switching frequency should be selected to avoid any switching spurs falling into the adjacent receiver band or switching noise lowering the PA adjacent channel power rejection in the transmitter. Although higher switching speed DC–DC converters can have major advantages in terms of both smaller passive devices and wider control bandwidth, the switching loss of gate driver stages and power switches will increase linearly with switching frequency. Thus, the switching speed of the converter has to be selected to provide a compromise between switching loss and closed-loop bandwidth.

8.2.1 Interleaved Topology with Negative Coupled Inductors

Multiphase interleaved DC–DC converters have become widely adopted to reduce current ripple and improve efficiency [2,13,14,24–28]. Compared to a single buck converter, the multiphase interleaved structure has more active components as well as more inductors that can increase the converter module size. Coupling the inductors has recently been proposed to improve the steady-state and transient responses of the converter by reducing the current ripple and the stabilization time [29]. While the analysis given in [29] has proceeded by developing an equivalent inductance under different coupling conditions for the steady-state and transient conditions separately, we will extend this analysis by quantifying the optimum coupling factor for ripple cancellation under different duty ratios, and use space state analysis to derive an expression for the converter's open-loop transfer function to study the bandwidth enhancement effect of coupled inductors [30–32].

8.2.1.1 Steady-State Analysis

Figure 8.5a shows the core of a two-stage interleaved structure with coupled inductors. Let us assume that the two branches have equal inductances L and the mutual inductance between the two phases is M, while $k = M/L$ is the coupling factor. The current and voltage waveforms at the input and output of the DC–DC converter for each phase, assuming ideal switching stages, are shown in Figure 8.5b. As can be seen from the figure, the circuit has four different states depending on the ON/OFF

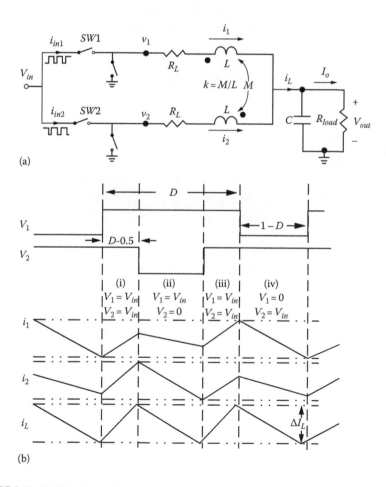

(a)

(b)

FIGURE 8.5 (a) Ideal two-phase interleaved topology with negative coupling between filter inductances. (b) Voltage and current waveforms at the terminals of the coupled inductors for $D \geq 0.5$.

condition of switches $SW1$ and $SW2$. To find the current ripple in each converter phase, let us derive the relationship between the currents i_1, i_2 and the voltages at the input and output of the DC–DC converter during each state.

State I: $0 \leq t < \left(D - \dfrac{1}{2} \right) T$,

$$L_1 \frac{di_1}{dt} + M \frac{di_2}{dt} = V_{in} - V_{out} = V_{in}\left(1 - D\right) \tag{8.5}$$

$$L_2 \frac{di_2}{dt} + M \frac{di_1}{dt} = V_{in} - V_{out} = V_{in}\left(1 - D\right) \tag{8.6}$$

The two phases are symmetrical, and thus the current ripple of each phase is identical. The current ripple in the first phase can be obtained as Δi_1 by substituting 8.6 into 8.5 such that

$$\Delta i_1 = \frac{V_{in}\left(1 - D\right)\left(1 - k\right)}{L\left(1 - k^2\right)}\left(D - \frac{1}{2} \right) T \tag{8.7}$$

State II: $\left(D - \dfrac{1}{2} \right) T \leq t < \dfrac{T}{2}$

$$L_1 \frac{di_1}{dt} + M \frac{di_2}{dt} = V_{in}\left(1 - D\right) \tag{8.8}$$

$$L_2 \frac{di_2}{dt} + M \frac{di_1}{dt} = -V_{out} = -DV_{in} \tag{8.9}$$

$$\Delta i_1 = \frac{V_{in}(1 - D)\left(1 + \dfrac{kD}{1 - D} \right)}{L(1 - k^2)}(1 - D)T \tag{8.10}$$

State III: $\dfrac{T}{2} \leq t \leq DT$

Since state III is identical to state I, we can rewrite Δi_1 as in Equation 8.7
Finally, state IV: $DT \leq t < T$

$$L_1 \frac{di_1}{dt} + M \frac{di_2}{dt} = -DV_{in} \tag{8.11}$$

$$L_2 \frac{di_2}{dt} + M \frac{di_1}{dt} = V_{in}(1-D) \tag{8.12}$$

$$\left|\Delta i_1\right| = \frac{V_{in}(1-D)\left(k+\dfrac{D}{1-D}\right)}{L(1-k^2)}(1-D)T \tag{8.13}$$

By inspecting Equations 8.7, 8.10, and 8.13, and as can be seen in Figure 8.5b, it is clear that the largest ripple occurs at state IV. Thus, Equation 8.13 can be used to represent the current ripple for $D \geq 0.5$. The same analysis can be repeated for $D < 0.5$ such that the current ripple can be derived as

$$\left|\Delta i_1\right| = \frac{V_{in}D\left(k+\dfrac{1-D}{D}\right)}{L\left(1-k^2\right)}DT \quad \text{for } D < 0.5 \tag{8.14}$$

To show the effect of coupling on the current ripple in each phase, we will define the term ripple reduction factor ξ as the ratio between the coupled to uncoupled current ripples, such that

$$\xi = \begin{cases} \dfrac{1+k\left(\dfrac{1}{1-D}-1\right)}{1-k^2} & \text{for } D < 0.5 \\[4ex] \dfrac{1+k\left(\dfrac{1}{D}-1\right)}{1-k^2} & \text{for } D \geq 0.5 \end{cases} \tag{8.15}$$

Figure 8.6a shows the ripple reduction factor ξ versus the duty cycle for different coupling factors. As can be seen, within the duty cycle range of 0.3–0.7, the current ripple is reduced in the coupled case compared to uncoupled case. It is also clear that there is an optimum coupling factor for maximum ripple cancellation at each duty cycle, which can be derived as

$$k_{opt} = \begin{cases} \dfrac{1-D}{D}\left(\sqrt{1-\left(\dfrac{D}{1-D}\right)^2}-1\right) & \text{for } D < 0.5 \\[4ex] \dfrac{D}{1-D}\left(\sqrt{1-\left(\dfrac{1-D}{D}\right)^2}-1\right) & \text{for } D \geq 0.5 \end{cases} \tag{8.16}$$

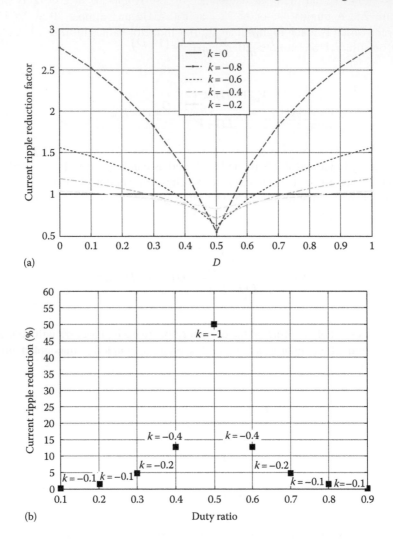

FIGURE 8.6 (a) Current ripple reduction as a function of duty cycle for different coupling factor, (b) ripple reduction percentage for two-phase interleaved converter as function of duty factor with optimum k listed.

Figure 8.6b shows the lowest achievable current ripple in each phase for different duty cycles, where the maximum reduction of 50% in current ripple occurs at $D = 0.5$ for $k = -1$. This improvement drops to 13% within 20% variation in duty cycle. The sensitivity of ripple reduction to the coupling factor is shown in Figure 8.7. For duty cycles other than 0.5, as the absolute value of the coupling factor increases beyond 0.4, the current ripple increases dramatically.

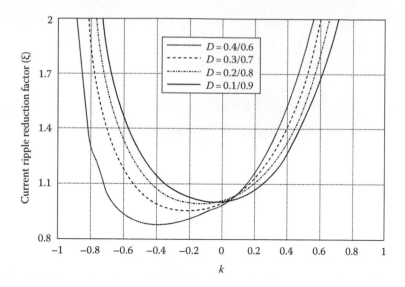

FIGURE 8.7 Sensitivity of current ripple reduction to the variation in coupling factor for different duty ratios.

8.2.1.2 System Stability and Transient Response

To study the effect of coupling on the transient response of the DC–DC converter, let us define $d(s)$ as the small signal function of the duty ratio D. The open-loop transfer function $V_{out}(s)/d(s)$ can be derived following the space state analysis described in [33]. The average space state equation is given in Equations 8.17 and 8.18, where i_{in1} and i_{in2} are the input currents to the two-phase converter and R_L is the DC resistance of the inductor, which is approximated as $8 \times L$ mΩ/nH. The transfer function of $V_{out}(s)/d(s)$

$$
\begin{bmatrix} \dfrac{di_1}{dt} \\[2mm] \dfrac{di_2}{dt} \\[2mm] \dfrac{dv_{out}}{dt} \end{bmatrix} = \begin{bmatrix} \dfrac{-LR_L}{L^2-M^2} & \dfrac{MR_L}{L^2-M^2} & \dfrac{-1}{L+M} \\[2mm] \dfrac{MR_L}{L^2-M^2} & \dfrac{-LR_L}{L^2-M^2} & \dfrac{-1}{L+M} \\[2mm] \dfrac{1}{C} & \dfrac{1}{C} & \dfrac{-1}{CR_{load}} \end{bmatrix} \times \begin{bmatrix} i_1(t) \\[2mm] i_2(t) \\[2mm] v_{out}(t) \end{bmatrix} + \begin{bmatrix} \dfrac{D}{L+M} \\[2mm] \dfrac{D}{L+M} \\[2mm] 0 \end{bmatrix} \times V_{in} + \begin{bmatrix} \dfrac{V_{in}}{L+M} \\[2mm] \dfrac{V_{in}}{L+M} \\[2mm] 0 \end{bmatrix} \times d(t)
$$

(8.17)

$$
\begin{bmatrix} i_{in1}(t) \\[2mm] i_{in2}(t) \end{bmatrix} = \begin{bmatrix} D & 0 \\ 0 & D \end{bmatrix} \times \begin{bmatrix} i_1(t) \\[2mm] i_2(t) \end{bmatrix} + \begin{bmatrix} \dfrac{I_O}{2} & 0 \\[2mm] 0 & \dfrac{I_O}{2} \end{bmatrix} \times d(t)
$$

(8.18)

$$\frac{V_{out}(s)}{d(s)} = \frac{\dfrac{2V_{in}}{C(L+M)}\left(s+\dfrac{R_L}{L-M}\right)}{s^3 + X1\cdot s^2 + X2\cdot s + X3}$$

$$X1 = \frac{1}{R_{load}C} + \frac{2LR_L}{L^2-M^2}$$

$$X2 = \frac{2LR_L}{R_{load}C(L^2-M^2)} + \frac{R_L^2}{L^2-M^2} + \frac{2}{C(L+M)}$$

$$X3 = \frac{R_L^2}{R_{load}C(L^2-M^2)} + \frac{2R_L}{C(L^2-M^2)}$$

(8.19)

is derived as in Equation 8.19 and plotted in Figure 8.8 for different coupling factors. The figure compares the bandwidth and phase margin between noncoupled ($k = 0$) positively coupled ($k = 0.4$) and negatively coupled ($k = -0.4$) two-phase interleaved DC–DC converters. The input voltage is selected as 4.5 V, $L = 8$ nH, $R_{load} = 3.3\ \Omega$, and the load capacitor $C = 20$ nF. As can be seen from the figure, for negative coupling factor of 0.4, the converter bandwidth increases by 29.2% compared to the noncoupled two-phase converter, and 52.7% compared to positively coupled converter with the same coupling factor. Figure 8.9 shows the improvement in bandwidth as a function of the coupling factor, where a strong indirect (negative) coupling translates to a wider bandwidth.

The DC–DC converter's transient behavior can also be explained using the equivalent inductor as follows: the transient response is measured by the time it takes the converter to stabilize when the input voltage or the duty cycle changes. The stabilization time is a function of the output filter network formed of the inductance and capacitance as well as their parasitic resistances. For faster transient response, the value of inductor should be small enough to allow a fast slew rate and prevent excessive voltage changes on the capacitor. The equivalent inductor for transient response is given by $L_{eq-trans} = L(1 - k)$ [29], which implies that higher coupling coefficients result in reduced rise and fall times. Hence, from the aforementioned discussion, negative coupling can be beneficial for both the steady-state and transient responses by reducing the current ripple and increasing the bandwidth of the DC–DC converter system. However, the improvement in the current ripple is very sensitive to duty cycle and at each duty cycle, there is an optimum coupling factor. Since the improvement is less for smaller k, the selection of coupling factor should mostly be based on the duty cycle of the system.

8.2.1.3 Circuit Implementation

To show the capabilities of GaAs power devices in a high-frequency power switching environment and demonstrate the effect of negatively coupled inductors in interleaving architectures, a prototype using 0.5 μm GaAs p-HEMT process is designed and fabricated. The circuit diagram for the proposed design is shown in Figure 8.10. The two-stage interleaved DC–DC converter is designed for 4.5–3.3 V, 1 A load current

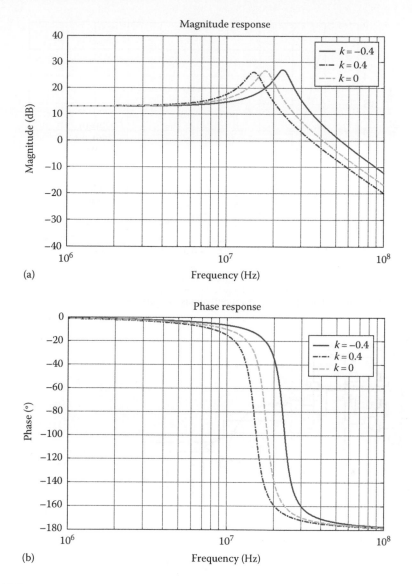

(a)

(b)

FIGURE 8.8   (a) Power stage gain and (b) phase versus coupling factor.

conversion with 150 MHz switching frequency. The coupled inductors are implemented on a separate GaAs die with 65 µm thick top copper layer.

8.2.1.3.1   Switching Stage Design

The output stage is implemented as a conventional buck converter with reversed diodes M_3 and M_4 as low-side switches and $SW1$ and $SW2$ as high-side switches. Two loss mechanisms are encountered in the switching stage: (1) the switching losses during the transition between on and off states, which increase with increasing the

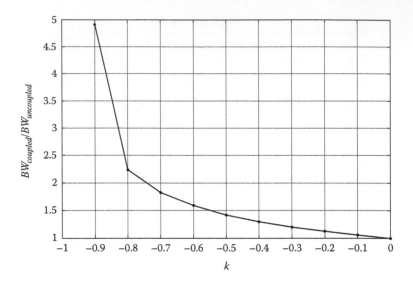

FIGURE 8.9   Power stage gain improvement versus coupling factor.

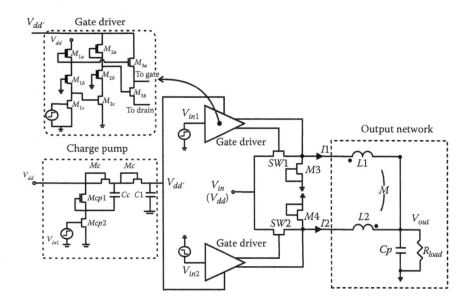

FIGURE 8.10   Schematic of GaAs pHEMT DC–DC converter.

transistors' sizes, and (2) conduction losses due to the finite on-resistance of the switching devices, which decreases with increasing the transistors' sizes. Figure 8.11 shows the variation in both losses as a function of the switching transistor width for 4.5–3.3 V conversion ratio. The optimum device widths are obtained by equating the conduction and switching losses of the switches to minimize their total power loss.

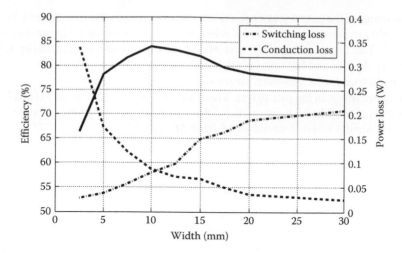

FIGURE 8.11 Variation of GaAs p-HEMT power losses and the corresponding output stage efficiency as a function of device width. The solid line represents the efficiency of the output stage.

It is important to consider the gate driver losses when sizing the switching transistors. This is mainly due to the fact that the switching losses increase as the gate driver stage fails to provide enough driving power. For a 1 A output current and 4.5/3.3 V voltage conversion ratio, the widths of $SW1$ and $SW2$ are chosen as 10 mm from Figure 8.11. They are implemented as 20 unit transistors in parallel, each with 500 μm width. This is mainly to satisfy a given aspect ratio for the converter die. The reverse connected diodes are realized by connecting the gate and source of the HEMT devices together. Transistors M_3 and M_4 have to provide a path for the current when $SW1$ and $SW2$ are off. In this design, they are sized at the same width of $SW1$ and $SW2$.

8.2.1.3.2 Gate Driver Design

Due to the lack of complementary transistors in the used GaAs p-HEMT process, the supply voltage for the gate driver has to be higher than the supply voltage of the output stage of the converter in order to drive the high-side switches. The minimum value for the gate driver supply is $V_{dd} + V_p$, where V_p is the pinch off voltage of the enhancement mode p-HEMT. A single-stage Dickson charge pump is adopted to generate $V_{dd} + V_p$ as the supply voltage of the gate driver circuit as shown in Figure 8.10.

The gate driver stage is a two-stage active inverter with the second stage referenced to the source of the high-side switches. This inverter stage is designed as pseudo-complementary switches with high-side depletion mode HEMT and low-side enhancement mode HEMT. The gate driver generates complementary control signals for M_{3a} and M_{3b}. Since the depletion mode HEMT has a negative pinch off voltage, the supply voltage for the first driver stage can be lowered to V_{dd} to reduce the power consumption. The second and third stages have a supply of $V_{dd} + V_p$, so that the output of second driver stage can swing between 0 and $V_{dd} + V_p$, and turn on switch $SW1$.

The sizing of the gate driver stages is a trade-off between the gate driver loss and its capability to drive the high-side switches, which affects the switching loss of $SW1$ and $SW2$. The larger sizes of M_{3a} and M_{3b} provide better driving capability while decreasing the rise and fall times, which will accordingly reduce the switching loss of the main transistors. However, larger sizes of M_{3a} and M_{3b} will also increase the power consumption in the gate driver. The width of enhancement mode pHEMT M_{3b} is chosen as 1/10 of the high-side switch $SW1$. M_{3a} is sized as 1/3 of M_{3b}. The first two stages are equally sized. Transistors M_{1c} and M_{2c} are selected as 1/4 of M_{3b} as well as M_{1b} and M_{2b} since they are pull-down transistors. Considering the same turn-on and turn-off time, pull-up transistors M_{1a} and M_{2a} are sized as 1/10 of M_{1c} and M_{2c}.

8.2.1.3.3   Coupled Inductors Design

While operating at high-switching frequency facilitates the monolithic integration of inductors, satisfying the requirements for low dc resistance and high current-handling capability makes the inductor design quite challenging. In this design, a 65 µm thick copper metal layer is used. The metal is electroplated on top of a field metal on semi-insulating GaAs substrate. The planar spirals fabricated on the process have a large cross-sectional area due to the large vertical sidewalls reducing the DC resistance and maximizing the quality factor. Contacts to the ends of the spiral inductor are made using solder bumps with a thickness of about 25 µm. The die is then flip-chipped to a laminate. The 25 µm separation between the copper metal and traces on the laminate ensures that less energy is coupled into the copper traces below, and thus, reduces eddy current losses.

A design example of a single inductor with 65 µm thick copper on GaAs is shown in Figure 8.12. The inductor has 3.5 turns, and the width and spacing between windings are 80 µm. As shown in Figure 8.12a, the total die area is 2.4 mm^2 and the real active region is much smaller. Seven spare bumps at the corner of the die are added for mechanical stability considerations. Two-port S-parameter measurement using a network analyzer is used to characterize the thick copper inductors once it is mounted on the laminate with signal and ground access. Picoprobes, with 50 Ω input impendence, and up to 3.5 GHz frequency response are used. The measured inductance value and quality factor is plotted in Figure 8.12b. The designed inductor has a dc resistance of 86.97 mΩ. At 100 MHz, inductance value is 11.94 nH and a peak inductance of 22 nH appears at 2 GHz. A quality factor of 22.1 at 100 MHz and almost 80 at 2 GHz is measured.

For the circuit shown in Figure 8.10, since the total current delivered to the load passes through the filter inductors, the size of the inductors must be properly selected to achieve an optimal balance between the required inductance value for a given current ripple and their respective series resistance that affects the converter efficiency. Thus, the inductance value can be determined based on either the required current ripple or to minimize the losses in the inductance. To find the minimum acceptable inductor value for a given current ripple, the maximum current ripple is defined at the boundary of continuous conduction mode (CCM) and discontinuous conduction mode (DCM) [1], such that

$$|\Delta i_1| = I_O \qquad (8.20)$$

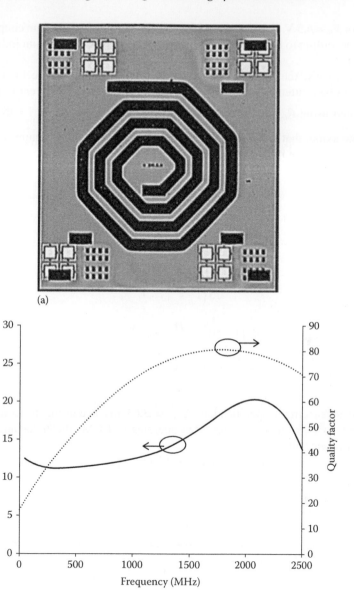

FIGURE 8.12 Design and measurement results of designed inductor in 65 μm thick copper layout: (a) layout; (b) measured inductance and quality factor.

where I_o is the current in the load resistance. Using Equations 8.13 and 8.20, the inductor required for minimum current ripple can be defined as

$$L_{min} = \frac{V_{in}(1-D)\left(k+\dfrac{D}{1-D}\right)(1-D)}{(1-k^2)f_{SW}I_o} \qquad (8.21)$$

Thus, for $V_{in} = 4.5$ V, $I_o = 1$ A, $f_{SW} = 150$ MHz, $D = 0.65$, the optimum coupling factor is -0.3 according to Equation 8.16, which corresponds to a minimum inductor value of 6.28 nH.

Alternatively, the inductor value can be found from the inductor power loss, which can be written as Equation 8.22 by adding the average current to the current ripple, and using R_{ind} as the resistance of the inductor, where $R_{ind} = R_L + \dfrac{2\pi f_{SW} L}{Q}$.

From the axiom that $a^2 + b^2 \geq 2ab$, the inductor value for a minimum inductor loss $P_{indloss}$ can be given by

$$P_{indloss} = \left(\frac{I_o^2}{2} + \frac{2}{3}\left(\frac{V_{in}\left(D + (1-D)k\right)}{L(1-k^2)}(1-D)T \right)^2 \right) R_{ind}$$

$$\geq \frac{2V_{in}I_o(1-D)^2\left(\dfrac{D}{1-D} + k \right)}{\sqrt{3}\left(1-k^2\right)f_{SW}L} R_{ind} \tag{8.22}$$

$$L_{min} = \frac{2V_{in}\left(1-D\right)^2\left(\dfrac{D}{1-D} + k \right)}{\sqrt{3}\left(1-k^2\right)f_{SW}I_o}$$

For the given circuit specification, $L_{min} = 3.63$ nH according to Equation 8.22. However, to make sure the converter operates at CCM, the inductance value of 6.28 nH is selected.

The coupled inductors are designed in 65 μm copper layer with interleaved square topology. ASITIC [34] is used to determine the number of turns and metal width for the given inductance and resistance values. Electromagnetic simulations using MOMENTUM are used to verify the inductance value and characterize the variation of Q versus frequency. Figure 8.13 shows the EM simulation results for the inductance and quality factor. For $L = 6.3$ nH, a quality factor of 25 at 150 MHz and a DC resistance = 55 mΩ is simulated assuming flip-chip packaging. The power loss of the designed inductor is only 2.4% of the total power loss in the converter.

8.2.1.3.4 Power Loss Analysis

The major power loss components in a DC–DC converter can be divided as (1) series resistance losses from the inductors $P_{indloss}$, (2) the conduction loss of the high-side switches P_{onloss}, (3) the switching losses of $SW1$ and $SW2$ P_{SWloss}, which depend on the turn-on and off time periods and the switching frequency, (4) the conduction loss of the diode-connected transistors M_3 and M_4 during $(1-D)T$ period $P_{diodeloss}$, and (5) the gate driver loss $P_{gateloss}$, which depends on the parasitic gate capacitances as well as the switching frequency.

Operating at high switching frequency reduces the inductor size, and its equivalent dc resistance. However, this comes at the expense of increasing the switching losses and

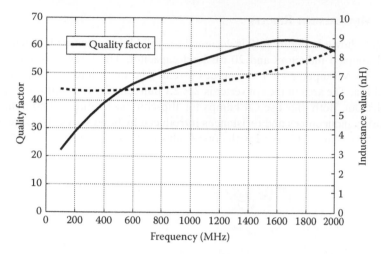

FIGURE 8.13 Electromagnetic simulations of coupled inductor.

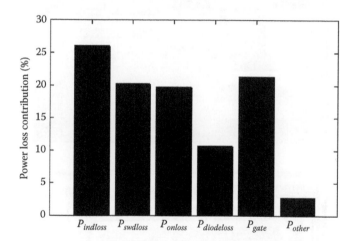

FIGURE 8.14 Power loss contribution of various elements in the two-phase interleaved GaAs converter with negative coupled inductors at 4.5/3.3 V, 1 A output and 150 MHz switching frequency.

gate driver losses. The distribution of power losses for the DC–DC converter, as shown in Figure 8.14, is based on simulation results. The conduction loss is almost the same as the switching loss, which contribute about 20% of the overall power losses. Since the duty cycle is 0.65, the diode loss is only half of the conduction loss. At 150 MHz, the inductor loss is still the dominant loss component, while the overall power efficiency is 84.5%.

The lack of complementary devices in the used technology contributed to the increase in gate driver losses. For the duty cycle of 0.65, the diode loss is almost 10% of the overall losses. The contribution of the diode loss is expected to increase for lower power level. Using synchronous rectifiers with dead-time control circuit as in [2,39,40] can be adopted to maintain the efficiency over a wide output power range.

8.2.2 MEASUREMENT RESULTS

The circuit shown in Figure 8.10 is designed for 150 MHz switching frequency with 8.77 nH coupled inductors and 20 nF load capacitor. The circuit converts 4.5 V input to 3.3 V output with 1 A output current. The die micrograph is shown in Figure 8.15a. The area of the converter is 2×2.1 mm^2 and 2.3×2.7 mm^2 for the coupled inductors. Both dies use C4 bumps for flip-chip packaging, which eliminates the parasitic inductances and resistances introduced by bond wires. The circuit test board is a four-layer PC board with copper plus OSP. An input decoupling capacitor of 22 µF is mounted close to the input supply voltage to avoid any oscillations.

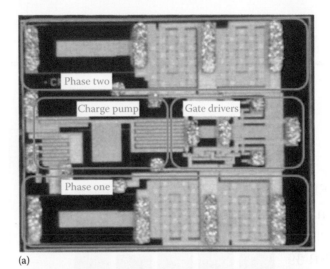

(a)

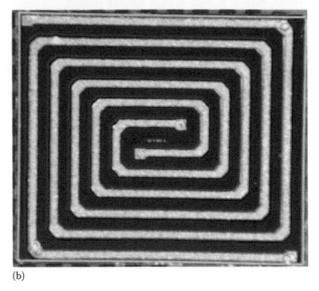

(b)

FIGURE 8.15   (a) GaAs p-HEMT DC–DC converter and (b) coupled inductors die photos.

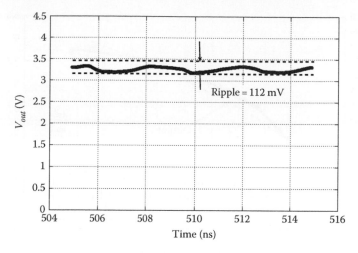

FIGURE 8.16 Output voltage ripple response.

The two-stage interleaved DC–DC converter is tested using an external pulse input provided by Agilent B1110A generator with the output measured using HP Infinium 1.5 GHz Oscilloscope. Figure 8.16 shows the signal at the output node with a measured output ripple of 116 mV. This is slightly higher than the simulated output ripple of 72 mA due to the deviation in the coupling factor of the implemented inductors from the optimum value for maximum ripple cancellation. The implemented inductors have an effective coupling factor of 0.46 compared to the optimum value of 0.3 since it is flip mounted on a four-layer PCB, which actually increases the distance between copper layer and ground. Figure 8.17 shows the two-phase operation waveform with the internal access at the input of coupled inductors. There is a 180° phase delay between two phases.

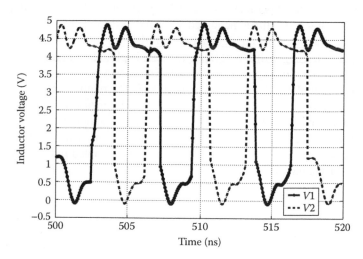

FIGURE 8.17 Two-phase operation of DC–DC converter.

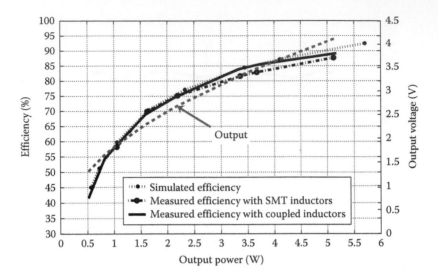

FIGURE 8.18 Efficiency comparisons for different output power at 150 MHz, 4.5 V input, and 1 A load current.

The efficiency using integrated coupled inductors as well as external uncoupled surface-mount inductance are compared to simulation results in Figure 8.18. The measured efficiency using integrated coupled inductors at the target conversion ratio of 4.5–3.3 V is 83.8%. As the duty ratio changes from 0.2 to 0.8, the efficiency drops about 30%. This can be reduced by changing the topology of the output switches and adding control technique to maintain constant efficiency. For the measurements using external uncoupled inductors, 7.5 nH 0603 SMT inductors by Coilcraft [37] with a quality factor of 28 at 150 MHz and dc resistance of 0.059 Ω are used. The simulation results are based on circuit models provided by the manufacturer and the extracted lumped element parameters from EM simulations of the coupled inductors. The measurement result using noncoupled SMT inductor is about 1.5% less efficient than coupled inductors at the target conversion ratio.

The output efficiency at different input voltages is plotted in Figure 8.19 at a constant load and duty ratio. Figure 8.20 shows the efficiency at different load resistors with constant input of 4.5 V and a duty ratio of 0.65. The relation between efficiency and switching frequency is shown in Figure 8.21. The measurement results show that the optimum point is at around 120 MHz, which is about 20% offset from the designed frequency. Table 8.2 summarizes the measurement results of the proposed two-phase interleaved DC–DC converter with negative coupled inductors in 0.5 μm GaAs p-HEMT technology.

8.2.2.1   Gate Driver Loss Analysis

In the design of high-frequency low-power DC–DC converters, the gate driver loss is a major power loss contribution that can significantly degrade overall efficiency, especially at lower power levels. From the measurement results, at $V_{out} = 3.21$ V, the overall efficiency is 83.17% and the power stage losses including power switch

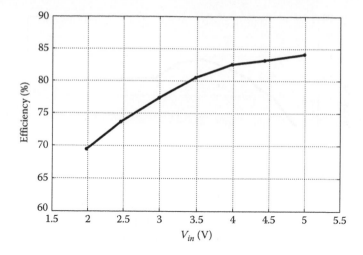

FIGURE 8.19 Efficiency versus input voltage for 150 MHz, duty ratio of 0.65 and 3.3 Ω load.

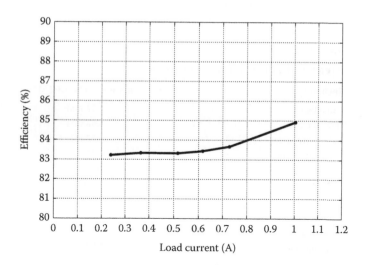

FIGURE 8.20 Efficiency versus load current at 150 MHz, 4.5/3.3 V conversion.

conduction loss, switching loss, diode conduction loss, and inductor loss contributed to 8.58% of the overall power. The gate driver loss is 8.25%, while at $V_{out} = 1.3$ V, the output efficiency drops to 41.16% and the power stage loss increases to 26.94% with the gate driver loss acting as the dominant loss component of 32%. Thus, in order to improve the light-load efficiency, the gate driver loss needed to be improved and synchronous rectifier topology is required. Figure 8.22 shows the simulated efficiency over output power for a single-phase synchronous rectifier buck converter with fixed dead time. The dotted line is calculated with gate driver loss excluded. By calculating the difference between these two lines, the power loss contribution of the gate driver circuit can be evaluated. The overall efficiency reaches 86% at full load

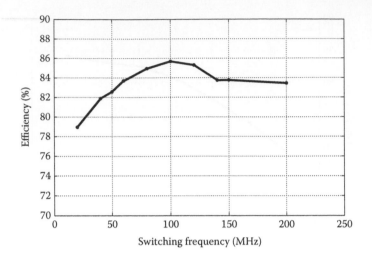

FIGURE 8.21   Efficiency versus switching frequency at 4.5/3.3 V conversion and 1 A load current.

TABLE 8.2
Measured Performance Summary

Technology	0.5 µm GaAs p-HEMT
Circuit area	4.22 mm²
Inductor area	5.94 mm²
Inductance value	8.77 nH
Coupling factor	0.46
Q at 150 MHz	26
Input voltage	4.5 V
Output voltage	3.3 V
Output current	1 A
Switching frequency	150 MHz
Peak efficiency	84%
Voltage ripple	116 mV

($I_{out} \sim 1$ A), drops to 83% at half load and 72% at light load. However, with gate drive loss excluded, the efficiency improves by 2.5% at peak load, 3.5% at half load, and 8% at light load. The efficiency curve (dotted line) is much flatter than the one with gate driver loss. Figure 8.23 also provides the power loss contribution over output power. At peak output power, the gate driver loss is less than 3% of the output power, while at low output power, it increases to 11% of the total output power. Hence, with synchronous rectifier buck converter, the efficiency range at different output power is limited to 15% compared to 26% for diode-connected buck converter. However, to further reduce gate driver power losses, some advanced gate driver topology needs to be developed, which will be discussed in the following section.

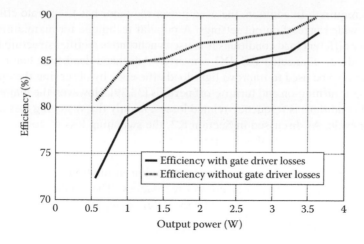

FIGURE 8.22 Simulated power efficiency over output power for single-phase synchronous buck converter with fixed dead time at 4.5 V input and 100 MHz switching frequency.

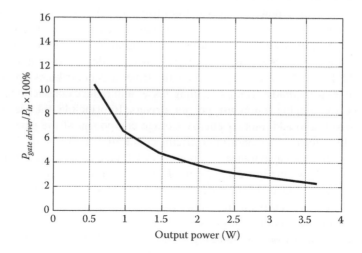

FIGURE 8.23 Simulated gate driver power loss contribution of single-phase synchronous buck converter with fixed dead time at 4.5 V input and 100 MHz switching frequency.

8.3 LIGHT LOAD EFFICIENCY IMPROVEMENT TECHNIQUES IN GaAs AND CMOS TECHNOLOGIES

High-efficiency and fast transient response of power regulators are becoming increasingly important for advanced electronic devices such as wireless portable systems, microprocessors, and other SOC, particularly when operating at different voltage and power settings. In radio transmitters employing envelop tracking applications, the variation of output voltage can change from zero volt up to the supply rails. For large SOC solutions, a variety of different output voltages and power levels are required. In addition, the used power regulator has to operate efficiently over such wide range of output settings. A popular technique for maintaining efficiency over different load conditions is to use synchronous rectifier structure instead of reverse diode-connected DC–DC converter. Advanced dead-time control techniques can also be used to improve light-load efficiency by observing the optimum dead time for turning-on and turning-off periods [38,39]. However, the major loss at light-load conditions is frequency-related losses, mainly the switching loss at every switching cycle. As discussed in Section 8.2, the switching loss of power switches and gate driver losses are the dominant loss components at light-load operation. There are three basic techniques to reduce such losses: (1) resonant gate drivers, (2) using phase shedding and switch segmentation in multiphase converters, and (3) variable switching frequency control techniques. The fundamentals of these techniques will be discussed here, while the detailed implementation and proposed architectures are given in Section 8.3.2.

8.3.1 Overview of Light-Load Efficiency Enhancement Techniques

8.3.1.1 Resonant Gate Driver

The development of a power converter requires careful design of gate drivers. At high-frequency, low-power applications, gate driver losses represent a significant portion of the overall power losses. For example, Figure 8.23 shows the gate driver losses of a single-phase DC–DC converter with fixed dead time control operating at 100 MHz and 4.5 V DC input. The gate driver loss contributes up to 11% at 1.0 V output. One method to limit switching losses and reduce conduction losses in gate drivers is to employ resonant-type gate drivers. In principle, the idea of resonant gate driver is to add another energy storage component at the gate of power switches, which forms a second-order circuit containing the resonant inductance L_r, power switch gate capacitance C_{in}, power switch gate resistance, and parasitic resistance R_g as shown in Figure 8.24. Here, L_r and C_{in} forms a resonant circuit with a characteristic impedance of $Z_o = \sqrt{L_r/C_{in}}$, and R_g is the only damping factor in the equivalent circuit and is much smaller than Z_o. The resonant inductor reduces the energy dissipation on the gate resistance and parasitic resistance during charging and discharging periods by changing the energy distribution. Ideally, a resonant gate driver can save as much as 40% energy during charging and discharging period according to [40]. Furthermore, with resonant inductors, less conduction time and

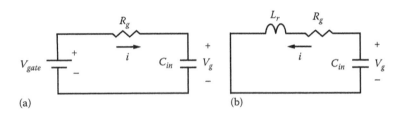

(a)   (b)

FIGURE 8.24 Equivalent circuit for resonant gate driver: (a) charging period and (b) discharging period.

zero voltage switching (ZVS) of gate driver switches can be achieved, which will further reduce the conduction and switching losses.

However, there are several disadvantages of resonant gate drivers. Adding resonant inductors will increase the chip or die size, which adds to the cost. Inherent resonant frequency is decided by L_r and C_g, and it determines the turning on/off transition of power switches. For a specified switching frequency, there will be a fixed and optimized L_r value. Hence, resonant gate drivers are not quite suitable for variable switching frequency applications unless a variable resonant inductance can be realized.

8.3.1.2   Phase Shedding and Switch Segmentation

When the output current level is high, larger switch and gate drivers are needed to drive the regulator output stage. However, at low-power settings, the large devices and drivers consume unneeded power that reduces the overall efficiency. In multi-phase converters, phase shedding can be employed to maintain peak efficiency trend even at low currents by disconnecting unnecessary phases [41,42]. Similarly, switch segmentation technique is the corresponding technique in single-phase operation. In this technique, the output switch and the corresponding gate driver are formed of a number of transistors and drivers in parallel. The peak efficiency is maintained by disconnecting some of the switching transistors and gate drivers as needed [43–46]. This approach is very effective with large variations in output current. One major disadvantage of the phase segmentation technique is that larger inductors and capacitors are required to keep the output ripple low during phase shedding/segmentations, which will increase passives size and their associated losses.

8.3.1.3   Variable Switching Frequency Control Techniques

Given that switching losses are the dominant losses at high-frequency converters, controlling the switching frequency can be a possible technique for improving efficiency at different load conditions. For example, sliding mode controllers for buck converter that switch the converter off for a given time period when the output is low have been employed in [47]. In this case, the PWM switching frequency is much higher than sliding mode control frequency. An auto-selectable frequency PWM modulator is used in [35], which is a combination of pulse frequency modulation (PFM) and pulse width modulation (PWM). A frequency selection block is inserted and PFM changes the switching frequency of buck converters with the load current which improves the light-load efficiency about 12% at 10 mA output current. In [48,49], dithering modulations have been utilized where some pulse periods are skipped at low-output current. However, this technique requires a very accurate current sensor. A buck converter topology based on cascode switch structure with a capacitor in between is proposed in [50]. The capacitor produces a voltage level other than 0 V and V_{dd}. The advantage of this topology is that the effective switching frequency at the inductor ripple is $2 \times f_{SW}$ while the switching loss–related frequency is still f_{SW}. In general, the presented topologies have attempted to reduce switching frequency in order to minimize the switching loss at light load. However, in order to sense the output and determine the frequency, a very fast analog to digital converter (ADC) is typically required, which makes this approach difficult to implement for power regulators running at hundreds of MHz.

8.3.2   Design of High-Frequency Resonant Gate Drivers

Conventional gate drivers are voltage-driven circuits that are typically formed of inverter chains with given fan-out factors according to the number of required stages, as well as input and output capacitors. The gate capacitor of the power switch is charged by the supply voltage through a parasitic gate resistance and interconnection resistance. The driver stage acquires duty ratio signal from the PWM generator, amplifies the signal, and sends it to the gate of the power switches. The loss in gate drivers consists of three parts. The first is conduction loss, which is related to the power dissipated through the parasitic gate resistor and interconnection resistor. The second is the cross-conduction loss caused by "shoot-through" between gate driver's two switches. The third part is the switching loss, which takes place at turn-on and turn-off transitions where there is a crossover between the voltage on the switch and the current through the switch. Among all three constituents, the conduction loss is the dominant one. The conduction loss takes place during charging and discharging of the main switch's gate. The conduction loss over the switching cycle ($P_{los,c}$) can be given as

$$P_{los,c} = V_{gate} \cdot Q_{gate} \cdot f_{SW} = C_{in} V_{gate}^2 f_{SW} \qquad (8.23)$$

where
V_{gate} is the gate voltage
Q_{gate} is the gate charge
f_{SW} is the switching frequency

To reduce the high-frequency gate driver losses, another energy storage component needs to be added to the gate driver to change the energy distribution and recover some of the energy back to the supply as a resonant gate driver.

8.3.2.1   Resonant Gate Driver Topologies

Conventional gate drivers are considered voltage-driven circuits while resonant gate drivers can be considered as current driven. Depending on the shape of the inductor current, there are basically three types of resonant gate drivers: (1) discontinuous resonant inductor current drivers, (2) continuous resonant inductor current drivers, and (3) coupled inductor resonant gate drivers. Figure 8.25a shows type I resonant gate driver topology with discontinuous inductor current, where diodes $D1$ and $D2$ are used to clamp the gate-source voltage [40]. When $M1$ turns on, the driver operates as an R–L–C circuit and the inductor current charges up as shown in Figure 8.25b. The diode $D1$ conducts once the gate source voltage V_{gs} of transistor M_s reaches $V_{gate} + V_{don}$ and prevents V_{gs} from increasing further, where V_{don} is the diode forward voltage drop. Transistor M_1 will turn off once $D1$ turns on and V_{gs} is clamped to $V_{gate} + V_{don}$. The current in L_r flows through the body diode of M_1 and back to V_{gate}. Once $M2$ is on, C_{in} discharges through R_g, L_r, and $S2$. Diode $D2$ conducts to prevent V_{gs} from having further negative voltage. In this structure, short conduction periods of $S1$ and $S2$ are preferred so that the energy stored in the inductor can return back to V_{gate}. Under this condition, M_1 and M_2 are turned off at the peak of inductor current, leading to an increase in the switching loss. However, in high-speed voltage regulators, it is difficult to

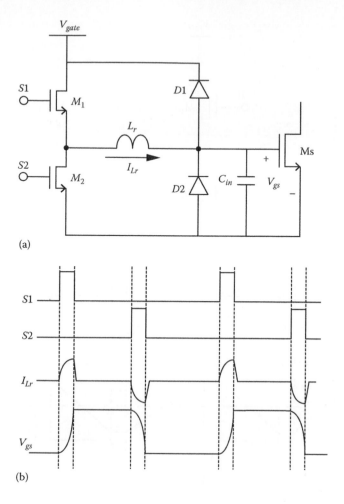

(a)

(b)

FIGURE 8.25 (a) Type I resonant gate driver with discontinuous inductor current and (b) operational waveform of type I resonant gate driver with discontinuous inductor current.

generate accurate drive pulses for $S1$ and $S2$. If the conduction time of driver switches becomes longer than charging/discharging intervals of power switch gate capacitance, it will cause higher conduction loss on the driver stages. Recent implementations of resonant gate driver with discontinuous inductor current use two additional transistors to replace the diodes and reduce the diode conduction loss [51–58].

An example of resonant gate driver with continuous inductor current is shown in Figure 8.26 [59–62]. This type of resonant gate driver requires a very large capacitor to block DC voltage across the resonant inductor. From Figure 8.26, the steady-state voltage across C_o equals to $D \times V_{gates}$, where D is the duty ratio and V_{gate} is the gate driver supply voltage. When both $M1$ and $M2$ are off, C_o starts charging C_{in} through L_r, with a constant value. When V_{gs} reaches V_{gate}, $M1$ will be turned on under zero voltage and clamp V_{gs} at V_{gate}. During turning off transition, C_{in} is first discharged through L_r at a constant value. Once V_{gs} is zero, $M2$ is turned on to clamp this voltage.

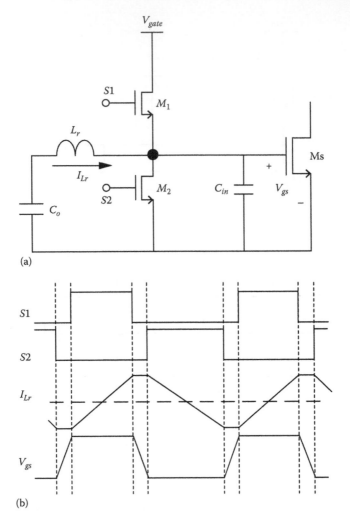

FIGURE 8.26 Type II resonant gate driver with continuous inductor current: (a) basic topology and (b) operational waveform of type II resonant gate driver with continuous inductor current.

At turning-on and -off periods, resonant inductor current is always constant at the peak value. Thus, a larger driving capability is typically needed compared to type I resonant gate driver. The dead time between $S1$ and $S2$ is determined by the charging and discharging intervals of L_r, C_{in}. A combined version of type II resonant gate drivers for both high-side and low-side switch utilizing only one resonant inductor L_r and one blocking capacitor C_r has been reported in [63–65].

The third type of resonant gate drivers is coupled inductor resonant gate drivers that adopts negative coupled inductors between the high-side resonant gate driver and the low-side resonant gate driver for synchronous rectifier buck converter as shown in Figure 8.27 [66,67]. In this topology, the energy from turning off Q_1 can be used to turn on the low-side switch Q_2. In order to transfer energy, an additional

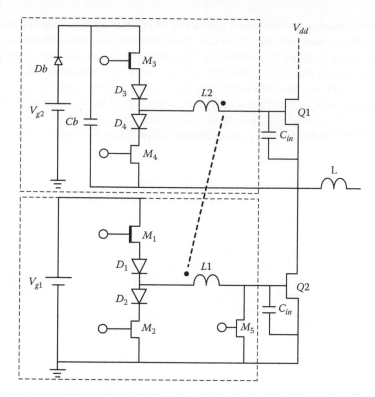

FIGURE 8.27 Type III resonant gate drivers with coupled inductors for synchronous buck converters.

switch M_5 is required. The added switch increases the control complexity, switching loss, and conduction loss, especially for ultra-high-speed power converters.

8.3.2.1.1 Resonant Gate Driver Design Challenges for Ultra-High-Frequency Voltage Regulators

As discussed before, the gate driver loss is the dominant loss component for voltage regulators at light-load operation. Since the gate driver loss is frequency dependent, it becomes more significant for high-frequency converters. While resonant gate drivers can provide a possible solution to this problem, they also present several design challenges at such high frequencies. First, resonant gate drivers require additional inductors, which directly affect the rising and falling transitions of power switches. Typically, the rising time is selected as 1/4 of resonant period [40,66]:

$$t_{on} = \frac{T_r}{4} = \frac{\pi}{2}\sqrt{L_r C_{in}}$$ (8.24)

When the required t_{on} is small (<1 ns) for ultra-high-speed voltage regulators, only few nH inductances are needed. Such small inductances are comparable to chip and board layout parasitics, and would require careful layout on the die and PCB laminate. Second, the control circuit design for resonant gate driver switches is a fundamental challenge at high frequencies. Resonant gate drivers are more efficient than conventional gate drivers because, on one hand, they reduce the charging and discharging loss of the power switch capacitance, and on the other hand, they realize zero voltage switching (ZVS) of gate driver switches and reduce their conduction losses. Hence, precise and accurate control signals are required for resonant gate drivers. For voltage regulators running at hundreds of MHz and above, the control circuits will generate additional losses. Thus, there is a design trade-off between resonant gate driver losses and digital control losses.

8.3.2.2 CMOS Resonant Gate Driver

In order to minimize the gate driver loss at medium- and low-output power ranges, and to reduce the switching loss of the power switches, a current-source resonant gate driver is proposed for synchronous rectifier DC–DC converters, implemented in CMOS technology, as shown in Figure 8.28. In Figure 8.28, $sw1$ and $sw2$ are the main power switch and synchronous rectifier, respectively, while C_{gsp} and C_{gsn} are the gate-source capacitors of $sw1$ and $sw2$. The two sets of drive circuits: high side (HS) and low side (LS) have a half-bridge topology. The topology consists of $s1$, $s2$ and $s3$, $s4$, respectively. With deliberated designed asymmetrical control, all the drive switches can achieve ZVS. Since in the PMOS-NMOS DC–DC converter circuit same phase driving signals are used for V_{ghi} and V_{gls}, drive LS can be considered as duplication of drive HS. L_{r1} is the resonant inductor for drive HS and $(L_{r1} + L_{r2})$ is the resonant inductor for drive LS. This is due to the fact that for PMOS-NMOS DC–DC converter design the PMOS switch is usually two to three times larger than the NMOS synchronous rectifier. In order to get the same rise and fall times for $sw1$ and $sw2$, larger resonant inductors are used for the LS driver. C_b is the blocking capacitor that removes the dc component of the voltage across the resonant inductors. Typically, C_b is relatively large. According to the charge balance over the resonant inductors, the steady-state voltage across C_b is DV_{dd}, where D is the duty ratio of the power switches. For C_1–C_4 being the intrinsic capacitors of $s1$–$s4$, respectively, there are six switching transitions in a switching period as marked in Figure 8.28. It is clear from the figure that the resonant inductor current remains at peak value during the turn-on and turn-off transitions. Based on the charge balance equation, the peak inductor current (I_{Lrpk}) can be approximately expressed as

$$I_{Lrpk} = I_{Lr1pk} + I_{Lr2pk} = \frac{V_{dd}D(1-D)}{2L_{r1}f_{SW}} + \frac{V_{dd}D(1-D)}{2L_{r2}f_{SW}} \qquad (8.25)$$

Here, we assume that the size of PMOS is 2× wider than the NMOS, 2× $L_{r1} = L_{r1} + L_{r2}$ and $L_{r1} = L_{r2}$. The total peak resonant current can also be derived as

$$I_{Lrpk} = \frac{3V_{dd}D(1-D)}{4L_{r1}f_{SW}} \qquad (8.26)$$

and

$$I_{Lr1pk} = \frac{C_{gsp}V_{dd}}{t_{rf}}$$

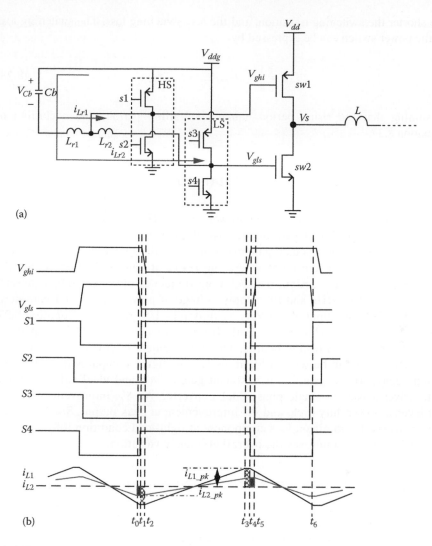

(a)

(b)

FIGURE 8.28 (a) CMOS resonant gate driver and (b) voltage and current waveforms of CMOS resonant gate driver.

Thus, the turn-on and turn-off time of V_{ghi} can be derived as

$$t_{r,f} = \frac{4C_{gsp}L_rf_{SW}}{3D(1-D)} \qquad (8.27)$$

8.3.2.2.1 Design Optimization

As can be seen from the main current waveforms in Figure 8.28, the peak current I_{Lrpk} of the resonant inductor L_r is regarded as a current source. The higher I_{Lrpk}, the shorter the switching transition, and the less switching loss. The switching loss of the power switch can be expressed by

$$P_{SW} = \frac{1}{2} V_{dd} I_{Lavg} (t_{on} + t_{off}) f_{SW}$$ (8.28)

Assuming that the rising period equals the falling period, and according to Equation 8.27,

$$P_{SW} = \frac{4 V_{dd} I_{Lavg} C_{gsp} L_r f_{SW}}{3D(1-D)}$$ (8.29)

Hence, the switching loss of the power switch (P_{SW}) is proportional to L_r. Meanwhile, the resonant gate driver loss $P_{resg} \propto (1/L_r)^2$. Therefore, there is a trade-off between the resonant gate driver loss and the switching loss. The minimum loss point is at the point where the resonant gate drive loss equals the switching loss. In this design, the switching frequency of the DC–DC converter is selected as 100 MHz and the supply voltage of power switch and gate drivers is 5 V with an output of 2.5 V. $t_{r,f}$ is chosen as $1/50\,f_{SW}$. From Equation 8.27, L_r is selected as 10 nH and C_r is 2 nF. Both components are SMT components mounted on the PCB testing boards. The improvement in efficiency using resonant gate driver can be seen in Figure 8.29. The figure compares the efficiency among conventional gate driver, resonant gate driver, and ideal values without gate driver losses for single-phase buck converter. An 8% improvement can be achieved at lower duty cycle and 1% improvement at peak output. Since the gate driver loss is the dominant loss component at light-load condition [68], resonant gate driver greatly improves the overall efficiency variation.

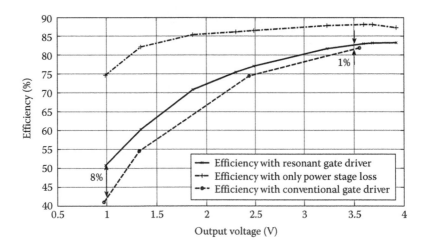

FIGURE 8.29 Efficiency versus output voltage for different gate driver topologies at 100 MHz switching frequency and $V_{in} = 5$ V.

8.3.2.3 Resonant Gate Driver Design in GaAs Technology

To evaluate the applicability of the resonant gate driver design technique in GaAs p-HEMT technology, a prototype of type I resonant gate driver circuit is implemented in 0.5 μm GaAs p-HEMT. Figure 8.30 shows a single-phase DC–DC converter with a resonant the low-side gate driver. Transistors M_{8a} and M_{8b} represent the high-side

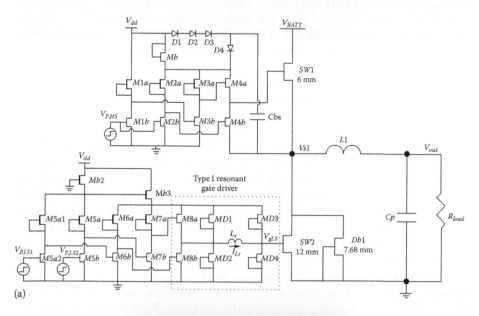

(a)

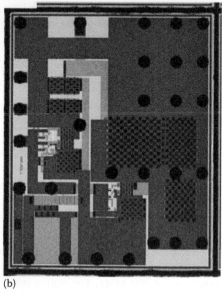

(b)

FIGURE 8.30 GaAs p-HEMT buck converter with low-side resonant gate driver: schematic and die micrograph.

switch and the low-side switch of the resonant gate driver, respectively, both sized at 0.5 mm. Due to the nature of GaAs p-HEMT devices, there is no parasitic body diode across the drain and source of HEMT devices. However, in order to realize ZVS for M_{8a} and M_{8b}, two diode-connected transistors with the gate shorted to the drain ($MD1$ and $MD2$) are connected in parallel with the two switches. Transistors $MD3$ and $MD4$ are used instead of $D1$ and $D2$ in Figure 8.25a to reduce the conduction loss and forward conduction voltage compared to regular diodes. The sizes of $MD1$–$MD4$ are 1 mm. To prevent the forward conduction of the Schottky gate of the synchronous rectifier switch $SW2$, the supply voltage for low-side gate drivers needs to be shifted down to a smaller value compared to V_{dd}. To down shift the supply voltage, two depletion mode p-HEMT M_{b2} and M_{b3} are used. The gate of M_{b2} is connected to ground so that the source voltage at M_{b2} is close to the pinch off voltage of the D-pHEMT. The gate of M_{b3} is connected to the source of M_{b2} and the final stages of gate drivers have a supply voltage close to twice the pinch off voltage of D-pHEMT. Two buffer stages are inserted to drive M_{8a} and M_{8b}, respectively. Ideally, the control signal waveform of $V_{p,LS1}$ and $V_{p,LS2}$ should have short conduction time that equals to the charging/discharging intervals of $SW2$ through L_r. However, in order to completely clamp V_{gls} at zero and reduce the complexity of the control signal generation, a larger period of $V_{p,LS2}$ is used. The first stage of the buffer stage of M_{5a1}/M_{5a} and M_{5b1}/M_{5b} are sized as 5 and 10 μm, respectively. The width of the second buffer stage of M_{6a}/M_{7a} and M_{6b}/M_{7b} are selected as 12 and 30 μm, which is three times that of the first stage.

The high-side gate drivers are conventional gate drivers. The high-side transistor $SW1$ requires a floating driver that can raise the gate potential above V_{dd}. Therefore, a bootstrapped gate driver structure is needed to boost the supply voltage to $V_{dd} + V_{pinchoff}$ to avoid the Schottky gate forward conduction. To control the boosted up voltage, three series diodes are inserted. When $SW1$ is off, bootstrap capacitor C_{bs} is charged up to $V_{dd} - 3V_D - V_{s1}$, where V_D is the forward conduction voltage drop on the diodes. Once V_{pHS} turns high, the gate of $SW1$ is pulled high by M_{4a} and V_{s1} starts rising. Capacitor C_{bs} starts discharging through D_4 and M_{4a}, raising the potential at the gate of $SW1$ above V_{dd}. In this design, C_{bs} is selected as 100 pF and M_{4a}, M_{4b} are selected as 0.25 mm. Two buffer stages are used to drive M_{4a} while one buffer stage is used to drive M_{4b}, considering the nonoverlapping control logics for M_{4a} and M_{4b}. Transistors M_{1a} and M_{1b} are sized as 5 and 10 μm, while M_{2a}/M_{3a} and M_{2b}/M_{3b} are selected as 6 and 25 μm, respectively.

To further improve the light-load efficiency, the sizes of power switches $SW1$ and $SW2$ are determined based on 4.5/1 V conversion ratio and 0.5 A optimum current. Since the duty ratio is about 0.25 at such output, the high-side switch $SW1$ is sized as 6 mm, which is about 1/2 of synchronous rectifier $SW2$ of 12 mm. A diode-connected transistor MD with 7.68 mm width is inserted in parallel with $SW2$ to provide a conduction path during dead-time period. In order to compare the performance of resonant gate driver, the same design with a conventional gate driver is included on the die. Two circuit prototypes are implemented in 0.5 μm GaAs p-HEMT technology with 4.5 V supply operating at 100 MHz. The output inductor is 0402 Coilcraft 11 nH inductor with $R_{dc} = 65$ mΩ and quality factor of 28 at 100 MHz.

The output efficiency at different output voltages is plotted in Figure 8.31. At $V_{out} = 1$ V, the converter employing a resonant gate driver reaches 53% efficiency

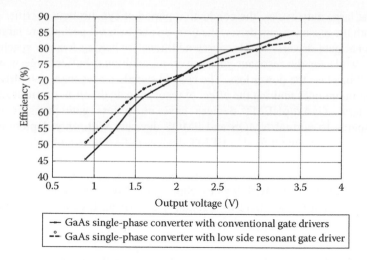

FIGURE 8.31 Measured output efficiency versus voltage for resonant gate driver buck converter and conventional gate driver buck converter.

while the circuit with conventional gate driver has an efficiency of 48% at 1 V output. At V_{out} = 2.1 V, the two circuits have almost the same efficiency of 73%. When the output level increases, the GaAs converter with the conventional gate driver achieves higher efficiency than the one with resonant gate driver. At V_{out} = 3.3 V, with conventional gate driver, the buck converter has 85% efficiency while with resonant gate driver has only 82.3% [69]. Figure 8.32 shows the gate driver loss contribution of the conventional gate driver and the resonant gate driver. When V_{out} > 2 V, the resonant gate driver has an even larger loss compared to the conventional gate driver. This is

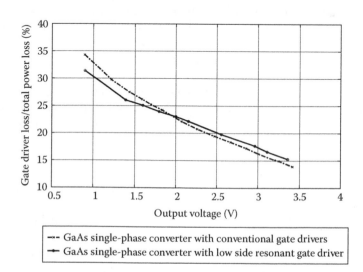

FIGURE 8.32   Measured gate driver power loss contribution versus output voltage.

because at large duty cycles, the conduction period of synchronous rectifier is very short. With the resonant gate driver, the turn-on and turn-off periods are larger than conventional gate drivers, which causes additional switching loss on synchronous rectifiers, while for the resonant gate driver, an additional buffer stage is needed, which will increase the power loss. The control signals for resonant gate drivers are not the optimum ones making the effect of resonant gate driver not significant.

In general, GaAs pHEMT devices have much less gate parasitic capacitors compared to normal NMOS and PMOS devices. With only depletion mode pHEMT and enhancement mode pHEMT devices in the GaAs technology, it is difficult to generate complex logic signals for resonant gate driver. Furthermore, the gate swing (V_{gs}) of pHEMT devices is between 0 and 0.85 V, which is much lower than the gate swing of CMOS devices. Since gate driver loss is greatly related to C_{gs} and V_{gs}, it makes the effect of resonant gate drivers less significant in GaAs pHEMT buck converter applications.

8.3.3 PHASE SHEDDING USING COUPLED INDUCTORS PASSIVE NETWORK

For high-frequency buck converters, the multiphase interleaved structure is beneficial for its low-current ripple, small inductor size, and fast transient response. However, the efficiency of multiphase converters tends to drop at medium and light-load conditions as each phase operates far from the optimal design conditions. While resonant gate driver discussed in the previous section is one possibility for improving efficiency, phase shedding technique can also be employed to eliminate unneeded phases during light-load operation [41,42]. For example, for the four-phase interleaved converter shown in Figure 8.33a, the efficiency curve ideally would follow the peak of the efficiency curves at different output currents as shown in Figure 8.34. A similar approach is called "phase segmentation," which divides each power switch and gate driver into several small segments and disconnect some of them during light-load operation [43,44]. Such technique does not require phase synchronization between phases, and thus, reduces the complexity of the control circuitry. However, at light load, the DC–DC converter for both phase shedding and phase segmentation can fall into DCM due to the smaller load inductor value and low current level, increasing the output voltage ripple. Higher load inductance and capacitance or a variable switching frequency should be adopted at light loads to avoid this effect. In this section, we present an output stage topology with coupled inductors network that changes the effective inductance values to reduce the ripple and increases the efficiency compared to the simple multiphase converters employing phase shedding or segmentation.

8.3.3.1 Proposed Architecture

The proposed architecture, shown in Figure 8.33b, is based on a two-phase interleaving structure. Each phase is divided into two equal segments having power switches and gate drivers. There are three operation schemes: two-phase four-segment at full load, two-phase two-segment at medium load, and one-phase one-segment at light load. To prevent the converter from operating in DCM, four additional inductors are added and positive coupling is introduced between segments to further improve the effective inductance value. Furthermore, negative-coupled inductors are used between the two

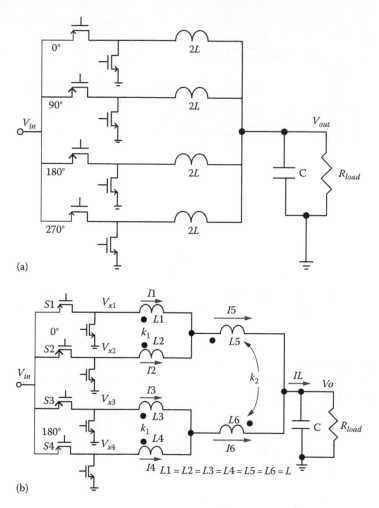

FIGURE 8.33 (a) Conventional four-phase buck converter and (b) proposed two-phase, four-segment buck converter.

phases to reduce the inductor current ripple and improve efficiency. The solid line in Figure 8.34 gives the efficiency trend of the proposed architecture when employing phase shedding and/or phase segmentation. At one-phase one-segment operation, the efficiency will be the same as conventional topology shown in Figure 8.33a. However, for two-phase two-segment and two-phase four-segment mode, the proposed architecture achieves higher efficiency due to the ripple reduction introduced by the multipolarity coupled inductors scheme analyzed in Section 8.3.3.2.

8.3.3.2 Steady-State Analysis

To get a better understanding of the proposed topology, the steady-state current ripple analysis when employing a coupling factor k_1 between inductor pairs ($L1$, $L2$) and ($L3$, $L4$) and coupling factor k_2 between inductors ($L5$, $L6$) is presented (Figure 8.33b),

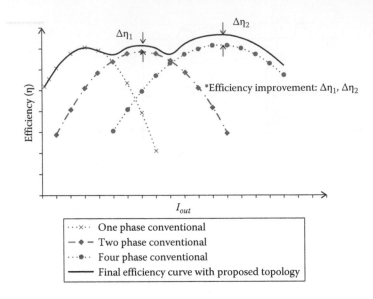

FIGURE 8.34 Efficiency trends for conventional four-phase buck converter and proposed topology.

where we assume k_1 and k_2 can be either positive or negative and all the inductors have the same value L. Based on the current charge balance equation, the inductor current can be derived as

$$
\begin{cases}
2L \cdot \dfrac{di_1}{dt} + L(1+k_1) \cdot \dfrac{di_2}{dt} + k_2 L \cdot \dfrac{di_3}{dt} + k_2 L \cdot \dfrac{di_4}{dt} = V_{x1} - V_o \\[2mm]
2L \cdot \dfrac{di_2}{dt} + L(1+k_1) \cdot \dfrac{di_1}{dt} + k_2 L \cdot \dfrac{di_3}{dt} + k_2 L \cdot \dfrac{di_4}{dt} = V_{x2} - V_o \\[2mm]
2L \cdot \dfrac{di_3}{dt} + L(1+k_1) \cdot \dfrac{di_4}{dt} + k_2 L \cdot \dfrac{di_1}{dt} + k_2 L \cdot \dfrac{di_2}{dt} = V_{x3} - V_o \\[2mm]
2L \cdot \dfrac{di_4}{dt} + L(1+k_1) \cdot \dfrac{di_3}{dt} + k_2 L \cdot \dfrac{di_1}{dt} + k_2 L \cdot \dfrac{di_2}{dt} = V_{x4} - V_o
\end{cases}
\tag{8.30}
$$

From the steady-state current ripple waveform shown in Figure 8.35, we can derive $\dfrac{di_1}{dt}$ for four-segment operation when $V_{x1} = V_{x2}$ and $V_{x3} = V_{x4}$ as

$$
\frac{di_1}{dt} = \frac{(V_{x1} - V_o)(3 - 2k_1 - k_1^2) + 2k_2(k_1 - 1)(V_{x3} - V_o)}{L(1 - k_1)(k_1^2 + 6k_1 - 4k_2^2 + 9)}
\tag{8.31}
$$

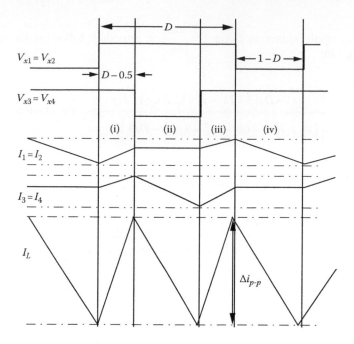

FIGURE 8.35   Steady-state waveform of the proposed architecture.

As shown in Figure 8.35, for a duty ratio larger than 0.5, during both states I and III, $\Delta t = (D - 1/2)T$, $V_{x1} = V_{x3} = V_{in}$; in state II, $\Delta t = (1 - D)T$, $V_{x1} = V_{in}, V_{x3} = 0$; in state IV, $\Delta t = (1 - D)T$, $V_{x1} = 0$, $V_{x3} = V_{in}$. By adding up the current ripple expressions over an entire period, the steady-state ripple per phase for $D > 0.5$ is as follows:

$$\left|\Delta i_1\right| = \frac{(1-D)(3-2k_1-k_1^2)+2(1-D)k_2(1-k_1)}{L(1-k_1)(k_1^2+6k_1-4k_2^2+9)}(1-D)V_{in}T \qquad (8.32)$$

A similar approach can be applied to duty ratio smaller than 0.5 such that the current ripple is

$$\left|\Delta i_1\right| = \frac{D(3-2k_1-k_1^2)+2Dk_2(1-k_1)}{L(1-k_1)(k_1^2+6k_1-4k_2^2+9)}DV_{in}T \qquad (8.33)$$

We can define the current ripple per segment for the uncoupled condition for comparison such that

$$\left|\Delta i_{1u}\right| = \frac{DV_{in}(1-D)}{3Lf_{SW}} \qquad (8.34)$$

The current ripple reduction factor $\xi = \dfrac{|\Delta i_1|}{|\Delta i_{1u}|}$ is provided, following the approach provided in [30]

$$\xi = \begin{cases} \dfrac{(1-D)\left(3-2k_1-k_1^2\right)+2Dk_2(1-k_1)}{\left(1-k_1\right)\left(k_1^2+6k_1-4k_2^2+9\right)\left(\dfrac{1}{3}-D/3\right)} & \text{for } D \le 0.5 \\[4mm] \dfrac{D\left(3-2k_1-k_1^2\right)+2(1-D)k_2(1-k_1)}{\left(1-k_1\right)\left(k_1^2+6k_1-4k_2^2+9\right)(D/3)} & \text{for } D > 0.5 \end{cases} \tag{8.35}$$

The effect of k_1 can be easily proved by examining the two segments in one phase and assuming $k_2 = 0$ in Equation 8.11. The current ripple reduction can be given as

$$\varepsilon = \frac{3}{3+k_1} \tag{8.36}$$

It is obvious that the larger k_1, the smaller the current ripple per segment. The maximum ripple reduction is 25% compared to noncoupled condition at $k_1 = 1$. However, given practical limitations for coupling coefficient, the highest coupling factor achieved for k_1 is about 0.8 and the ripple reduction factor is 0.79. Table 8.3 compares the simulated current ripple per segment and converter output stage efficiency for three different coupling conditions. For small duty ratios, the effect of negative-coupled inductors (k_2) is not significant. However, for 50% duty ratios, the ripple reduction percentage is 45% for $k_2 = -0.4$ compared to 21% for $k_2 = 0$ case. The efficiency improvement is mostly related to the positive coupled inductors k_1 with maximum efficiency improvement around 1.2% at 50% duty ratio for the case of tight positively coupled inductors at $k_1 = 0.8$ and tight negative coupling $k_2 = -0.8$.

TABLE 8.3

Ripple Reduction and Efficiency Comparison for Different Optimum k_1 and k_2

	$D = 0.25$		
	$k_1 = k_2 = 0$	$k_1 = 0.8, k_2 = 0$	$k_1 = 0.8, k_2 = -0.4$
Ripple reduction factor (ξ)	1	0.79	0.75
Efficiency (%)	83.5	84.2	84.4
	$D = 0.5$		
	$k_1 = k_2 = 0$	$k_1 = 0.8, k_2 = 0$	$k_1 = 0.8, k_2 = -0.4$
Ripple reduction factor (ξ)	1	0.79	0.55
Efficiency (%)	86.6	87.2	87.8

8.3.4 COMPLETE DC–DC CONVERTER DESIGN

In the previous two subsections, the design of resonant gate drivers and the introduction of phase shedding and its passive network implementation have been presented. In this section, these techniques are applied to the design of a 100 MHz CMOS multi-phase DC–DC converter. The complete schematic of the proposed DC–DC converter is shown in Figure 8.36. The circuit is implemented in 0.18 μm CMOS technology with 5 V CMOS transistors used for the power stage [70]. The design targets a 5 V input, 1–4 V output, and 2 A total current. The output switching stage utilizes PMOS as high-side switch and NMOS as low-side switch. The power switches are optimized at 5 V input to 2.5 V output. Power switches have both conduction loss and switching loss. Sizing of the transistors is selected at the point where the switching loss is equal to the conduction loss. Another important issue for multiphase DC–DC converter with phase shedding/segmentation is to determine the optimum current value per phase. The efficiency of the power stage and gate driver stages is plotted at different load currents with 5 V input and 2.5 V output, and the current per phase is selected at the point where the efficiency is maximized [70]. The peak current point is at 1 A load current. The power switches were implemented in 0.18 μm CMOS technology with six metal layers plus distribution metal for flip-chip bumps using 5 V devices. The physical layout adopts a butted source topology, where substrate contacts are placed next to every source diffusion area to ensure uniform voltage distribution and avoid device breakdown. The breakdown voltage of the 5 V devices is 6 V. For driving the power switches, the resonant gate driver, discussed in Section 8.3.2, is employed. The schematic of the driver is shown in Figure 8.37. The size of the resonant gate driver is determined by balancing the gate driver loss and switching loss of power switch. Typically, the larger the gate driver, the higher the driving capability required to drive the power switch and the less switching loss encountered due to shorter turn-on and turn-off periods. However, this also means higher gate driver loss. Considering that the main power switch is sized as 40 mm for high-side PMOS switch and 15 mm for low-side NMOS switch, the high-side resonant gate driver is selected as 7.5 and 2.5 mm and the low-side resonant gate driver is sized as 3 and 1 mm, respectively. The ratio between power switch and resonant gate driver is around 5:1. Four nonoverlapping control signals are required for resonant gate drivers. The optimum solution is to generate these control signals by examining the internal switching node V_x, and V_{gs} of both high-side and low-side switches. This approach would require very high-speed and low-propagation delay comparators. However, since the converter operates at 100 MHz switching speed, the dead time between signals is less than 1 ns. This complicates the design of the comparator and might lead to increased delay or power consumption due to possible overlapping between high-side and low-side switches, which will drive increased current from the power supply.

In this design, a fixed dead-time circuit has been selected. The circuit is composed of three clock tree logic gates. The first generates nonoverlapping signals for the high-side and the low-side resonant gate driver, while the second generates control signals for $s1$ and $s2$ and the third generates control signals for $s3$ and $s4$. Each clock tree circuit is a cascade of inverters with specific fan-out factor. The inverter scaling

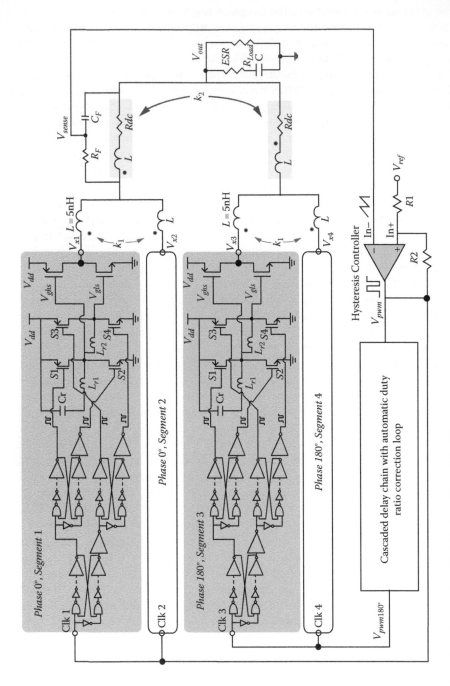

FIGURE 8.36   Schematic of CMOS closed-loop DC–DC converter.

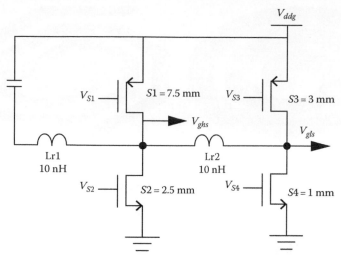

FIGURE 8.37   Resonant gate driver.

factor (F) for the inverter chains is determined based on both load capacitance and input capacitance $F = \left(\dfrac{C_{gs,n,p}}{C_{in,I1}}\right)^{1/N}$, where I_1 is the first stage of inverter chain and N is the optimum number of inverter stages, determined by $N = \ln \dfrac{C_{gs,n,p}}{C_{in,I1}}$. The minimum total propagation delay occurs at the optimum number of inverter stages. In this design, the optimum number of stages is equal to four for all clock tree circuits.

8.3.4.1   Coupled Inductor Design

To implement the coupled inductor network, copper layers on the printed circuit board (PCB) are utilized for planar inductors. The thickness of the PCB copper layer is 20 μm with 1 oz copper. This is much thicker than most of the metal layers on CMOS process as those reported in [14,16,17] and can provide higher-quality factor at hundreds of MHz. Furthermore, the spacing between different layers can be selected as low as 2.7 mil with Micro Via technology. It can significantly improve the coupling coefficient between inductors for multiphase interleaved DC–DC converter applications. Figure 8.38 shows the layouts of designed positive and negative coupled inductors. The minimum size of the inductor is determined by $|\Delta i_1| = 2xI_{1avg}$, when the converter is operating at the boundary of CCM and DCM [1]. I_{1avg} is the average current per segment. The minimum inductance can be found as

$$L = \frac{(1-D)(3-2k_1-k_1^2)+2Dk_2(1-k_1)}{2I_{avg}(1-k_1)(k_1^2+6k_1-4k_2^2+9)} \cdot DV_{in}T \tag{8.37}$$

where k_1 is the positive coupling between $L1$ and $L2$ as shown in Figure 8.33a, and k_2 is the negative coupling factor between $L5$ and $L6$. For $V_{in} = 5$ V, $T = 10$ ns, $D = 0.25$, $I_{1avg} = 0.5$ A, and k_1 & k_2 selected as 0.8 and −0.4, respectively, the minimum inductance value can be calculated as 3.9 nH. However, for smaller inductor ripple and

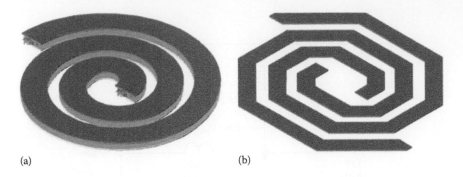

(a)   (b)

FIGURE 8.38 Coupled inductors layout: (a) positive coupled inductors and (b) negative coupled inductors.

TABLE 8.4
PCB Inductor Parameters

	$L_{Positive,Coupled}$	$L_{Negative,Coupled}$
Area	2.58×2.37 mm^2	3.4×4.35 mm^2
Width	300 μm	300 μm
Turns	2.5	1.5
Spacing	120 μm	200 μm
L	6.27 nH	6.48 nH
Rds	0.032 Ω	0.032 Ω
K	0.8	0.4
Q at 100 MHz	37	45

to consider the inductor area limitation of the PCB laminate, 6 nH inductance is selected for this design. To achieve the highest coupling factor, the positive coupled inductors are built on the top and second PCB layers with 68.58 μm spacing between layers. The negative coupled inductors are designed on the top copper layer with the spacing between windings adjusted to achieve the specific coupling factor of 0.4. Design parameters and electromagnetic simulation results for the two coupled inductors are given in Table 8.4.

8.3.4.2 Hysteresis Controller Design

For a high-speed transient response, a hysteresis controller is selected in this design as shown in Figure 8.39. However, the main disadvantage for the hysteresis controller is the weak regulation of switching frequency [30], where an overall 40% variation of switching frequency is encountered when the duty ratio sweeps from 0.2 to 0.8. To synchronize the two phases, an open-loop delay chain with duty ratio corrector circuits is employed. As in multiphase interleaved DC–DC converters, it is important to maintain the same duty ratio among phases to balance the current and avoid current-sharing problems. Hence, in this design, an automatic duty ratio correction is adopted. The function of phase delay is realized using series-connected voltage-controlled delay cells.

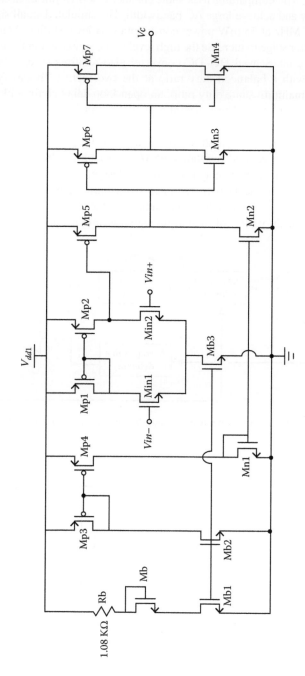

FIGURE 8.39 Hysteresis comparator circuit schematic.

The hysteresis comparator circuit uses a two-stage differential structure with a bias current of 70 μA. The comparator uses short channel $L = 0.18$ μm devices to reduce propagation delay and achieve large AC bandwidth. The simulated small signal bandwidth equals 278 MHz at 55 mW power consumption. A level shifter circuit is added after the comparator stage to increase the high level of the signal swing from 1.8 to 5 V.

The two-phase interleaved DC–DC converter also requires 0° and 180° phase-delayed signals with a balanced duty ratio at the two inputs. To satisfy the 180° phase delay and maintain same duty ratio, an open-loop delay chain with duty ratio corrector loop is employed. The schematic of the circuit is shown in Figure 8.40. Conventional voltage-controlled delay cells are used to generate the required delay. The schematic of the delay cell is shown in Figure 8.41. V_{dcont} controls the delay time by adjusting the effective load resistance of $Mn2$ through the biasing current. The larger V_{dcont} leads to a smaller delay time. In this design, eight cascaded delay cells are used for a total 5 ns delay with each cell contributing about 0.625 ns delay at $V_{dcont} = 1.7$ V. Since the delay is achieved by adjusting the turning on/off transition of each delay cells, the duty ratio of the output signal is smaller than the input signal.

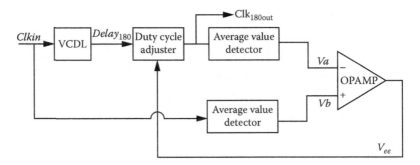

FIGURE 8.40 Block diagram of delay chain and duty ratio corrector circuits.

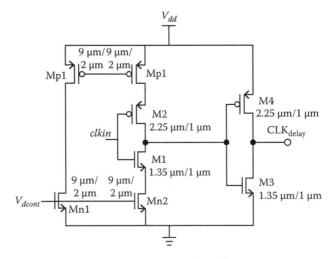

FIGURE 8.41 Schematic of voltage-controlled delay cell.

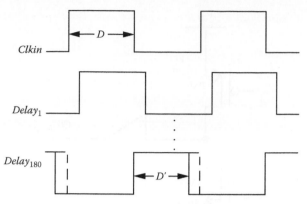

FIGURE 8.42 Diagram of the delay and duty ratio reduction cycles.

At the end of the delay chain, the variation of the duty ratio compared to *Clkin* can be as high as 25%, as shown in Figure 8.42. If this signal is sent to the second phase of the DC–DC converter, severe current-sharing and distribution problems will be encountered. Hence, the duty ratio information at the output of the delay chain needs to be adjusted. The duty ratio corrector block shown in Figure 8.40 compares the average voltage of both *Clkin* and *Delay*$_{180}$ signals and adjusts the duty ratio of *Delay*$_{180}$ accordingly. The circuit is composed of an average value detector, op-amp, and a duty ratio corrector. The schematic of the average value detector is shown in Figure 8.43, which is basically a charge pump circuit. The average output voltage is $V_{outa} = \dfrac{Ib \cdot T_{off}}{C_1}$, where T_{off} is the off time of the input signal V_{in}. Thus, the larger the duty ratio, the smaller the output voltage V_{outa}. In this design, R_1 is chosen as 1 Ω and is implemented using the fifth metal layer resistance provided in the

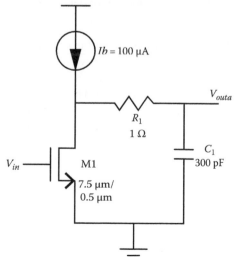

FIGURE 8.43 Circuit schematic of the average value detector.

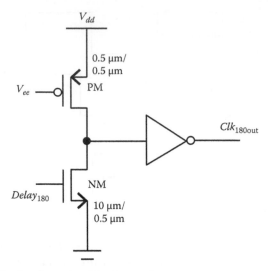

FIGURE 8.44 Circuit schematic of the duty ratio corrector.

0.18 μm CMOS process on the chip where there is a total of six metal layers with a distribution metal layer on top and C_1 is selected as 300 pF and implemented on chip using poly capacitors. The Op-amp circuit is used to compare the two average values from the input clock signal and delayed signal and generate V_{ee} for the duty ratio corrector. The output of the Op-amp is sent to the duty ratio corrector circuit as depicted in Figure 8.44. The voltage V_{ee} is connected to the gate of the load PMOS device to adjust the load resistance of the inverter. By changing the value of V_{ee}, the duty ratio of $Delay_{180}$ signal can be modified. The open-loop delay chain with duty ratio corrector satisfies the requirements of the same duty ratio and phase delay for multiphase interleaved DC–DC converter systems. It is a simplified design with less risk of system instability compared to the use of delayed locked loop.

8.3.4.3 Measurement Results

The circuit shown in Figure 8.36 is implemented in 0.18 μm CMOS technology with 5 V devices available in the technology. The high-voltage devices were used for power switch, gate drivers, and delay chains, while the 1.8 V devices are used for the hysteresis comparator. The die photo of the chip is shown in Figure 8.45. The chip contains two segments of the converter power switches and gate drivers, and all the control blocks. The positive and negative coupled inductors are implemented on the PCB laminate with 1 oz copper of 20 mm thickness to generate the required 6 nH inductance. The output capacitor is 47 nF and uses discrete 0402 components. The converter operates at peak 4 V power supply with 1–3.2 V and 0.1–1.86 A output. The die area is 2.5 mm × 3.0 mm with 30 C4 solder bumps for flip-chip packaging. The die size is mainly limited by the solder bumps. The converter die is mounted on a four-layer PCB laminate using Micro Via design rules. Input decoupling capacitor of 22 and 0.1 μF are mounted close to the V_{dd} supply pads.

The output of the hysteresis comparator is monitored as shown in Figure 8.46 using an external reference signal running at 100 MHz. The lower and upper limits

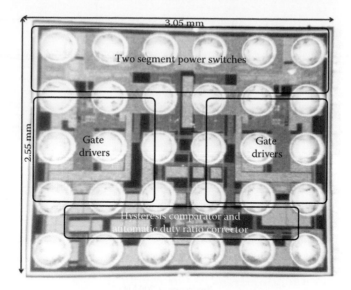

FIGURE 8.45   Die photo of flip-clip CMOS DC–DC converter.

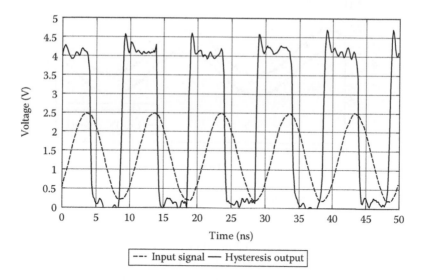

FIGURE 8.46   Measurement results of hysteresis comparator.

of the hysteresis window is set at 0.5 and 2.2 V, respectively. The propagation delay at turning-on transition is about 0.5 and 0.7 ns at turning-off transition. The two-phase four-segment interleaved DC–DC converter is tested for close-loop operation and the output is measured using HP Infinium 1.5 GHz Oscilloscope. R_f and C_f in the feedback network are selected at 1 kΩ and 2.7 pF, respectively, for 100 MHz switching frequency. Figure 8.47 shows the measured output transient response at $V_{in} = 4$ V and $V_{out} = 2.3$ V. The measured output ripple is 50 mV.

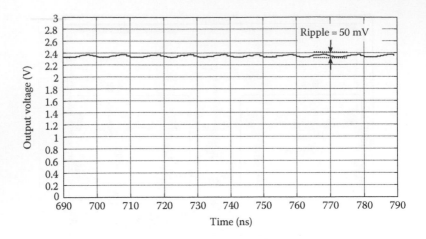

FIGURE 8.47 Transient response of output voltage.

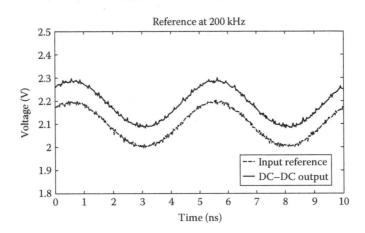

FIGURE 8.48 Closed-loop circuit tracking mode performance with a reference changing at 200 kHz.

Figure 8.48 demonstrates the tracking mode transient response when the input reference signal is running at 200 kHz, where it can be seen that there is no phase delay at this reference frequency. The closed-loop efficiency at different output voltages with input supply of 4 V is provided in Figure 8.49. For a conventional gate driver at two-phase two-segment operation with 4.3 Ω resistor, the efficiency is about 73.8% at 3 V output and drops to 46.4% at 1 V output, which is about 27.8% reduction. Better efficiency at lower output can be achieved with more advanced dead-time control circuits. The dotted line provides the efficiency value excluding the gate driver loss. At peak output, the gate driver loss is about 4.5% of the overall input power, while at V_{out} = 1 V, it is around 30% higher in efficiency.

The effect of resonant gate driver topology is demonstrated by adding resonant inductors L_r and DC blocking capacitor C_b on the evaluation PCB laminate.

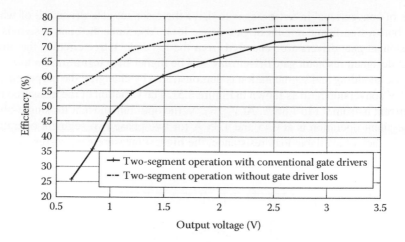

FIGURE 8.49 Efficiency versus output voltage at two-phase, two- segment operation with conventional gate drivers.

The calculated and simulated optimum L_r is about 10 nH. However, considering the PCB trace inductance, which is about 1 nH/mm, the external discrete L_r is selected as 2 nH with Coilcraft inductors and C_b of 2.7 nF. Figure 8.50 shows the efficiency value with the resonant gate driver at two-phase two-segment operation with the same 4.3 Ω resistor. A 5% efficiency improvement is achieved at V_{out} = 0.8 V and 1.4% improvement at 3 V output. By using resonant gate drivers, the total driver loss reduces from 33.3% to 27.32% at V_{out} = 0.7 V. However, there are still a large proportion of losses coming from other driver stages. That is because we can only apply resonant gate driver at the last stage of gate drivers and there are still several

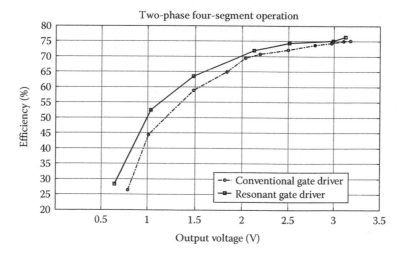

FIGURE 8.50 Efficiency versus output voltage at two-phase, four-segment operation with resonant gate drivers.

stages of conventional inverter stages to drive the resonant gate driver, all of which contribute a significant amount of gate driver loss. Also since the control signals for the resonant gate drivers are generated from the fixed clock tree circuits, the power loss of designed resonant gate driver may still be higher than its minimum loss.

Phase shedding and segmentation is demonstrated by changing the output current at 4–2 V conversion ratio as shown in Figure 8.51. The variation of the measured output current is from 0.1 to 1.86 A. At $V_{out} = 2$ V, the optimum current for single-phase one-segment operation is at 0.75 and 1.35 A for two-phase two-segment operation. With the lack of available load resistance, the highest measurement current is about 1.86 A. It is estimated that for two-phase four-segment operation the peak efficiency

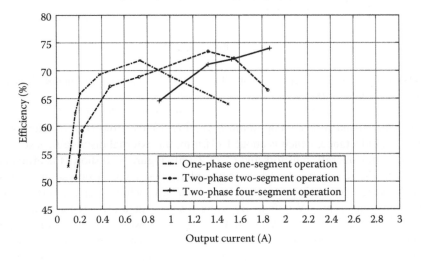

FIGURE 8.51 Efficiency versus output voltage at $V_{in} = 4$ V and $V_{out} = 2$ V.

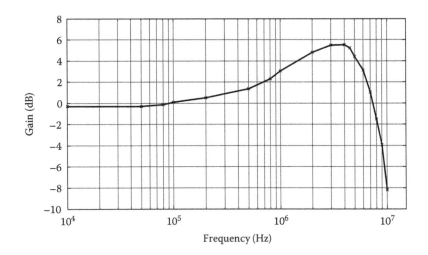

FIGURE 8.52 Gain characteristic of the tracking mode frequency response.

TABLE 8.5
Measured Performance Summary

Technology	0.18 μm CMOS
Circuit area	7.78 mm²
Positive coupled inductor area	6.11 mm²
Negative coupled inductor area	15.3 mm²
Input voltage	4 V
Peak output voltage	3 V
Peak output current	1.8 A
Switching frequency	100 MHz
Peak efficiency	77.4%
Voltage ripple	50 mV

is at I_{out} = 2.7 ~ 3A. By reducing phases and segmentations, the converter will be operating at peak efficiency as the load current changes. The closed-loop frequency response is measured by feeding the reference node with sinusoidal signals at different frequencies and monitoring the time-domain reference and output signals. Fourier transformation is applied to both signals to generate the required frequency response. Figure 8.52 shows the gain response of the hysteresis-controlled two-phase converter. The 3 dB bandwidth is about 8 MHz. Table 8.5 summarizes the measurement results of the presented two-phase four-segment interleaved DC–DC converter with positive and negative coupled inductors in 0.18 μm CMOS technology.

8.3.5 SUMMARY

Two techniques for light-load efficiency improvement are addressed. The resonant gate driver topology employs a resonant inductor between the gate driver output stage and the gate of the power switch to save the energy dissipated on the gate resistor and parasitic interconnection resistor. A two-stage resonant gate driver topology is used for a CMOS high-frequency DC–DC converter. An 8% efficiency improvement is achieved at 1 V output. Type I resonant gate driver is also implemented in GaAs p-HEMT technology at low-side gate drivers. However, due to the nature of GaAs devices, the effect of resonant gate driver circuit is not significant. Another technique discussed is phase shedding/segmentation, which is effective when the output current varies over a wide range. A novel inductor network with positive coupled inductors and negative coupled inductors is proposed, which reduces the inductor current ripple, increases the effective inductance value, and improves the efficiency compared to conventional multiphase buck converter. The proposed topology is particularly effective for high-power DC–DC converters when the inductor power loss is the dominant component. To demonstrate the two proposed techniques, a prototype of a two-phase, four-segment DC–DC converter with the proposed inductor network is built in 0.18 μm CMOS technology operating at 100 MHz with 5 V power supply. Complete circuit analysis, including power switch design, gate driver design, inductor design, and control loop design, is addressed. The peak measured efficiency at two-phase four-segment operation with 4 V supply and 3.2 V/1.86 A output is 77.4%. The closed-loop bandwidth with hysteresis controller is 8 MHz. The architecture improves light-load efficiency by 5% with resonant gate driver and the output efficiency follows the peak value at 2 V output when the output current changes from 0.1 mA to 2 A with phase shedding/segmentation technique and proposed inductor output network.

REFERENCES

1. R. W. Erickson and D. Maksimovic, *Fundamentals of Power Electronics*, 2nd edn. Springer, 2001.
2. J. Wibben and R. Harjani, A high-efficiency DC-DC converter using 2 nH integrated inductors, *IEEE J. Solid-State Circuits*, 43, 844–854, April 2008.
3. R. Severns and G. Bloom, *Modern DC-DC Switch Mode Power Converter Circuits*. Van Nostrand Reinhold Electrical/Computer Science and Engineering Series, Springer, New York, 1985.
4. S. Ajram and G. Salmer, Ultrahigh frequency DC-to-DC converters using GaAs power switches, *IEEE Trans. Power Electron.*, 16(5), 594–602, September 2001.
5. J. M. Rivas, D. Jackson, O. Leitermann, A. D. Sagneri, Y. Han, and D. J. Perreault, Design considerations for very high frequency DC–DC converters, in *Proceedings of IEEE Power Electronics Specialists Conference (PESC'06)*, Jeju, South Korea, June 2006, pp. 1–11.
6. T.-W. Kwak, M.-C. Lee, and G.-H. Cho, A 2W CMOS hybrid switching amplitude modulator for EDGE polar transmitters, *IEEE J. Solid-State Circuits*, 42(12), 2666–2676, December 2007.
7. W.-Y. Chu, B. Bakkaloglu, and S. Kiaei, A 10 MHz-bandwidth 2 mV-ripple PA-supply regulator for CDMA transmitters, in *Proceedings of IEEE International Solid-State Circuits Conference*, San Francisco, CA, February 2008, pp. 448–626.
8. C. Schaef, B. Reese, C. R. Sullivan, and J. T. Stauth, Design aspects of multi-phase interleaved resonant switched-capacitor converters with mm-scale air-core inductors, *2015 IEEE 16th Workshop on Control and Modeling for Power Electronics, COMPEL*, 2015.
9. C. Huang and P. K. T. Mok, A 100 MHz 82.4% efficiency package-Bondwire based four-phase fully-integrated buck converter with flying capacitor for area reduction, *IEEE J. Solid-State Circuits*, 48(12), 2977–2988, 2013.
10. M. K. Song, M. F. Dehghanpour, J. Sankman, and D. Ma, A VHF-level fully integrated multi-phase switching converter using bond-wire inductors, on-chip decoupling capacitors and DLL phase synchronization. *Applied Power Electronics Conference and Exposition (APEC), 2014 29th Annual IEEE*, IEEE, Long Beach, CA, 2014.
11. J. Sun, J. Lu, D. Giuliano, T. Chow, and R. Gutmann, 3D power delivery for microprocessors and high-performance ASICs, in *Proceedings of IEEE Applied Power Electronics Conference (APEC'07)*, Anaheim, CA, February 2007, pp. 127–133.
12. S. Abedinpour, B. Bakkaloglu, and S. Kiaei, A multistage interleaved synchronous buck converter with integrated output filter in 0.18 μm SiGe process, *IEEE Trans. Power Electron.*, 22(6), 2164–2175, November 2007.
13. M. Wens and M. Steyaert, A fully-integrated 130 nm CMOS DC-DC step-down converter, regulated by a constant on/off-time control system, in *European Solid-State Circuits Conference (ESSCIRC)*, Edinburgh, Scotland, September 2008, pp. 62–65.
14. H. Bergveld, K. Nowak, R. Karadi, S. Iochem, J. Ferreira, S. Ledain, E. Pieraerts, and M. Pommier, A 65-nm-CMOS 100-MHz 87%-efficient DC-DC down converter based on dual-die system-in-package integration, in *IEEE Energy Conversion Congress and Exposition (ECCE)*, September 2009, pp. 3698–3705.

15. K. Onizuka, K. Inagaki, H. Kawaguchi, M. Takamiya, and T. Sakurai, Stacked-chip implementation of on-chip buck converter for distributed power supply system in SiPs, *IEEE J. Solid-State Circuits*, 42(11), 2404–2410, November 2007.

16. M. Lee, Y. Choi, and J. Kim, A 0.76 W/mm^2 on-chip fully-integrated buck converter with negatively-coupled, stacked-LC filter in 65 nm CMOS, in *IEEE Energy Conversion Congress and Exposition (ECCE)*, 2014.

17. H. K. Krishnamurthy, V. A. Vaidya, P. Kumar, G. E. Matthew, S. Weng, B. Thiruvengadam, W. Proefrock, K. Ravichandran, and V. De, A 500 MHz, 68% efficient, fully on-die digitally controlled buck Voltage Regulator on 22 nm Tri-Gate CMOS, in *VLSI Circuits Digest of Technical Papers, 2014 Symposium on IEEE*, 2014.

18. B. Minnis, P. Moore, P. Whatmough, P. Blanken, and M. van der Heijden, System-efficiency analysis of power amplifier supply-tracking regimes in mobile transmitters, *IEEE Trans. Circuits Syst. I*, 56(1), 268–279, January 2009.

19. J. N. Kitchen, I. Deligoz, S. Kiaei, and B. Bakkaloglu, Polar SiGe class E and F amplifiers using switch-mode supply modulation, *IEEE Trans. Microw. Theory Tech.*, 55(5), 845–856, May 2007.

20. J. Qin, R. Guo, J. Park, and A. Huang, A low noise, high efficiency two stage envelope modulator structure for EDGE polar modulation, in *Proceedings of IEEE International Symposium on Circuits and Systems*, 2009, pp. 1089–1092.

21. G. Hanington, P.-F. Chen, P. Asbeck, and L. Larson, High-efficiency power amplifier using dynamic power-supply voltage for CDMA applications, *IEEE Trans. Microw. Theory Tech.*, 47(8), 1471–1476, August 1999.

22. P. Hazucha, G. Schrom, J. Hahn, B. Bloechel, P. Hack, G. Dermer, S. Narendra, D. Gardner, T. Karnik, V. De, and S. Borkar, A 233-MHz 80%–87% efficient four-phase DC-DC converter utilizing air-core inductors on package, *IEEE J. Solid-State Circuits*, 40, 838–845, April 2005.

23. G. Schrom, P. Hazucha, J. Hahn, D. Gardner, B. A. Bloechel, G. Dermer, S. Narendra, T. Karnik, and V. De, A 480-MHz, multi-phase interleaved buck DC-DC converter with hysteretic control, in *Proceedings of IEEE Power Electronics Specialists Conference (PESC'04)*, Aachen, Germany, June 2004, pp. 4702–4707.

24. P. Li, D. Bhatia, L. Xue, and R. Bashirullah, A 90–240 MHz hysteretic controlled DC-DC buck converter with digital PLL frequency locking, in *Proceedings of Custom Integrated Circuits Conference (CICC)*, San Jose, CA, September 2008, pp. 21–24.

25. S. J. Kim, R. K. Nandwana, Q. Khan, R. C. N. Pilawa-Podgurski, and P. K. Hanumolu, A 4-Phase 30–70 MHz switching frequency buck converter using a time-based compensator, *IEEE J. Solid-State Circuits*, December 2015.

26. C. K. Teh, A. Suzuki, M. Yamada, M. Hamada, and Y. Unekawa, 4.1 A 3-phase digitally controlled DC-DC converter with 88% ripple reduced 1-cycle phase adding/dropping scheme and 28% power saving CT/DT hybrid current control, in *IEEE International Solid-State Circuits Conference Digest of Technical Papers (ISSCC)*, San Francisco, CA, February 2014.

27. P.-L. Wong, P. Xu, P. Yang, and F. Lee, Performance improvements of interleaving VRMs with coupling inductors, *IEEE Trans. Power Electron.*, 16(5), 499–507, July 2001.

28. H. Peng, V. Pala, P. Wright, T. P. Chow, and M. M. Hella, High efficiency, high switching speed, AlGaAs/GaAs P-HEMT DC-DC converter for integrated power amplifier modules. *J. Analog Integr. Circ. Sig. Process.*, 66(3), 331–348, March 2011.

29. H. Peng, V. Pala, T. P. Chow, and M. M. Hella, A 150 MHz, 84% efficiency, two phase interleaved DC-DC converter in AlGaAs/GaAs P-HEMT technology for integrated power amplifier modules. *IEEE Radio Frequency Integrated Circuits Symposium RFIC*, Anaheim, CA, May 2010, pp. 259–262.

30. H. Peng, T. P. Chow, and M. M. Hella, A highly efficient interleaved DC-DC converter using coupled inductors in GaAs technology. *IEEE International Symposium on Circuits and Systems ISCAS*, Taiwan, May 24–27, 2009, pp. 2037–2040.

31. J. Abu-Qahouq, M. Batarseh, L. Huang, and I. Batarseh, Analysis and small signal modeling of a non-uniform multiphase buck converter, in *Proceedings of IEEE Power Electronics Specialists Conference (PESC'07)*, Orlando, FL, June 2007, pp. 961–967.

32. ASITIC. [Online]. Available: http://rfic.eecs.berkeley.edu/~niknejad/asitic.html. Accessed on January 1, 2000.

33. T. Man, P. Mok, and M. Chan, An auto-selectable-frequency pulse-width modulator for buck converters with improved light-load efficiency, in *Proceedings of IEEE International Solid-State Circuits Conference*, San Francisco, CA, February 2008, pp. 440–626.

34. V. Yousefzadeh and D. Maksimovic, Sensorless optimization of dead times in DC-DC converters with synchronous rectifiers, *IEEE Trans. Power Electron.*, 21(4), 994–1002, July 2007.

35. Coilcraft. [Online]. Available: http://www.coilcraft.com. Accessed on 04.11.2011.

36. M. Mulligan, B. Broach, and T. Lee, A constant-frequency method for improving light-load efficiency in synchronous buck converters, *IEEE Power Electron. Lett.*, 3(1), 24–29, March 2005.

37. S. W. Lee et al., Accurate dead-time control for synchronous buck converter with fast error sensing circuits, *IEEE Trans. Circuits and Syst. Regul. Pap.*, 60(11), 3080–3089, 2013.

38. Y. Chen, Resonant gate drive techniques for power MOSFETs, Master's thesis, Virginia Polytechnic Institute and State University, May 2000.

39. J. Costabeber, P. Mattavelli, and S. Saggini, FPGA implementation of phase shedding with time-optimal controller in multi-phase buck converters, in *Proceedings of IEEE 35th Annual Conference of Industrial Electronics*, Porto, Portugal, 2009, pp. 2919–2924.

40. J. Su, C. Hung, T. Chang, and C. Liu, A novel phase shedding scheme for improved light load efficiency of interleaved DC/DC converters, in *Proceedings of IEEE Applied Power Electronics Conference and Exposition*, Palm Springs, CA, 2010, pp. 1482–1487.

41. V. R. H. Lorentz et al., Light-load efficiency increase in high-frequency integrated DC DC converters by parallel dynamic width controlling, *Analog Integr. Circ. Signal Process.*, 62(1), 1–8, January 2010.

42. S. Musunuri and P. Chapman, Improvement of light-load efficiency using width-switching scheme for CMOS transistors, *IEEE Power Electron. Lett.*, 3(3), 105–110, September 2005.

43. K.-H. Chen, C.-C. Chien, C.-H. Hsu, and L.-R. Huang, Optimum power-saving method for power MOSFET width of DC/DC converters, *IET Circuits, Devices Syst.*, 1(1), 57–62, February 2007.

44. O. Trescases, A digitally controlled DC-DC converter module with a segmented output stage for optimized efficiency, in *International Symposium on Power Semiconductor Devices and IC's (ISPSD)*, May 2007.

45. M. Hoyerby, M. Andersen, and P. Andreani, A 0.35 µm 50V CMOS sliding-mode control IC for buck converters, in *European Solid State Circuits Conference (ESSCIRC)*, Muenchen, Germany, September 2007, pp. 182–185.

46. H.-W. Huang, H.-H. Ho, C.-C. Chien, K.-H. Chen, G.-K. Ma, and S.-Y. Kuo, Ditherng skip modulator with a width controller for ultra-wide-load high-efficiency DC-DC converters, in *IEEE Custom Integrated Circuits Conference (CICC)*, San Jose, CA, September 2006, 643–646.

47. H.-W. Huang, K.-H. Chen, and S.-Y. Kuo, Dithering skip modulation, width and dead time controllers in highly efficient DC-DC converters for system-on-chip applications, *IEEE J. Solid-State Circuits*, 42(11), 2451–2465, November 2007.

48. V. Yousefzadeh, E. Alarcon, and D. Maksimovic, Three-level buck converter for envelope tracking applications, *IEEE Trans. Power Electron.*, 21(2), 549–552, March 2006.

49. M. Bathily, B. Allard, and F. Hasbani, A 200 MHz integrated buck converter with resonant gate drivers for an RF power amplifier, *IEEE Trans. Power Electron.*, 27(2), 610–613, 2012.

50. W. Eberle, Y.-F. Liu, and P. Sen, A new resonant gate-drive circuit with efficient energy recovery and low conduction loss, *IEEE Trans. Ind. Electron.*, 55(5), 2213–2221, May 2008.

51. X. Zhou, Z. Liang, and A. Huang, A new resonant gate driver for switching loss reduction of high-side switch in buck converter, in *Applied Power Electronics Conference and Exposition (APEC)*, Palm Springs, CA, February 2010, pp. 1477–1481.

52. W. Eberle, Y.-F. Liu, and P. Sen, A resonant gate drive circuit with reduced MOSFET switching and gate losses, in *IEEE Industrial Electronics, IECON 2006*, November 2006, pp. 1745–1750.

53. M. Bathily, B. Allard, J. Verdier, and F. Hasbani, Resonant gate drive for silicon integrated DC/DC converters, in *Energy Conversion Congress and Exposition (ECCE) 2009*, IEEE, September 2009, pp. 3876–3880.

54. W. Eberle, Z. Zhang, Y.-F. Liu, and P. Sen, A current source gate driver achieving switching loss savings and gate energy recovery at 1-MHz, *IEEE Trans. Power Electron.*, 23(2), 678–691, March 2008.

55. Z. Zhang, F.-F. Li, and Y.-F. Liu, A high-frequency dual-channel isolated resonant ate driver with low gate drive loss for ZVS full-bridge converters, *IEEE Trans. Power Electron.*, 29(6), 3077–3091, June 2014.

56. Alihossein Sepahvand, Yuanzhe Zhang and Dragan Maksimovic, 100 MHz isolated DC-DC resonant converter using spiral planar PCB transformer, in *IEEE 16th Workshop on Control and Modeling for Power Electronics (COMPEL)*, Vancouver, British Columbia, Canada, July 2015.

57. D. Maksimovic, A MOS gate drive with resonant transitions, in *Proceedings of IEEE Power Electronics Specialists Conference*, Cambridge, MA, June 1991, pp. 527–532.

58. Z. Zhang, W. Eberle, Y.-F. Liu, and P. Sen, A new current-source gate driver for a buck voltage regulator, in *IEEE Applied Power Electronics Conference and Exposition (APEC)*, Austin, TX, February 2008, pp. 1433–1439.

59. M. Mulligan, B. Broach, and T. Lee, A 3MHz low-voltage buck converter with improved light load efficiency, in *IEEE International Solid-State Circuits Conference (ISSCC)*, San Francisco, CA, February 2007, pp. 528–620.

60. T. Lopez, G. Sauerlaender, T. Duerbaum, and T. Tolle, A detailed analysis of a resonant gate driver for PWM applications, in *IEEE Applied Power Electronics Conference and Exposition (APEC)*, vol. 2, Miami Beach, FL, February 2003, pp. 873–878.

61. Z. Yang, S. Ye, and Y.-F. Liu, A new resonant gate drive circuit for synchronous buck converter, *IEEE Trans. Power Electron.*, 22(4), 1311–1320, July 2007.

62. Z. Zhang, W. Eberle, P. Lin, Y.-F. Liu, and P. Sen, A 1-MHz high-efficiency 12-V buck voltage regulator with a new current-source gate driver, *IEEE Trans. Power Electron.*, 2817–2827, 2008.

63. Z. Zhang, Z. Yang, S. Ye, and Y.-F. Liu, Topology and analysis of a new resonant gate driver, in *IEEE Power Electronics Specialists Conference*, Dallas, TX, June 2006, pp. 1–7.

64. K. Yao and F. Lee, A novel resonant gate driver for high frequency synchronous buck converters, *IEEE Trans. Power Electron.*, 23(6), 180–186, November 2002.

65. L. Liu, Y. Ma, X. Xie, C. Zhao, W. Yao, and Z. Qian, A new resonant gate driver for low voltage synchronous buck converter based on topologies optimization, in *IEEE Applied Power Electronics Conference and Exposition*, Austin, TX, February 2008, pp. 1067–1072.

66. H. Peng, I. D-Anderson, and M. Hella, A 100 MHz two-phase four segment DC-DC converter with light load efficiency enhancement in 0.18 µm CMOS technology. *IEEE Custom Integrated Circuits Conference CICC*, San Jose, CA, September 2012.
67. H. Peng, V. Pala, T. P. Chow, and M. M. Hella, 100 MHz, 85% efficient integrated AlGaAs/GaAs supply modulator for RF power amplifier modules. *The Applied Power Electronics Conference and Exposition APEC 2013*, Long Beach, CA, March 2013.
68. H. Peng, and M. M. Hella, 100 MHz CMOS DC-DC converter with light load efficiency enhancement. *IEEE Trans. Circuits Syst. I*, 60(8), 2213–2224, August 2013.

Index

Printed and bound by CPI Group (UK) Ltd, Croydon, CR0 4YY

22/10/2024

01777645-0001